T0145215

# Statistics and Computing

**Series editor**

Wolfgang Karl Härdle, Humboldt University of Berlin, Berlin, Germany

Statistics and Computing (SC) includes monographs and advanced texts on statistical computing and statistical packages.

More information about this series at http://www.springer.com/series/3022

Luca Martino • David Luengo • Joaquín Míguez

# Independent Random Sampling Methods

Luca Martino
Department of Signal Theory
and Communications
Carlos III University of Madrid
Madrid, Spain

David Luengo
Department of Signal Theory
and Communications
Technical University of Madrid
Madrid, Spain

Joaquín Míguez
Department of Signal Theory
and Communications
Carlos III University of Madrid
Madrid, Spain

The code can be found on GitHub: https://github.com/BookIndRandSamplingMethods/code.
Links to all material are available at http://www.springer.com/.

ISSN 1431-8784 ISSN 2197-1706 (electronic)
Statistics and Computing
ISBN 978-3-030-10241-8 ISBN 978-3-319-72634-2 (eBook)
https://doi.org/10.1007/978-3-319-72634-2

Mathematics Subject Classification (2010): 65C10, 65C05



Printed on acid-free paper

This Springer imprint is published by the registered company Springer International Publishing AG part of Springer Nature.
The registered company address is: Gewerbestrasse 11, 6330 Cham, Switzerland

# Contents

# List of Tables

# Chapter 1
# Introduction

**Abstract** This chapter provides an introduction to the different approaches available for sampling from a given probability distribution. We start with a brief history of the Monte Carlo (MC) method, one of the most influential algorithms of the twentieth century and the main driver for the current widespread use of random samples in many scientific fields. Then we discuss the need for MC approaches through a few selected examples, starting with two important classical applications (numerical integration and importance sampling), and finishing with two more recent developments (inverse Monte Carlo and quasi Monte Carlo). This is followed by a review of the three types of "random numbers" which can be generated ("truly" random, pseudo-random, and quasi-random), a brief description of some pseudo-random number generators and an overview of the different classes of random sampling methods available in the literature: direct, accept/reject, MCMC, importance sampling, and hybrid. Finally, the chapter concludes with an exposition of the motivation, goals, and organization of the book.

> Anyone who considers arithmetical methods of producing random digits is, of course, in a state of sin.
>
> John von Neumann (1940)[1]

## 1.1 The Monte Carlo Method: A Brief History

Several techniques to perform random experiments have been used since, at least, the beginning of the nineteenth century. For example, Georges-Louis Leclerc (comte de Buffon) formulated (and solved) in 1777 what has come to be known as *Buffon's needle problem*, the earliest problem in geometric probability [73]. He found

[1]This sentence simply means that there are no true random "numbers," just means to produce them, and that "a strict arithmetic procedure is not such a method" [34].



L. Martino et al., *Independent Random Sampling Methods*, Statistics and Computing, https://doi.org/10.1007/978-3-319-72634-2_1

analytically that, given an infinite plane ruled with parallel lines uniformly separated a distance $d$, the probability that a needle of length $\ell < d$ cast at random intersects one of the lines is $\frac{2\ell}{\pi d}$. Much later on, in 1812, Laplace noted that Buffon's needle experiment could be used to approximate $\pi$ [70], and during the second half of the nineteenth century a number of people performed experiments in which they threw a needle in a haphazard manner (i.e., "randomly") onto a board ruled with parallel straight lines and tried to infer the value of $\pi$ [51].[2]

In spite of other isolated and undeveloped instances during the nineteenth and the beginning of the twentieth centuries, widespread interest and development of statistical sampling techniques did not start until the advent of the first electronic computers in 1945. Apparently, Stanislaw Ulam had the original idea in 1946, while convalescing from an illness and playing solitaires [33], of calculating probabilities by repeatedly performing an experiment and counting the number of outcomes of each type. Then, in the spring of 1946, after attending the review of the results of a preliminary computational model of a thermonuclear reaction on the ENIAC (the world's first electronic digital computer, built at the University of Pennsylvania), he realized the potential of the digital computers for the implementation of this approach and discussed the idea with John von Neumann [89]. Von Neumann noticed immediately the potential of Ulam's idea and, on March 11, 1947, he sent a letter to Robert Richtmyer, the Theoretical Division Leader at Los Alamos, where he provided a detailed outline of the first Monte Carlo method[3]: an approach to solving the problem of neutron diffusion in fissile material [33, 89]. This was followed by the publication of Metropolis and Ulam's classical paper [90], which provided an open name for the new method[4] and sparked a large interest on it, with a first symposium devoted exclusively to it celebrated that same year [56].

The main requirement to use the Monte Carlo method for simulation of a physical system is that it must be possible to describe the system in terms of a probability density function (or a cumulative distribution function). Once the density function of a system is known, then the simulation begins to generate random numbers from this density. There must be a rule available, based on some reasonable mathematical and/or physical theory, to decide the outcome of such a trial. Many trials are conducted and outcomes of all of these trials are recorded. The final step in the Monte Carlo method is estimating the behavior of the overall system by computing the average of the outcomes of the trials.

After the development of the generic Monte Carlo framework, the next milestone came in 1953, when Metropolis et al. developed an efficient algorithm to calculate the properties of equilibrium systems [91]. The Metropolis algorithm, which has

[2]Interestingly, the most accurate experiment, performed by Lazzarini in 1901 [72], has been recently discredited as a fraud [6].

[3]According to Emilio Segré, Enrico Fermi's student and collaborator, Fermi had already invented the Monte Carlo method nearly fifteen years earlier, without naming it or publishing anything about it, and used it to perform remarkably accurate predictions of experimental results [89, 102].

[4]It seems that it was Nick Metropolis who suggested the obvious name for the new method [89], which was "used in meetings" at Los Alamos [56].

**Table 1.1** Some milestones in Monte Carlo research

| | |
|---|---|
| 1949 | The Monte Carlo method (N. Metropolis and S. Ulam [90]) |
| 1951 | Rejection sampling (John von Neumann [113]) |
| 1953 | The Metropolis algorithm (N. Metropolis et al. [91]) |
| 1954–55 | Sequential sampling (J. Harmersly and K. Morton [52]; A. Rosenbluth and M. Rosenbluth [100]) |
| 1956 | Importance sampling (A. Marshall [85]) |
| 1970 | Metropolis-Hastings algorithm (W. K. Hastings [53]) |
| 1987 | Sampling importance resampling (D. B. Rubin [101]) |
| 1993 | Particle filters (Gordon et al. [48]) |
| 1995 | Reversible jump MCMC (Peter J. Green [49]) |
| 1996–2004 | Adaptive importance sampling (P. Zhang [120], O. Cappé et al. [17]) |

become the basic building block of many Monte Carlo algorithms, was cited in *Computing in Science and Engineering* as being among the top ten algorithms having the "greatest influence on the development and practice of science and engineering in the twentieth century." Since then, the Monte Carlo method has found widespread use in several scientific fields, and many other milestones have followed. Table 1.1 summarizes the most important steps in the development of the Monte Carlo approach.

Furthermore, because of the great potential of this methodology, various techniques are still actively being developed by researchers. Recent advances on Monte Carlo algorithms include rejection control [77, 79], umbrella sampling [118], density-scaling Monte Carlo [110], multigrid Monte Carlo [47], hybrid Monte Carlo [31], simulated annealing [62, 82], simulated tempering [84], parallel tempering [42, 106], multiple try Metropolis [75, 77, 80], adaptive and sequential MCMC [2, 43], adaptive importance sampling [97], sequential Monte Carlo [10, 27], particle filtering [5, 11, 29, 30, 37, 48, 67, 78, 98], etc.

There is also a current trend in studying and analyzing *population-based* methods [17, 43, 59, 77]. The underlying idea is to generate a collection of random variables in parallel and then incorporate an additional step of information exchange. For instance, population-based Markov chain Monte Carlo operates by embedding the target into a sequence of related probability measures and simulating $N$ parallel chains (the population). In addition, the chains are allowed to interact via various crossover moves.

Finally, let us remark that many algorithms developed in different fields are related, and sometimes they are even identical. For instance, the configuration bias Monte Carlo [103, 104] is equivalent to a sequential importance sampler combined with the Metropolis-Hastings algorithm with independent proposal (transition) density. Similarly, the multiple try Metropolis [75, 77, 80] is an extension of a technique described in [39], the exchange Monte Carlo [55] recalls the parallel tempering [42, 106] approach, and sequential Monte Carlo and particle filtering are often used as synonymous.

## 1.2 The Need for Monte Carlo

The range of applications of Monte Carlo (MC) algorithms is enormous, from statistical physics problems [100, 103, 104] (e.g., simulating galaxy formation [95]) to nuclear medicine applications [81] (e.g., predicting the exact path of photons, electrons, or $\alpha$-particles that traverse different regions of the body [105]). Other examples of applications arise in finance [58], genetics [50], state space models (e.g., in epidemiology [116] and meteorology [119]), time series analysis [7], mixture models for inference [38], or operations research [64] (e.g., in traffic control, quality control, and production optimization). Monte Carlo methods are also employed in many areas of engineering, e.g., to simulate the turbulent combustion of a diesel engine spray injection [96] or to track moving targets and estimate their positions [5].

Monte Carlo techniques have been applied to all of these problems in order to calculate complicated integrals, to simulate a complex phenomenon or to reduce the amount of computation. The resulting algorithms are often concurrent and well suited to implementation on parallel computers. In the sequel, in order to introduce the basic MC approach we focus our attention on the problem of approximating integrals numerically.

### *1.2.1 Numerical Integration*

The best known (and maybe the most important) applications of Monte Carlo techniques involve the approximation of complicated integrals. Given an $m$-dimensional variable, $\mathbf{x} \in \mathbb{R}^m$, a crucial part of many scientific problems is the computation of integrals of the form, e.g.,

$$I = \frac{1}{|\mathcal{D}|} \int_{\mathcal{D}} f(\mathbf{x}) d\mathbf{x}, \tag{1.1}$$

where $\mathcal{D} \subseteq \mathbb{R}^m$ is some domain of interest and $|\mathcal{D}| < \infty$ indicates the measure of $\mathcal{D}$. Let us assume that we are able to generate $N$ random samples $\mathbf{x}^{(1)}, \dots, \mathbf{x}^{(N)}$ uniformly distributed inside $\mathcal{D}$. Then we can approximate the integral $I$ as

$$\hat{I}_N = \frac{1}{N} \left( f(\mathbf{x}^{(1)}) + \ldots + f(\mathbf{x}^{(N)}) \right). \tag{1.2}$$

Furthermore, given a collection of independent random variables $X_1, \dots, X_N$ with common mean $\mu$ and finite variances, the *strong law of large numbers* [25, 61, 117] states that

$$\frac{X_1 + \ldots + X_N}{N} \to \mu$$

almost surely when $N \to +\infty$. Hence, based on this result we can claim that

$$\lim_{N \to +\infty} \hat{I}_N \to I, \tag{1.3}$$

with probability 1. This is the basic formulation of the Monte Carlo technique for numerical integration. We can provide further theoretical ground for the methodology by invoking the *central limit theorem* [25, 61, 117], which ensures that the error of the approximation converges in distribution to a normal random variable, i.e.,[5]

$$\sqrt{N}(\hat{I}_N - I) \to \mathcal{N}(0, \sigma^2) \tag{1.4}$$

as $N \to +\infty$, where $\sigma^2 = \text{var}[f(\mathbf{x})]$. It is important to remark that this variance $\sigma^2$ measures how the random variable $f(\mathbf{x})$ is distributed over $\mathcal{D}$ (e.g., if $f(\mathbf{x})$ is constant in $\mathcal{D}$ the variance is zero, i.e., $\sigma^2 = 0$). Moreover, given Eq. (1.4), we can state that the Monte Carlo approximation error $(\hat{I}_N - I) \to 0$ decays to zero with a rate proportional to $1/\sqrt{N}$, regardless of the dimensionality[6] of $\mathbf{x} \in \mathbb{R}^m$. This is probably the main advantage of the Monte Carlo techniques when we compare them with their deterministic counterparts (see, e.g., [16, 69]).

To see how the Monte Carlo methodology works in multidimensional problems, let us first consider the simplest case, i.e., $m = 1$. In this situation, we can carry out a deterministic approximation of $I$, such as the *Riemann approximation* [69], obtaining an error rate that decays as $1/N$, better than the Monte Carlo method. Moreover, using more sophisticated deterministic algorithms, such as *Simpson's rule* or *Newton-Cote's rule*, we can improve the approximation [69]. However, these deterministic methods are computationally expensive when the dimension $m$ increases, since the number of points required is typically $O(N^m)$. For instance, for $m = 20$ we need to evaluate $N^{20}$ grid points in the Riemann approximation to obtain an accuracy $O(1/N)$. On the contrary, drawing $N$ points $\mathbf{x}^{(1)}, \ldots, \mathbf{x}^{(N)}$ uniformly in $\mathcal{D}$, the Monte Carlo scheme has a theoretical accuracy $O\big(1/\sqrt{N}\big)$, regardless of the dimension of $\mathcal{D}$. From this point of view, some researches have argued that the Monte Carlo approach beats the "curse of dimensionality" [77, 99].

In practice, however, there are relevant drawbacks to the Monte Carlo methodology. For example, one should bear in mind that the errors of the deterministic algorithms (trapezoidal rule, Simpson's rule, Newton-Cote's rule, etc.) are deterministic, whereas the error of the Monte Carlo approach is random, so we can only characterize it through its variance. Moreover, when the dimensionality increases, the "crude" Monte Carlo technique has further problems [77, 99]:

- The variance $\sigma^2$ can be very large.
- The accuracy $O\big(1/\sqrt{N}\big)$ is only a probabilistic bound.

---

[5] In the sequel, $\mathcal{N}(\mu, \sigma^2)$ denotes a Gaussian distribution with mean $\mu$ and variance $\sigma^2$.

[6] Specifically, the approximation error can be proved to have an upper bound of the form $\frac{c}{\sqrt{N}}$, where $c$ is a constant independent of $N$, but possibly dependent on the dimension $m$.

- Specific features of the integrand $f(\mathbf{x})$ are not exploited.
- We may not be able to draw samples uniformly inside $\mathcal{D}$.

The second issue implies that there is no guarantee that the expected accuracy is achieved in a specific calculation. Moreover, the third point remarks that the probabilistic bound $O(1/\sqrt{N})$ is obtained under very weak regularity conditions, but we do not make any improvements from any additional properties of $f(\mathbf{x})$. For instance, $f(\mathbf{x})$ could have a sharp peak, so that drawing samples $\mathbf{x}^{(1)}, \ldots, \mathbf{x}^{(N)}$ uniformly inside $\mathcal{D}$ would yield an inefficient estimator, i.e., an estimator with high variance.

### 1.2.2 Importance Sampling

In order to overcome some of the aforementioned difficulties, the concept of *importance sampling* (IS)[29, 30, 85, 99] has been introduced in the literature. This approach consists in drawing $N$ samples $\mathbf{x}^{(1)}, \ldots, \mathbf{x}^{(N)}$ from a non-uniform probability density function (pdf), $\pi(\mathbf{x})$, that concentrates more probability mass on the "important" parts of the region $\mathcal{D}$ in order to save the computational resources. In this case, the approximation of $I$ is given by[7]

$$\hat{I}_N = \frac{1}{N|\mathcal{D}|}\left(\frac{f(\mathbf{x}^{(1)})}{\pi(\mathbf{x}^{(1)})} + \ldots + \frac{f(\mathbf{x}^{(N)})}{\pi(\mathbf{x}^{(N)})}\right). \tag{1.5}$$

This estimator is unbiased and it has a variance $\sigma^2 = \text{var}_\pi[f(\mathbf{x})/\pi(\mathbf{x})]$. In a "fortunate" case, if $f(\mathbf{x})$ is non-negative and $I$ is finite, we may select $\pi(\mathbf{x}) \propto f(\mathbf{x})$ so that $\hat{I}_N = I$. However, in most practical problems we can only try to find a good proposal density $\pi(\mathbf{x})$ that is reasonably close in shape to $f(\mathbf{x})$. Therefore, the challenge in this case is being able to find a good proposal from which random samples can be efficiently drawn.

If the measure $|\mathcal{D}| < \infty$ of the support domain $\mathcal{D}$ is unknown, or if we know the analytic form of the proposal density $\pi(\mathbf{x})$ only up to a multiplicative constant (i.e., we know a non-negative unnormalized function $\pi_u(\mathbf{x}) \propto \pi(\mathbf{x})$), we can construct a *biased* estimator,

$$\hat{I}_N = \frac{1}{w^{(1)} + \ldots + w^{(N)}}\left(w^{(1)}f(\mathbf{x}^{(1)}) + \ldots + w^{(N)}f(\mathbf{x}^{(N)})\right), \tag{1.6}$$

where $w^{(i)} = 1/\pi_u(\mathbf{x}^{(i)})$, $i = 1, \ldots, N$. With this approach, we can effectively use an unnormalized function $\pi_u(\mathbf{x})$ and, moreover, $\hat{I}_N$ often has a smaller mean square error than the unbiased estimator of Eq. (1.5). However, we remark that the choice of

[7] If different proposal pdfs are jointly used, alternative IS approximations of $I$ are possible and provide more robust estimations, albeit at the expense of an increased computational cost [35, 36]

$\pi$ (or $\pi_u$) is crucial to the performance of both estimators, i.e., a good choice of the proposal can reduce drastically the variance of the estimate [77, 99].

### Bayesian Inference via IS

Bayesian methods have become very popular in statistics, machine learning, and signal processing during the past decades [77, 99]. Monte Carlo techniques are often required for the implementation of optimal a posteriori estimators. More specifically, in Bayesian inference it is often necessary to compute integrals of the type

$$I = E_{p_o}[f(\mathbf{X})] = \int_{\mathcal{D}} f(\mathbf{x})p_o(\mathbf{x}|\mathbf{y})d\mathbf{x}, \tag{1.7}$$

$$= \frac{1}{Z}\int_{\mathcal{D}} f(\mathbf{x})p(\mathbf{x}|\mathbf{y})d\mathbf{x} \tag{1.8}$$

where $\mathbf{X} \sim p_o(\mathbf{x}|\mathbf{y}) = \frac{1}{Z}p(\mathbf{x}|\mathbf{y})$ and $p_o(\mathbf{x}|\mathbf{y})$ represents the posterior pdf of the variable of interest $\mathbf{x}$ given the observed data $\mathbf{y}$. Moreover, $f(\mathbf{x})$ is an integrable function w.r.t. $p_o(\mathbf{x}|\mathbf{y})$. A direct Monte Carlo approach follows these two steps: (a) draw $N$ independent samples $\mathbf{x}^{(i)}$, $i = 1, \ldots, N$ from the posterior $p_o(\mathbf{x}|\mathbf{y})$ and (b) compute $\hat{I}_N = \frac{1}{N}\sum_{i=1}^{N} f(\mathbf{x}^{(i)})$. However, in many applications, it is not possible to draw directly from $p_o(\mathbf{x}|\mathbf{y})$ so that a direct Monte Carlo approach cannot be applied. An alternative strategy is suggested by the following equality

$$I = E_{p_o}[f(\mathbf{X})] = \frac{1}{Z}\int_{\mathcal{D}} f(\mathbf{x})p(\mathbf{x}|\mathbf{y})d\mathbf{x} \tag{1.9}$$

$$= \frac{1}{Z}\int_{\mathcal{D}} f(\mathbf{x})\frac{p(\mathbf{x}|\mathbf{y})}{\pi(\mathbf{x})}\pi(\mathbf{x})d\mathbf{x}, \tag{1.10}$$

$$= E_{\pi}[f(\mathbf{X})w(\mathbf{X})], \tag{1.11}$$

where $w(\mathbf{X}) = \frac{p(\mathbf{X}|\mathbf{y})}{\pi(\mathbf{X})}$ and $\pi(\mathbf{x})$ is a suitable proposal density. Therefore, if the normalizing constant $Z = \int_{\mathcal{D}} p(\mathbf{x}|\mathbf{y})d\mathbf{x}$ of the posterior is known, a possible IS estimator is

$$\hat{I}_N = \frac{1}{NZ}\sum_{i=1}^{N} w^{(i)} f(\mathbf{x}^{(i)}),$$

where $\mathbf{x}^{(i)} \sim \pi(\mathbf{x})$ and $w^{(i)} = \frac{p(\mathbf{x}^{(i)}|\mathbf{y})}{\pi(\mathbf{x}^{(i)})}$. Otherwise, if $Z$ is unknown, an alternative IS approximation is given by

$$\hat{I}_N = \frac{1}{\sum_{n=1}^{N} w^{(n)}}\sum_{i=1}^{N} w^{(i)} f(\mathbf{x}^{(i)}).$$

### 1.2.3 Quasi-Monte Carlo

Another important class of algorithms introduced to overcome the drawbacks of the basic Monte Carlo methodology includes the so-called *quasi-Monte Carlo methods* [41, 92, 114]. The basic idea of a quasi-Monte Carlo technique is replacing the random samples in a Monte Carlo method by *well-chosen* deterministic points, often termed *nodes* or *quasi-random* numbers. These nodes are chosen judiciously in order to guarantee a small error in the numerical approximation of the integral $I$. The selection criterion is based on the concepts of *uniformly distributed sequence* and *discrepancy*, which is a measure of the deviation from the uniform distribution [92]. For a suitable choice of $N$ nodes, quasi-Monte Carlo methods can obtain a deterministic error bound $O(N^{-1}\log(N)^{m-1})$, where $m$ is the dimension of the space.

### 1.2.4 Inverse Monte Carlo

The term inverse Monte Carlo (IMC) is often used to identify a class of Monte Carlo algorithms designed to solve inverse problems [32, Chap. 7]. For instance, consider the integral

$$I^{(k)}(\mathbf{y}_k, \boldsymbol{\theta}) = \frac{1}{|\mathcal{D}|}\int_{\mathcal{D}} f(\mathbf{x}, \mathbf{y}_k, \boldsymbol{\theta})d\mathbf{x}, \tag{1.12}$$

where $\mathbf{y}_k$ and $\boldsymbol{\theta}$ are different parameters. In a direct Monte Carlo problem, the goal is estimating the value of the integral $I^{(k)}$ assuming that the parameters are known in advance. On the other hand, in some inverse problems, we are interested in approximating $\boldsymbol{\theta}$ given the outcomes of several integrals, $I^{(k)}$ for $k = 1, \ldots, K$. Namely, we are essentially trying to solve a parameter estimation problem: we know the value of the integral $I^{(k)}$ for different values of $\mathbf{y}_k$ and we want to infer the parameter vector $\boldsymbol{\theta}$. A general and simple Monte Carlo scheme to solve such inverse problem consists of the following steps [32, Chap. 7]:

1. Choose an initial value $\hat{\boldsymbol{\theta}}_0$.
2. Given the current estimate of the parameter vector, $\hat{\boldsymbol{\theta}}_t$, compute the approximations, $\hat{I}^{(k)}(\mathbf{y}_k, \hat{\boldsymbol{\theta}}_t)$ for $k = 1, \ldots, K$, using a suitable Monte Carlo approach (recall that all the values $\mathbf{y}_k$ are known).
3. Compare the estimates $\hat{I}^{(k)}(\mathbf{y}_k, \hat{\boldsymbol{\theta}}_t)$ with the true values, $I_k(\mathbf{y}_k, \boldsymbol{\theta})$ for $k = 1, \ldots, K$:
   - If they are "good enough," then set $\hat{\boldsymbol{\theta}} = \hat{\boldsymbol{\theta}}_t$ and stop.
   - Otherwise, set $t = t + 1$, propose a new value $\hat{\boldsymbol{\theta}}_{t+1}$ (following some predefined procedure), and go back to step 2.

## 1.3 Random Number Generation

Random number generation is the core issue of Monte Carlo simulations. A reliable random number generator is critical for the success of a Monte Carlo method. Indeed, the accuracy of Monte Carlo calculations depends on the pertinence and suitability of the underlying stochastic model, but also on the "quality" of the random numbers that simulate the random variables in the model.

Random numbers are required in a variety of areas. We have emphasized their application in Monte Carlo methods, but they also play a crucial role in many other simulation problems in different areas, such as computational statistics, VLSI testing, finance, cryptography, bioinformatics, computational chemistry and physics, etc. [24, 40, 92]. In all these scientific fields we need to draw samples from $p_o(\mathbf{x})$, the non-uniform pdf associated to the complex system of interest, where $\mathbf{x}$ denotes the (random) configuration of the system. For instance, in the analysis of a macromolecule, $\mathbf{x}$ could represent the three-dimensional coordinates of all the atoms in the molecule. Hence, the target density in this example would correspond to the so-called Boltzmann distribution [45, 75, 77],

$$p_o(\mathbf{x}) = \frac{1}{Z(T)} \exp\left(-\frac{V(\mathbf{x})}{kT}\right),$$

where $k \approx 1.38\times10^{-23}$ J/K is the Boltzmann constant, $T$ is the system's temperature (in Kelvin degrees), $V(\mathbf{x})$ is the energy function, and $Z(T)$ is the *partition function*, which is difficult to calculate in general and plays the role of a normalizing constant, since it does not depend on $\mathbf{x}$. In Bayesian statistical inference, $\mathbf{x}$ usually represents the missing data jointly with the unknown parameter values and $p_o(\mathbf{x})$ often denotes the joint posterior pdf of these variables.

There is another (broad) class of applications where we come across optimization problems that can be conveniently tackled by generating random samples [4, 82]. Indeed, let us consider the problem of minimizing a generic cost function $V(\mathbf{x})$. This is equivalent to maximizing another function $p(x) = \exp(-V(\mathbf{x})/T)$ with $T > 0$. If $p(x)$ is integrable for all $T > 0$, we can define the target density

$$p_o(x) \propto p(x) = \exp(-V(\mathbf{x})/T).$$

Thus, if we are able to draw samples from $p_o(x)$ when $T$ is small enough, the generated samples are located (with high probability) in a region close to the global minimum of $V(\mathbf{x})$ [77, 99].

### 1.3.1 *Random, Pseudo-Random, Quasi-Random*

There are several types of "random numbers":

- "Truly" random numbers are generated using a physical device as the source of randomness. Examples found in the literature include coin flipping, roulette wheels, white noise, and the count of particles emitted by a radioactive source [115].

- Pseudo-random numbers correspond to a deterministic sequence that passes tests of randomness. The standard procedures for generating sequences of pseudo-random numbers are based on recursive methods and yield sequences that can be periodic (with a very large period) [40, Chap. 1], [92, Chap. 7] or *chaotic*[8] [1, 13].
- Quasi-random numbers refer to a deterministic number sequence that presents a low discrepancy with respect to (w.r.t.) a given distribution. Deterministic constructions have been studied to build low-discrepancy point sets and sequences (using, for example, digit and fractional expansions) [92, Chap. 3].

The general requirements that are usually placed on random number generation methods can be divided into four categories [24, 40, 92]:

1. *Computational requirements*: they refer to the computational cost and the resources needed to generate the random numbers.
2. *Structural requirements*: they include, e.g., the period length (in periodic sequences of pseudo-random numbers) and the lattice structure.
3. *Statistical requirements*: the produced numbers are expected to pass statistical tests related essentially to their distribution and certain statistical independence properties.
4. *Complexity-theoretic requirements*: some definitions of "randomness" for a finite string of random digits have been presented in the literature [20, 21, 65, 66, 86]. These represent a collection of conditions that a finite sequence has to satisfy in order to be considered "random."

In computational statistics, random variate generation is usually divided into two steps:

(1) generating "imitations" of independent and identically distributed (i.i.d.) random numbers having a uniform distribution and
(2) applying some transformation and/or selection techniques such that these i.i.d. uniform samples are converted into variates from the target probability distribution.

These two steps are essentially independent. The expression *pseudo-random number generator* usually refers to an algorithm used for the first step, while the term *sampling method* is usually associated to an algorithm used in the second step. In particular, a sampling technique assumes that some random or pseudo-random number generator with a known distribution (typically uniform) is available. Random sampling algorithms are also termed *non-uniform random variate generators*. Figure 1.1 summarizes this classification.

[8]In fact, we note that most chaotic random number generators actually yield *pseudo-chaotic* sequences, since they are implemented using finite precision arithmetic.

| Real – Random Numbers | Pseudo – Random Numbers | | Quasi – Random Numbers |
|---|---|---|---|
| | Pseudo – Random generators | Sampling methods | |

**Fig. 1.1** General scheme of random number generation

## 1.4 Pseudo-Random Number Generators

The focus of this book is on random sampling methods. However, it is instructive to pay some attention at this point to pseudo random number generators. As shown in the previous section, several applications require "random-like" sequences that must also be *reproducible*. These reproducible random-like sequences are called *pseudo-random number* (PRN) streams. In order to ensure reproducibility, PRNs are usually generated using deterministic recursive equations. In this section, we provide a brief overview of some of the most relevant pseudo-random number generators (PRNGs).

### *1.4.1 Nonlinear Recursions*

Most PRNGs can be formally represented as discrete-time dynamical systems. If we intend to produce a one-dimensional PRN stream, such systems consist of a nonlinear recursion of the form

$$y_{n+1} = f(y_n), \quad n = 0, 1, 2, \dots \tag{1.13}$$

where $y_0 \in \mathbb{R}$ is a user-defined initial condition and $f$ is a non-linear function or *map*. The sequence $\{y_n\}_{n=0}^{+\infty}$, known as *orbit* or *trajectory*, is produced by applying Eq. (1.13) recursively.

#### Invariant Sets

For any starting value $y_0 \in \mathbb{R}$, and after a transient period, the dynamical system described by Eq. (1.13) converges to an *attractive invariant set*, i.e., for a sufficiently large value of $n$, the recursion (1.13) produces iterates that take values from an invariant set. To be specific, the asymptotic behavior of the sequence $\{y_n\}_{n=0}^{+\infty}$ generated by system (1.13) can be classified as follows [1]:

- **Fixed point:** For $n \geq n^*$ the system reaches a single value $y^*$ which is maintained for the rest of the trajectory, i.e., $y_n = y^*$ for $n \geq n^*$.
- **Periodic sequence or limit cycle**: After $n^*$ iterations the system reaches a periodic solution, i.e., $y_n = y_{n-T}$ for $n \geq n^*$, where $T \in \mathbb{N}^+$ is the period of the

corresponding cyclic trajectory. In the frequency domain the signal is composed of a finite collection of harmonically related frequencies.

- **Quasi-periodic sequence:** After $n^*$ iterations the system reaches a "periodic-like" solution, i.e., a sequence which looks periodic but does not meet exactly the definition of periodicity. In the frequency domain these signals are composed of two or more *incommensurate* natural frequencies.
- **Chaotic sequence:** The sequence $y_n$ behaves in a "random-like" manner for $n \geq n^*$ in spite of having been generated by a deterministic system (i.e., the system does not diverge, but never reaches a fixed point, a limit cycle or a quasi-periodic solution). In the frequency domain they are characterized by a continuous wide-band spectrum.

The asymptotic behavior of the system can be different depending on the choice of the starting point $y_0$. Consequently, a map can have different invariant sets which can be *stable* (i.e., attractive) or *unstable* (i.e., repulsive).

### 1.4.2 Chaotic Pseudo-Random Number Generators

Chaotic dynamical systems are particularly appealing, both intuitively and theoretically, as PRNGs, since the sequences they generate naturally enjoy certain properties that greatly resemble what we perceive as randomness. Formally, it is possible to associate an invariant probability density function, denoted $\rho(y)$, to a chaotic attractor, based on the fact that some regions of the state space are visited more frequently than others by the sequence $\{y_n\}_{n \geq n^*}^{+\infty}$ [13, 71]. Given a generic map, $f(y) : \mathcal{D} \to \mathcal{D}$, an invariant density $\rho(y)$ is defined as a function that satisfies

$$\int_{\mathcal{A}} \rho(y)dy = \int_{f^{-1}(\mathcal{A})} \rho(y)dy, \tag{1.14}$$

for any subset $\mathcal{A} \subseteq \mathcal{D}$, where $f^{-1}(\mathcal{A})$ represents the preimage of all points contained in $\mathcal{A}$. Below we provide some specific examples of uniform and non-uniform invariant densities.

*Example 1.1 (Logistic Map)* The best known discrete-time one-dimensional chaotic system is probably the *logistic map*[9]

$$y_{n+1} = \lambda y_n(1 - y_n), \tag{1.15}$$

[9]The Belgian mathematician P.F. Verhulst proposed in 1838 the logistic equation [111], $dN(t)/dt = rN(t)(1-N(t)/K)$ with $r, K > 0$, whose exact solution is $N(t) = K/[1+CK\exp(-rt)]$ with $C = 1/N(0) - 1/K$ [112], as a continuous-time alternative to the Malthusian model of population growth [83]. Much later on, the Australian biologist Robert May showed that the discrete-time version of the logistic equation, $f(x) = \lambda x(1 - x)$ with $0 < \lambda \leq 4$, so-called logistic map since then, could exhibit fixed points, limit cycles and chaotic behavior depending on the value of the bifurcation parameter $\lambda$ [87, 88].

for some $\lambda \in (0, 4]$, and $y_0 \in (0, 1)$. When $\lambda = 4$, almost the whole domain of the logistic map, $[0, 1]$,[10] is a chaotic attractor with an associated invariant density

$$\rho(y) = \frac{1}{\pi \sqrt{y(1-y)}}, \qquad 0 < y < 1, \tag{1.16}$$

which is a Beta pdf, $p(y) \propto y^{\gamma-1}(1-y)^{\eta-1}$, with $\gamma = \eta = 0.5$ [46, 93].

*Example 1.2 (Tent and Bernoulli Shift Maps)* Two simple maps with uniform invariant density in $(0, 1)$, i.e.,

$$\rho(y) = 1, \qquad 0 < y < 1, \tag{1.17}$$

are the *tent map*,

$$y_{n+1} = \begin{cases} 2y_n, & 0 \leq y_n \leq 0.5; \\ 2(1-y_n), & 0.5 < y_n \leq 1, \end{cases} \tag{1.18}$$

and the *Bernoulli shift map* [1],

$$y_{n+1} = 2y_n \bmod 1 = \begin{cases} 2y_n, & 0 \leq y_n \leq 0.5; \\ 2y_n - 1, & 0.5 < y_n \leq 1, \end{cases} \tag{1.19}$$

with a starting point $y_0 \in (0, 1)$. More generally, the class of Bernoulli shift maps (also known as sawtooth maps), which also have a uniform invariant density, is often expressed as

$$y_{n+1} = a y_n \bmod 1, \tag{1.20}$$

where $a = 1/M$, with $M \in \mathbb{N}^+$ being the number of intervals of the map. This type of recursions are related to the linear congruential generators described in Sect. 1.4.4. below.

Besides the collection of well-studied discrete-time chaotic systems that can be found in the literature, it is also possible, given a prescribed invariant pdf $\rho(y)$, to design a piece-wise map displaying chaotic behavior and with its trajectories distributed over the state space according to $\rho(y)$ [13].

Although chaotic maps can be used to produce sequences with good statistical properties, they also suffer from some significant limitations:

- The produced sequences still present certain structure.

[10]Note that scattered within this interval we may find fixed points (e.g., $y = 0$ and $y = 0.75$) and limit cycles with arbitrarily large periods (e.g., $y = (5 - \sqrt{5})/8$ and $y = (5 + \sqrt{5})/8$ form a period two limit cycle).

- The finite precision of computing devices causes the generated sequence to be pseudo-chaotic. Hence, it can get stuck in a low period limit cycle or even a stable fixed point.

For these reasons, other approaches are often used to produce PRN streams.

### 1.4.3 The Middle-Square Generator

Probably the first PRNG ever employed to generate uniform numbers in a computer is the *middle-square method*, proposed by John von Neumann already in 1946 [63], although it was not published until 1951 [113]. The procedure starts from an initial number. Then, it iteratively takes the square of the previous number and extracts the middle digits to generate a new element. More precisely, the algorithm consists of the following steps[11]:

1. *Initialization.* Select an initial four-digit state, $x_0 = 0.d_1^0 d_2^0 d_3^0 d_4^0$ with $d_i^0$ representing natural numbers uniformly distributed[12] within $(0, 1)$. Set $n = 1$.
2. *Iteration.* Assume $x_{n-1}$ is available. The next number in the sequence, $x_n$, is obtained by taking the square of $x_{n-1}$ and extracting the middle four digits, i.e., let $y_n = x_{n-1}^2 = 0.\tilde{d}_1^n \tilde{d}_2^n \tilde{d}_3^n \tilde{d}_4^n \tilde{d}_5^n \tilde{d}_6^n \tilde{d}_7^n \tilde{d}_8^n$ and set $x_n = 0.\tilde{d}_3^n \tilde{d}_4^n \tilde{d}_5^n \tilde{d}_6^n$.

The middle-square method was developed at a time when computers were just starting and was extensively used, since it was simple and much faster than any other method available at that time. However, by current standards its quality is quite poor, since its output can get stuck at a fixed point or a short limit cycle, or degenerate quickly to zero [63].

### 1.4.4 Linear Congruential Generators

The linear congruential method (LCG) was proposed by Lehmer in 1951 [74]. It is often known also as the *multiplicative* LCG, Lehmer generator or, more formally, as the prime modulus multiplicative linear congruential generator (PMMLCG) [94]. It has become a classical and still very popular approach for the generation of uniform PRNs within the $[0, 1)$ interval [58, 63, 92].

[11] In the sequel, the notation $x_n = 0.d_1^n d_2^n d_3^n d_4^n$ is used to indicate that $x_n = d_1^n \times 10^{-1} + d_2^n \times 10^{-2} + d_3^n \times 10^{-3} + d_4^n \times 10^{-4}$.

[12] The initial state was originally obtained from a mechanical device that generated truly random numbers, from a table of random digits or manually by the user (e.g., rolling out a ten-sided dice or dealing out cards).

The generator has the following parameters:

- The *modulus*, $M > 0$, which should be large positive integer.
- The *multiplier*, $a \in \{1, 2, \ldots, M - 1\}$, which is a positive integer such that $\gcd(a, M) = 1$, where $\gcd(x, y)$ denotes the greatest common divisor of the natural numbers $x$ and $y$.
- The *increment*, $c \in \{0, 1, \ldots, M - 1\}$, which is a non-negative integer.
- An initial value, $y_0 \in \{0, 1, \ldots, M - 1\}$, usually called the *seed*.

A sequence of non-negative integers, $\{y_n\}_{n=0}^{\infty} \in \{0, 1, \ldots, M - 1\}$, is generated by the recursion

$$y_{n+1} = (ay_n + c) \bmod M, \quad n = 0, 1, 2, \ldots \tag{1.21}$$

and the linear congruential pseudo-random numbers are

$$x_n = \frac{y_n}{M} \in [0, 1), \quad n = 0, 1, 2, \ldots \tag{1.22}$$

Although the sequence $\{x_n\}_{n=0}^{\infty}$ may seem random, it is actually periodic with period $T \leq M$ (see [58], [92, Chap. 7]). The parameters $a$ and $c$ should be appropriately chosen in order produce a sequence with the largest possible period, i.e.., $T = M$, in which case the generator is said to have a full period.

The choice of the modulus $M$ is usually made according to the word length of the machine, e.g., we might choose $M \approx 2^{32}$ for single-precision, $M \approx 2^{48}$ for extended-precision and even $M \approx 2^{64}$ for high-precision calculations.[13]

## General LCGs

Higher order linear recursive equations have been used in order to improve the "random features" of the sequences generated by the LCG. Let $M$ be a large prime number, let $K \geq 2$ be the order of the recursion and choose coefficients $a_j \in \{0, 1, \ldots, M-1\}$, with $j = 1, \ldots, K$ and $a_1 \neq 0$. We can generate a sequence of non-negative integers, $\{y_n\}_{n=0}^{\infty}$ with $y_n \in \{0, 1, \ldots, M - 1\}$, by way of the recursion

$$y_n = \sum_{j=1}^{K} a_j y_{n-j} \bmod M, \qquad n \in \mathbb{N}, \tag{1.23}$$

where the initial values $y_1, \ldots, y_K$ should not all be equal to zero. The pseudo uniform random numbers are again obtained using Eq. (1.22). In this case the

[13] In fact, we must choose $M \neq 2^b$, since the resulting generator would not have a full period otherwise [94]. Hence, the word length of the machine only provides the order of magnitude of $M$.

two sequences, $\{x_n\}_{n=0}^{\infty}$ and $\{y_n\}_{n=0}^{\infty}$, are both periodic with the same period $T \leq M^K - 1$ [40].

*Example 1.3 (Fibonacci Generator)* A classical example is obtained using $K = 2$ and $a_0 = a_1 = 1$, which leads to the recursion

$$y_n = (y_{n-1} + y_{n-2}) \bmod M, \qquad n \in \mathbb{N}, \tag{1.24}$$

known as the basic Fibonacci generator, because it replicates the construction of Fibonacci numbers [14].

## 1.5 Random Sampling Methods

Non-uniform random numbers are also referred to simply as *random variates*. The field of non-uniform random variate generation is an area in the crossroad of mathematics, statistics, and computer science. It is often considered a subarea of statistical computing and simulation methodology [40, 45, 54, 75, 77, 109].

Sampling techniques can be classified in three large categories that we describe briefly in the following sections: direct methods, accept/reject methods, and Markov chain Monte Carlo (MCMC) methods. This classification is summarized graphically in Fig. 1.2. In all cases, the random sampler can be interpreted as a device or algorithm to transform a random number produced by the available source (typically uniform) into a sample from some desired probability distribution. This is illustrated by the simple scheme shown in Fig. 1.3.

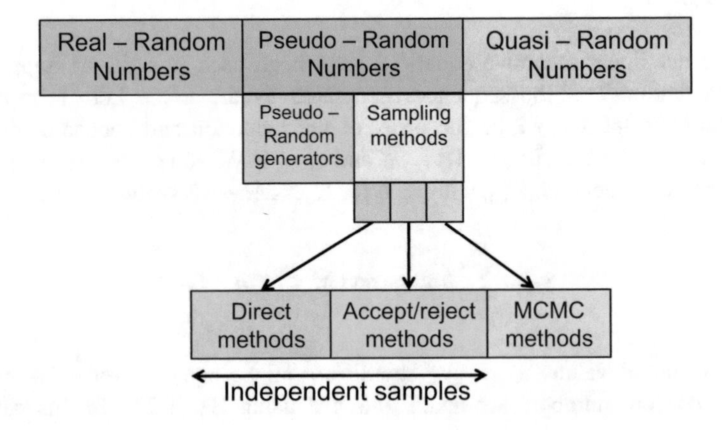


**Fig. 1.2** General scheme of random number generation and random sampling categories

**Fig. 1.3** Schematic representation of a general sampling method, which converts a sample $\xi'$ from an available random source and distributed according to $\pi(\xi)$ (often termed the *proposal density*) into another random sample $x'$ distributed according to the target pdf, $p_o(x)$

### 1.5.1 Direct Methods

These techniques apply an appropriate transformation or a direct random mechanism to convert the samples provided by an available random source into samples with the desired statistical properties [54, Chap. 2]. These methods use known relationships among random variables, transformations, mixtures, convolutions or, more generally, different mathematical representations of the target distribution.

In general, direct methods are fast and the samples generated are independent. For this reason, when a direct method is available, it is often the best random sampling algorithm. However, in many practical situations a suitable transformation or, more generally, a direct connection is unknown, and alternative approaches must be considered. Chapter 2 is devoted to describe this class of techniques.

### 1.5.2 Accept/Reject Methods

Given samples from an available random source, these algorithms accept or discard those samples by performing an adequate test. The main benefit of this class of techniques is that their range of applicability is greater than that of direct methods. Indeed, they are often termed *universal samplers*, since they can be applied to draw samples from virtually any kind of target distribution. Moreover, the generated samples are still independent, as in the case of direct methods.

One inconvenience of this approach is that an analytic construction of a proposal distribution or the knowledge of an upper bound of the target pdf is required. However, their main drawback is that the acceptance rate can be very low, implying that the computational cost can be very high. For these two reasons, many adaptive schemes have been proposed in order to:

(a) build automatically a good proposal distribution and
(b) improve the probability of accepting candidate samples [44], [54, Chap. 4].

An in-depth description of this class of methods is provided in Chaps. 3 and 4.

### 1.5.3 Markov Chain Monte Carlo (MCMC)

The techniques in this class rely on the ability to construct a Markov chain that converges to a prescribed stationary distribution [45, 99]. On the one hand, the main advantage of this methodology is that it can be applied almost universally. On the other hand, its main drawback is that MCMC algorithms produce sequences of correlated variates. Due to this correlation among samples, the transient (or burn-in) period of the chain can be very long. Hence, a considerable computational effort is invested in generating a set of samples that have to be discarded later. Furthermore, if the target distribution is multimodal, the chain can get trapped in a local mode and fail to adequately explore the complete space where the target density has support. As a consequence, MCMC estimators tend to have greater variance than those based on independent samples. A particular sub-class of these techniques that produces *asymptotically independent* samples is described in Chap. 6.

### 1.5.4 Importance Sampling

The importance sampling technique, that we have briefly introduced in Sect. 1.2.1, is another class of broadly used Monte Carlo methods. However, importance sampling cannot be considered as a random sampling algorithm. Indeed, the importance sampling procedure approximates a probability distribution with a set of weighted samples, but it does not produce random numbers from the target pdf. Some authors classify importance sampling within the category of *variance reduction techniques* [99].

### 1.5.5 Hybrid Techniques

Many mixed strategies have been proposed that combine the methods from the different categories previously discussed [15, 18, 23, 26, 77]. For instance, many authors have tried to integrate the MCMC techniques within particle filters [9, 45, 60]. One example is the *resample-move algorithm* [8], which combines sequential importance resampling (SIR) with MCMC sampling. Recently, a technique termed particle MCMC [3] that employs particle filters to construct efficient MCMC kernels has become very popular because of its broad applicability.

Another interesting mix of different classes is provided by algorithms that combine rejection and importance sampling. Some examples are the so-called *weighted rejection sampling* and *rejection control* algorithms [15, 19, 77, 79]. Other methods, such as *partial rejection control* [77] and *rejection particle filters* [12, 57, 68, 107, 108], combine rejection sampling with sequential importance sampling [79]. All of these techniques are accept/reject methods that re-incorporate (in different ways) the discarded samples into the computed estimators. For a detailed comparison of the performance of rejection and importance sampling estimators, see [18, 22, 76].

Last but not least, the *ratio-of-uniforms* (RoU) technique combines features from the direct and accept/reject approaches in an appealing manner. In Chap. 5 we describe the RoU method and its generalizations in detail.

## 1.6 Goal and Organization of This Book

### *1.6.1 Motivation and Goals*

This monograph is concerned with the theory and practice of pseudo-random variate generation, an issue which is at the core of Monte Carlo simulations and, hence, of practical importance for a large number of applications in various fields, including computational statistics, cryptography, computer modeling, games, etc. The focus will be placed on independent and exact sampling methods, as opposed to techniques that produce approximate (e.g., importance sampling) and/or correlated populations (e.g., MCMC).

A number of relevant references can be found in the literature related to these topics [24, 28, 40, 54]. In this monograph, we distinctly aim at:

- Presenting a comprehensive and unified view of independent random samplers, that includes the most relevant classes of methods and emphasizes their generality—as opposed to the common trend of investigating algorithms "tailored" to specific problems.
- Exploring in depth the connections, relationships, and relative merits of the different families of techniques, including systematic comparisons with non-independent samplers, such as MCMC methods and importance samplers.
- Consolidating the considerable body of knowledge that has been generated during the last decade, concerning principally a broad class of new and flexible adaptive samplers.

Although our main interests are in the theory and methodologies for independent random sampling, we have made a special effort in the choice of application examples that enjoy a clear practical interest. In this respect, we expect that the materials included in this book may be of interest to engineers working in signal processing and statisticians interested in computational methods, but they should also be useful to scientists working in the fields of biology, quantitative finance, or physics, where complex models that demand Monte Carlo computations are needed. Computer code of the examples and the main algorithms is provided in a companion website.

### 1.6.2 Organization of the Book

The rest of the book is organized as follows. Chapter 2 describes the so-called *direct methods*: a collection of classical and modern techniques used for random sampling based on suitable transformations and/or specific connections among random variables. All of them assume the availability of a random source with known distribution, and all of them are aimed to produce independent and identically distributed (i.i.d.) samples. Many of them are intrinsically connected, and we make a special effort to remark the relationships among different techniques or different categories of methods. Indeed, we note that some techniques can be classified within more than one category of algorithms and can be derived in different ways (e.g., the Box-Muller method). These different points of view are explored and the connections among categories are highlighted.

Chapters 3 and 4 are devoted to *accept/reject methods*, also known as *rejection sampling* (RS) algorithms. The basic RS approach, which is described in Chap. 3, was suggested by John von Neumann as early as in 1946, although it was not published until 1951, and it is a classical technique for *universal sampling*. In an accept/reject method, each sample is either accepted or rejected by an adequate test of the ratio of the proposal and the target pdfs, and it can be proved that accepted samples are actually distributed according to the target density. The fundamental figure of merit of a rejection sampler is the mean acceptance rate, i.e., the expected number of accepted samples over the total number of proposed candidates. To attain good acceptance rates, adaptive rejection sampling (ARS) schemes, which are the focus of Chap. 4, have been proposed in the literature. These techniques aim at sequentially building proposal functions that become closer and closer to the target pdf as the algorithm is iterated (i.e., as more samples are drawn and more accept/reject tests are carried out). In these two chapters we present the standard RS and ARS schemes, together with various recent developments, and including some original material. Again, the connections and relationships among different methods are highlighted, and we also explore some of the main applications of the methodology.

In Chap. 5 we focus on the ratio of uniforms (RoU) method, which is a classical technique that combines both of the approaches of the previous chapters: suitable transformations and rejection sampling. Assume that $p(x)$ is the target density from which it is needed to generate samples. The RoU technique aims at calculating a bounded region $A$ such that points drawn independently and uniformly inside $A$ yield i.i.d. samples from $p(x)$ in a very straightforward manner. Since uniform samples within $A$ cannot be obtained usually by direct methods, in practice the RoU is often combined with the RS approach. In this chapter, we present the standard RoU method and some extensions first. Then we focus on adaptive implementations of the RoU technique, including some original contributions. The connection of the RoU approach, both with transformation methods and with the accept/reject class of techniques, is highlighted and several application examples are also provided.

The theory and applications in the previous chapters deal with sampling from scalar (i.e., one-dimensional) random variables. While relatively simple for a few special and well-known cases (e.g., Gaussian random variables), the design of general methods to draw samples from multidimensional distributions (i.e., to generate random vector-samples) which can be efficiently implemented is a hard problem. Some methods have been proposed, though, and we review them in Chap. 6. Most of them are only *partially general*, as they involve a number of constraints on the target distributions. Others are more general, e.g., the multivariate extension of the RoU and ARS techniques discussed in Chaps. 5 and 4, but the computational complexity can be prohibitive. Both theoretical constraints and computational limitations are explored, and illustrated by way of a few examples.

The goal of Chap. 7 is introducing another family of samplers which is widely used in the literature: MCMC algorithms. MCMC methods [75, 99] are Monte Carlo techniques that produce a Markov chain of correlated samples whose stationary distribution is known. From the perspective of this work (the generation of i.i.d. samples), they present two drawbacks: the correlation of the samples drawn and the fact that the samples only come exactly from the desired distribution when the chain is in its stationary distribution, which is not straightforward to determine. In this chapter we describe special classes of MCMC approaches that produce "asymptotically" independent samples. This means that the MCMC sampler tends to become an exact sampler as the number of iterations grows or as a parameter of the algorithm is increased, thus ensuring that the correlation among samples quickly vanishes to zero and the samples generated eventually become i.i.d. Finally, Chap. 8 provides the conclusions and a discussion about future research directions in the field.

Matlab code, related to the algorithms and numerical examples presented in the different chapters, is provided at https://github.com/BookIndRandSamplingMethods/code.

## References

1. K. Alligood, T. Sauer, J.A. York, *Chaos: An Introduction to Dynamical Systems* (Springer, New York, 1997)
2. C. Andrieu, N. de Freitas, A. Doucet, Sequential MCMC for Bayesian model selection, in *Proceedings of the IEEE HOS Workshop* (1999)
3. C. Andrieu, A. Doucet, R. Holenstein, Particle Markov chain Monte Carlo methods. J. R. Stat. Soc. B **72**(3), 269–342 (2010)
4. M.J. Appel, R. Labarre, D. Radulovic, On accelerated random search. SIAM J. Optim. **14**(3), 708–730 (2003)
5. M.S. Arulumpalam, S. Maskell, N. Gordon, T. Klapp, A tutorial on particle filters for online nonlinear/non-Gaussian Bayesian tracking. IEEE Trans. Signal Process. **50**(2), 174–188 (2002)
6. L. Badger, Lazzarini's lucky approximation of $\pi$. Math. Mag. **67**(2), 83–91 (1994)

7. R. Bellazzi, P. Magni, G. De Nicolao, Bayesian analysis of blood glucose time series from diabetes home monitoring. IEEE Trans. Biomed. Eng. **47**(7), 971–975 (2000)
8. C. Berzuini, W. Gilks, Resample-move filtering with cross-model jumps, in *Sequential Monte Carlo Methods in Practice*, ed. by A. Doucet, N. de Freitas, N. Gordon, Chap. 6 (Springer, New York, 2001)
9. C. Berzuini, N.G. Best, W. Gilks, C. Larizza, Dynamic conditional independence models and Markov chain Monte Carlo methods. J. Am. Stat. Assoc. **92**, 1403–1412 (1996)
10. A. Beskos, D. Crisan, A. Jasra, On the stability of sequential Monte Carlo methods in high dimensions. Ann. Appl. Probab. **24**(4), 1396–1445 (2014)
11. E. Bolviken, G. Storvik, Deterministic and stochastic particle filters in state-space models, in *Sequential Monte Carlo Methods in Practice*, ed. by A. Doucet, N. de Freitas, N. Gordon, Chap. 5 (Springer, New York, 2001), pp. 97–116
12. E. Bolviken, P.J. Acklam, N. Christophersen, J.M. Stordal, Monte Carlo filters for non-linear state estimation. Automatica **37**(2), 177–183 (2001)
13. A. Boyarsky, P. Góra, *Laws of Chaos: Invariant Measures and Dynamical Systems in One Dimension* (Birkhäuser, Boston, 1997)
14. R.P. Brent, Uniform random number generators for supercomputers, in *Proceedings of the 5th Australian Supercomputer Conference*, Melbourne (1992), pp. 95–104
15. B.S. Caffo, J.G. Booth, A.C. Davison, Empirical supremum rejection sampling. Biometrika **89**(4), 745–754 (2002)
16. J. Candy, *Bayesian Signal Processing: Classical, Modern and Particle Filtering Methods* (Wiley, Hoboken, 2009)
17. O. Cappé, A. Gullin, J.M. Marin, C.P. Robert, Population Monte Carlo. J. Comput. Graph. Stat. **13**(4), 907–929 (2004)
18. G. Casella, C.P. Robert, Rao-Blackwellisation of sampling schemes. Biometrika **83**(1), 81–94 (1996)
19. G. Casella, C.P. Robert, Post-processing accept-reject samples: recycling and rescaling. J. Comput. Graph. Stat. **7**(2), 139–157 (1998)
20. G. Chaitin, On the length of programs for computing finite binary sequences. J. ACM **13**, 547–569 (1966)
21. G. Chaitin, On the length of programs for computing finite binary sequences: statistical considerations. J. ACM **16**, 145–159 (1969)
22. R. Chen, Another look at rejection sampling through importance sampling. Stat. Probab. Lett. **72**, 277–283 (2005)
23. R. Chen, J.S. Liu, Mixture Kalman filters. J. R. Stat. Soc. B **62**, 493–508 (2000)
24. J. Dagpunar, *Principles of Random Variate Generation* (Clarendon Press, Oxford/New York, 1988)
25. M.H. DeGroot, M.J. Schervish, *Probability and Statistics*, 3rd edn. (Addison-Wesley, New York, 2002)
26. P. Del Moral, *Feynman-Kac Formulae: Genealogical and Interacting Particle Systems with Applications* (Springer, New York, 2004)
27. P. Del Moral, A. Doucet, A. Jasra, Sequential Monte Carlo samplers. J. R. Stat. Soc. Ser. B Stat. Methodol. **68**(3), 411–436 (2006)
28. L. Devroye, Random variate generation for unimodal and monotone densities. Computing **32**, 43–68 (1984)
29. P.M. Djurić, S.J. Godsill (eds.), Special issue on Monte Carlo methods for statistical signal processing. IEEE Trans. Signal Process. **50**(3), 173 (2002)
30. A. Doucet, N. de Freitas, N. Gordon (eds.), *Sequential Monte Carlo Methods in Practice* (Springer, New York, 2001)
31. S. Duane, A.D. Kennedy, B.J. Pendleton, D. Roweth, Hybrid Monte Carlo. Phys. Lett. B **195**(2), 216–222 (1987)
32. W.L. Dunn, J.K. Shultis, *Exploring Monte Carlo Methods* (Elsevier, Amsterdam, 2011)
33. R. Eckhardt, Stan Ulam, John von Neumann, and the Monte Carlo method. Los Alamos Sci. **15**, 131–137 (1987). Special Issue: Stanislaw Ulam 1909–1984

34. I. Elishakoff, Notes on philosophy of the Monte Carlo method. Int. Appl. Mech. **39**(7), 753–764 (2003)
35. V. Elvira, L. Martino, D. Luengo, M. Bugallo, Efficient multiple importance sampling estimators. IEEE Signal Process. Lett. **22**(10), 1757–1761 (2015)
36. V. Elvira, L. Martino, D. Luengo, M.F. Bugallo, Generalized multiple importance sampling (2015). arXiv:1511.03095
37. P. Fearnhead, Sequential Monte Carlo methods in Filter Theory. Ph.D. Thesis, Merton College, University of Oxford (1998)
38. Y. Fong, J. Wakefield, K. Rice, An efficient Markov chain Monte Carlo method for mixture models by neighborhood pruning. J. Comput. Graph. Stat. **21**, 197–216 (2012)
39. D. Frenkel, B. Smit, *Understanding Molecular Simulation: From Algorithms to Applications* (Academic Press, San Diego, 1996)
40. J.E. Gentle, *Random Number Generation and Monte Carlo Methods* (Springer, New York, 2004)
41. M. Gerber, N. Chopin, Sequential quasi Monte Carlo. J. R. Stat. Soc. Ser. B Stat. Methodol. **77**(3), 509–579 (2015)
42. C.J. Geyer, Markov Chain Monte Carlo maximum likelihood, in *Computing Science and Statistics: Proceedings of the 23rd Symposium on the Interface* (1991), pp. 156–163
43. W.R. Gilks, N.G.O. Robert, E.I. George, Adaptive direction sampling. Statistician **43**(1), 179–189 (1994)
44. W.R. Gilks, P. Wild, Adaptive rejection sampling for Gibbs sampling. Appl. Stat. **41**(2), 337–348 (1992)
45. W.R. Gilks, S. Richardson, D. Spiegelhalter, *Markov Chain Monte Carlo in Practice: Interdisciplinary Statistics* (Taylor & Francis, London, 1995)
46. B.V. Gnedenko, *The Theory of Probability*, 6th ed. (Gordon and Breach, Amsterdam, 1997)
47. J. Goodman, A.D. Sokal, Multigrid Monte Carlo method for lattice field theories. Phys. Rev. Lett. **56**(10), 1015–1018 (1986)
48. N. Gordon, D. Salmond, A.F.M. Smith, Novel approach to nonlinear and non-Gaussian Bayesian state estimation. IEE Proc. F Radar Signal Process. **140**, 107–113 (1993)
49. P.J. Green, Reversible jump Markov chain Monte Carlo computation and Bayesian model determination. Biometrika **82**(4), 711–732 (1995)
50. R.C. Griths, S. Tavaré, Monte Carlo inference methods in population genetics. Math. Comput. Model. **23**(8–9), 141–158 (1996)
51. A. Hall, On an experimental determination of Pi. J. Messenger Math. **2**, 113–114 (1873)
52. J.M. Hammersley, K.W. Morton, Poor man's Monte Carlo. J. R. Stat. Soc. Ser. B Methodol. **16**(1), 23–38 (1954)
53. W.K. Hastings, Monte Carlo sampling methods using Markov chains and their applications. Biometrika **57**(1), 97–109 (1970)
54. W. Hörmann, J. Leydold, G. Derflinger, *Automatic Nonuniform Random Variate Generation* (Springer, New York, 2003)
55. K. Hukushima, K. Nemoto, Exchange Monte Carlo method and application to spin glass simulations. J. Phys. Soc. Jpn. **65**, 1604–1608 (1996)
56. C.C. Hurd, A note on early Monte Carlo computations and scientific meetings. Ann. Hist. Comput. **7**(2), 141–155 (1985)
57. M. Hürzeler, H.R. Künsch, Monte Carlo approximations for general state-space models. J. Comput. Graph. Stat. **7**(2), 175–193 (1998)
58. P. Jaeckel, *Monte Carlo Methods in Finance* (Wiley, New York, 2002)
59. A. Jasra, D.A. Stephens, C.C. Holmes, Population-based reversible jump Markov chain Monte Carlo. Biometrika **94**(4), 787–807 (2007)
60. L. Jing, P. Vadakkepat, Interacting MCMC particle filter for tracking maneuvering target. Digit. Signal Process. **20**, 561–574 (2010)
61. S. Karlin, H.M. Taylor, *A First Course on Stochastic Processes* (Academic, New York, 1975)
62. S.K. Kirkpatrick, C.D. Gelatt Jr., M.P. Vecchi, Optimization by simulated annealing. Science **220**(4598), 671–680 (1983)

63. D.E. Knuth, *The Art of Computer Programming. Volume 2: Seminumerical Algorithms*, 2nd edn. (Addison-Wesley, Reading, MA, 1981)
64. J. Kohlas, *Monte Carlo Simulation in Operations Research* (Springer, Berlin, 1972)
65. A.N. Kolmogorov, On tables of random numbers. Sankhya Indian J. Stat. Ser. A **25**, 369–376 (1963)
66. A.N. Kolmogorov, Three approaches to the quantitative definition of information. Probl. Inf. Transm. **1**(1), 1–7 (1965)
67. J. Kotecha, P.M. Djurić, Gaussian sum particle filtering. IEEE Trans. Signal Process. **51**(10), 2602–2612 (2003)
68. H.R. Künsch, Recursive Monte Carlo filters: algorithms and theoretical bounds. Ann. Stat. **33**(5), 1983–2021 (2005)
69. P.K. Kythe, M.R. Schaferkotter, *Handbook of Computational Methods for Integration* (Chapman and Hall/CRC, Boca Raton, 2004)
70. P.-S. Laplace, *Théorie Analytique des Probabilités* (Mme Ve Courcier, Paris, 1812)
71. A. Lasota, M.C. Mackey, *Chaos, Fractals and Noise: Stochastic Aspects of Dynamics*, 2nd edn. (Springer, New York, NY, 1994)
72. M. Lazzarini, Un' applicazione del calcolo della probabilità alla ricerca sperimentale di un valore approssimato di $\pi$. Periodico di Matematica **4**, 140–143 (1901)
73. G.-L. Leclerc (Comte Buffon), Essai d'arithmétique morale. *Supplément à l'Histoire Naturelle*, 4 (1777)
74. D.H. Lehmer, Mathematical methods in large-scale computing units. Ann. Comput. Lab. Harv. Univ. **26**, 141–146 (1951)
75. F. Liang, C. Liu, R. Caroll, *Advanced Markov Chain Monte Carlo Methods: Learning from Past Samples*. Wiley Series in Computational Statistics (Wiley, London, 2010)
76. J.S. Liu, Metropolized independent sampling with comparisons to rejection sampling and importance sampling. Stat. Comput. **6**(2), 113–119 (1996)
77. J.S. Liu, *Monte Carlo Strategies in Scientific Computing* (Springer, New York, 2004)
78. J.S. Liu, R. Chen, Sequential Monte Carlo methods for dynamic systems. J. Am. Stat. Assoc. **93**(443), 1032–1044 (1998)
79. J.S. Liu, R. Chen, W.H. Wong, Rejection control and sequential importance sampling. J. Am. Stat. Assoc. **93**(443), 1022–1031 (1998)
80. J.S. Liu, F. Liang, W.H. Wong, The multiple-try method and local optimization in Metropolis sampling. J. Am. Stat. Assoc. **95**(449), 121–134 (2000)
81. M. Ljungberg, S.E. Strand, M.A. King, *Monte Carlo Calculations in Nuclear Medicine* (Taylor & Francis, Boca Raton, 1998)
82. M. Locatelli, Convergence of a simulated annealing algorithm for continuous global optimization. J. Glob. Optim. **18**, 219–234 (2000)
83. T.R. Malthus, *An Essay On The Principle Of Population* (Electronic Scholarly Publishing Project, London, 1998)
84. E. Marinari, G. Parisi, Simulated tempering: a new Monte Carlo scheme. Europhys. Lett. **19**(6), 451–458 (1992)
85. A. Marshall, The use of multistage sampling schemes in Monte Carlo computations, in *Symposium on Monte Carlo* (Wiley, New York, 1956), pp. 123–140
86. P. Martin-löf, Complexity of oscillations in infinite binary sequences. Z. Wahrscheinlichkeitstheorie verw. Geb. **19**, 225–230 (1971)
87. R.M. May, Biological populations with nonoverlapping generations: stable points, stable cycles and chaos. Science **186**, 645–647 (1974)
88. R.M. May, Simple mathematical models with very complicated dynamics. Nature **261**(5560), 459–467 (1976)
89. N. Metropolis, The beginning of the Monte Carlo method. Los Alamos Sci. **15**, 125–130 (1987). Special Issue: Stanislaw Ulam 1909–1984
90. N. Metropolis, S. Ulam, The Monte Carlo method. J. Am. Stat. Assoc. **44**, 335–341 (1949)
91. N. Metropolis, A. Rosenbluth, M. Rosenbluth, A. Teller, E. Teller, Equations of state calculations by fast computing machines. J. Chem. Phys. **21**, 1087–1091 (1953)

92. H. Niederreiter, *Random Number Generation and Quasi-Monte Carlo Methods* (Society for Industrial and Applied Mathematics, Philadelphia, 1992)
93. A. Papoulis, *Probability, Random Variables and Stochastic Processes*. McGraw-Hill Series in Electrical Engineering (McGraw-Hill, New York, 1984)
94. S.K. Park, K.W. Miller, Random number generators: good ones are hard to find. Commun. ACM **31**(10), 1192–1201 (1988)
95. M.M. Pieri, H. Martel, C. Grenón, Anisotropic galactic outflows and enrichment of the intergalactic Medium. I. Monte Carlo simulations. Astrophys. J. **658**(1), 36–51 (2007)
96. S.B. Pope, A Monte Carlo method for the PDF equations of turbolent reactive flow. Combust. Sci. Technol. **25**, 159–174 (1981)
97. D. Remondo, R. Srinivasan, V.F. Nicola, W.C. van Etten, H.E.P. Tattje, Adaptive importance sampling for performance evaluation and parameter optimization of communication systems. IEEE Trans. Commun. **48**(4), 557–565 (2000)
98. B. Ristic, S. Arulampalam, N. Gordon, *Beyond the Kalman Filter* (Artech House, Boston, 2004)
99. C.P. Robert, G. Casella, *Monte Carlo Statistical Methods* (Springer, New York, 2004)
100. M. Rosenbluth, A. Rosenbluth, Monte Carlo calculation of average extension of molecular chains. J. Chem. Phys. **23**, 356–359 (1955)
101. D.B. Rubin, A noniterative sampling/importance resampling alternative to the data augmentation algorithm for creating a few imputations when fractions of missing information are modest: the SIR algorithm. J. Am. Stat. Assoc. **82**, 543–546 (1987)
102. E. Segré, *From X-Rays to Quarks: Modern Physicists and Their Discoveries* (Freeman, New York, 1980)
103. J.I. Siepmann, A method for the direct calculation of chemical potentials for dense chain systems. Mol. Phys. **70**(6), 1145–1158 (1990)
104. J.I. Siepmann, D. Frenkel, Configurational bias Monte Carlo: a new sampling scheme for flexible chains. Mol. Phys. **75**(1), 59–70 (1992)
105. T. Siiskonen, R. Pollanen, Alpha-electron and alpha-photon coincidences in high-resolution alpha spectrometry. Nucl. Instrum. Methods Phys. Res. Sect. A Accel. Spectrom. Detect. Assoc. Equip. **558**(2), 437–440 (2006)
106. R.H. Swendsen, J.S. Wang, Replica Monte Carlo simulation of spin glasses. Phys. Rev. Lett. **57**(21), 2607–2609 (1986)
107. H. Tanizaki, On the nonlinear and non-normal filter using rejection sampling. IEEE Trans. Autom. Control **44**(3), 314–319 (1999)
108. H. Tanizaki, Nonlinear and non-Gaussian state space modeling using sampling techniques. Ann. Inst. Stat. Math. **53**(1), 63–81 (2001)
109. M.D. Troutt, W.K. Pang, S.H. Hou, *Vertical Density Representation and Its Applications* (World Scientific, Singapore, 2004)
110. J.P. Valleau, Density-scaling: a new Monte Carlo technique in statistical mechanics. J. Comput. Phys. **96**(1), 193–216 (1991)
111. P.F. Verhulst, Notice sur la loi que la population poursuit dans son accroissement. Correspondance Mathématique et Physique **10**, 113–121 (1838)
112. P.F. Verhulst, Recherches mathématiques sur la loi d'accroissement de la population. Nouveaux Mémoires de l'Académie Royale des Sciences et Belles-Lettres de Bruxelles **18**, 1–42 (1845)
113. J. von Neumann, Various techniques in connection with random digits, in *Monte Carlo Methods*, ed. by A.S. Householder, G.E. Forsythe, H.H. Germond. National Bureau of Standards Applied Mathematics Series (U.S. Government Printing Office, Washington, DC, 1951), pp. 36–38
114. X. Wang, Improving the rejection sampling method in quasi-Monte Carlo methods. J. Comput. Appl. Math. **114**(2), 231–246 (2000)
115. T. Warnock, Random-number generators. Los Alamos Sci. **15**, 137–141 (1987). Special Issue: Stanislaw Ulam 1909–1984

116. E.M. Wijsman, Monte Carlo Markov chain methods and model selection in genetic epidemiology. Comput. Stat. Data Anal. **32**(3–4), 349–360 (2000)
117. D. Williams, *Probability with Martingales* (Cambridge University Press, Cambridge, 1991)
118. S.R. Williams, D.J. Evans, Nonequilibrium dynamics and umbrella sampling. Phys. Rev. Lett. **105**(11), 1–26 (2010)
119. P. Zanetti, New Monte Carlo scheme for simulating Lagrangian particle diffusion with wind shear effects. Appl. Math. Model. **8**(3), 188–192 (1984)
120. P. Zhang, Nonparametric importance sampling. J. Am. Stat. Assoc. **91**(435), 1245–1253 (1996)

# Chapter 2
# Direct Methods

**Abstract** In this chapter we look into a collection of direct methods for random sampling. The term direct is used here to indicate that i.i.d. random draws with exactly the desired probability distribution are produced by applying a transformation that maps one (or many) realizations from the available random source into a realization from the target random variable. Most often, this transformation can be described as a deterministic map. However, we also include here techniques that rely on either discrete or continuous mixtures of densities and which can be interpreted as (pseudo)stochastic transformations of the random source.

Furthermore, many of the techniques proposed in the literature are found to be closely related when studied in sufficient detail. We pay here specific attention to some of these links, as they can be later exploited for the design of more efficient samplers, or simply to attain a better understanding of the field.

## 2.1 Introduction

In this chapter we describe in detail various classes of methods for direct random sampling, i.e., techniques that map one or more samples from a (pseudo)random source into one or more samples of the target probability distribution without requiring any approximations, convergence periods, or (stochastic) acceptance tests. We have classified these methodologies into three broad classes: transformation techniques, universal sampling schemes, and tailored (distribution-specific) procedures.

The chapter is organized as follows. After briefly introducing the main notational conventions to be used throughout the book in Sect. 2.2, we devote Sect. 2.3 to the study of transformation-based techniques. This term refers to schemes that take one or more random variates as input and apply a deterministic mapping to produce a collection of output variates. We study various forms of transformations, including invertible and non-invertible functions and many-to-one maps. We also include mixture densities (both continuous and discrete) in this section. Note that a discrete mixture can be viewed as a deterministic map that takes a collection of random variates (from the distributions in the mixture) and a uniform sample in the (0,1)



L. Martino et al., *Independent Random Sampling Methods*, Statistics and Computing, https://doi.org/10.1007/978-3-319-72634-2_2

interval (from where the mixture component is chosen) into a single output. Sorting transformations (i.e., order statistics) are also reviewed.

Section 2.4 contains different general strategies to build a transformation that relates the target density to another pdf from which samples can be easily drawn. In this section a collection of universal sampling techniques is presented. They rely on the ability to evaluate the target pdf and often demand additional assumptions. These methods are termed "universal" because, *theoretically*, they allow sampling from any target distribution. In practice, they can be used to generate a broad class of random variates but are subject to significant limitations too.

Section 2.5 includes sampling techniques that can be applied when the target density satisfies further (and more demanding) assumptions, such as convexity or a certain recursive structure. Therefore, they are tailored to generating samples from more specific classes of distributions.

In Sect. 2.6 we present a collection of examples that we consider relevant, not only to illustrate the techniques described earlier but also to introduce a number of specific relationships, derivations, and "tricks" that will be useful in subsequent chapters. Finally, we conclude with a brief summary (in Sect. 2.7), including some tables for quick reference to frequently used transformations.

## 2.2 Notation

### 2.2.1 Vectors, Points, and Intervals

Scalar magnitudes are denoted using regular face letters, e.g., $x$ and $X$, while vectors are displayed as boldface letters, e.g., $\mathbf{x}$ and $\mathbf{X}$. The scalar coordinates of a column vector in $n$-dimensional space are denoted with square brackets, e.g., $\mathbf{x} = [x_1, \ldots, x_n]^\top$. Often, it is more convenient to interpret $\mathbf{x}$ as a point in an $n$-dimensional space. When needed, we emphasize this representation with the alternative notation $\mathbf{x} = (x_1, \ldots, x_n)$.

We use a similar notation for the intervals in the real line. Specifically, for two boundary values $a \leq b$, we denote a closed interval as $[a, b] = \{x \in \mathbb{R} : a \leq x \leq b\}$, while

$$(a, b] = \{x \in \mathbb{R} : a < x \leq b\},$$

and

$$[a, b) = \{x \in \mathbb{R} : a \leq x < b\},$$

are used to indicate half-open intervals and $(a, b) = \{x \in \mathbb{R} : a < x < b\}$ is an open interval.

### 2.2.2 Random Variables, Distributions, and Densities

We indicate random variables (r.v.'s) with uppercase letters, e.g., $X$ and $\mathbf{X}$, while lowercase letters are used to denote the corresponding realizations, e.g., $x$ and $\mathbf{x}$. Often, when we draw a collection of samples from a r.v., we use the superscript notation $x^{(i)}$ or $\mathbf{x}^{(i)}$, where $i$ indicates the sample number.

We use lowercase letters, e.g., $q(\cdot)$, to denote the probability density function (pdf) of a random variable or vector, e.g., $q(y)$ is the pdf of $Y$. The conditional pdf of $X$ given $Y = y$ is written as $p(x|y)$. The cumulative distribution function (cdf) of a r.v. $X$ is written as $F_X(\cdot)$. The probability of an event, e.g., $X \leq x$, is indicated as $\text{Prob}\{X \leq x\}$. In particular, $F_X(a) = \text{Prob}\{X \leq a\}$. The target pdf from which we wish to draw samples is denoted as $p_o(x)$ while $p(x)$ is a function proportional to $p_o(x)$, i.e., $p(x) \propto p_o(x)$. We often denote as $c = \int_{\mathcal{D}} p(x)dx$ the normalizing constant of p(x).

The uniform distribution in an interval $[a, b]$ is indicated as $\mathcal{U}([a, b])$. The Gaussian distribution with mean $\mu$ and variance $\sigma^2$ is denoted as $\mathcal{N}(\mu, \sigma^2)$, while $\mathcal{N}(x; \mu, \sigma^2)$ represents a Gaussian density. The symbol $\sim$ means that a r.v. $X$ or a sample $x'$ have the indicated distribution, e.g., $X \sim \mathcal{U}([a, b])$ or $x' \sim \mathcal{N}(\mu, \sigma^2)$, or a given pdf, e.g., $X \sim p_o(x)$. Finally, the expression $X \stackrel{d}{=} Z$ denotes that the two r.v.'s $X$ and $Z$ are "equal in distribution," i.e., they have the same cdf.

### 2.2.3 Sets

Sets are denoted with calligraphic uppercase letters, e.g., $\mathcal{R}$. The support of the r.v. of interest $X$ is denoted as $\mathcal{D} \subseteq \mathbb{R}$, i.e., $\mathcal{D}$ is the domain of the target pdf $p_o(x)$. In some cases, without loss of generality, we may consider $\mathcal{D} = \mathbb{R}$ for convenience. When needed, we denote with $\mathcal{C}$ the support of auxiliary variables.

Finally, we write the indicator function on the set $\mathcal{S}$ as $\mathbb{I}_{\mathcal{S}}(x)$. This function takes a value equal to one if $x \in \mathcal{S}$ and zero otherwise, i.e.,

$$\mathbb{I}_{\mathcal{S}}(x) = \begin{cases} 1, & \text{if } x \in \mathcal{S}, \\ 0, & \text{if } x \notin \mathcal{S}. \end{cases} \tag{2.1}$$

## 2.3 Transformations of Random Variables

Sampling methods rely on the assumption that a random source is available that produces samples from a known distribution which, in general, differs from the target distribution. Fortunately, in some cases of interest we can find adequate transformations that convert the samples provided by the available random source into samples distributed according to the target pdf. In this section, we review several such transformations. The resulting connections between various families of random variables will be useful throughout the rest of the book.

### 2.3.1 *One-to-One Transformations*

In this section, we handle transformations that map a random variable into another one. We consider the monotonic (invertible) case and the non-monotonic (non-invertible) case separately.

#### Invertible Transformations

Consider two random vectors $\mathbf{Y} = [Y_1, Y_2, \ldots, Y_m]^\top \in \mathbb{R}^m$ and $\mathbf{Z} = [Z_1, Z_2, \ldots, Z_m]^\top \in \mathbb{R}^m$ with joint pdfs $p(y_1, y_2, \ldots y_m)$ and $q(z_1, z_2, \ldots, z_m)$, respectively, related through an invertible transformation $\boldsymbol{\psi} = [\psi_1, \ldots, \psi_m]^\top$, i.e., $\mathbf{Y} = \boldsymbol{\psi}(\mathbf{Z})$, such that

$$\begin{cases} Y_1 = \psi_1(Z_1, Z_2, \ldots, Z_m), \\ \quad \vdots \\ Y_m = \psi_m(Z_1, Z_2, \ldots, Z_m), \end{cases}$$

and the inverse transformation is given by

$$\begin{cases} Z_1 = \psi_1^{-1}(Y_1, Y_2, \ldots, Y_m), \\ \quad \vdots \\ Z_m = \psi_m^{-1}(Y_1, Y_2, \ldots, Y_m), \end{cases}$$

i.e., $\mathbf{Z} = \boldsymbol{\psi}^{-1}(\mathbf{Y})$, where $\boldsymbol{\psi}^{-1} = [\psi_1^{-1}, \ldots, \psi_m^{-1}]^\top$. The two joint pdfs, $p$ and $q$, are linked by the relationship [8, 10, 24]

$$p(y_1, \ldots y_m) = q\left(\psi_1^{-1}(y_1, \ldots, y_m), \ldots, \psi_m^{-1}(y_1, \ldots, y_m)\right) \left|\det \mathbf{J}^{-1}\right|, \quad (2.2)$$

where $\mathbf{J}^{-1}$ is the Jacobian matrix of the inverse transformation, i.e.,

$$\mathbf{J}^{-1} = \begin{bmatrix} \frac{\partial \psi_1^{-1}}{\partial y_1} & \frac{\partial \psi_1^{-1}}{\partial y_2} & \cdots & \frac{\partial \psi_1^{-1}}{\partial y_m} \\ \vdots & \vdots & \ddots & \vdots \\ \frac{\partial \psi_m^{-1}}{\partial y_1} & \frac{\partial \psi_m^{-1}}{\partial y_2} & \cdots & \frac{\partial \psi_m^{-1}}{\partial y_m} \end{bmatrix}.$$

Hence, if we are able to draw samples $\mathbf{z}' = (z_1', z_2', \ldots, z_m')$ from $q(z_1, z_2, \ldots, z_m)$, then we can also obtain samples $\mathbf{y}' = (y_1', y_2', \ldots, y_m')$ from $p(y_1, y_2, \ldots, y_m)$

simply by computing $\mathbf{y}' = \boldsymbol{\psi}(\mathbf{z}')$. Similarly, if we are able to draw samples $\mathbf{y}' = (y'_1, y'_2, \ldots, y'_m)$ from $p(y_1, y_2, \ldots, y_m)$, then we can generate $\mathbf{z}' = (z'_1, z'_2, \ldots, z'_m)$ from $q(z_1, z_2, \ldots, z_m)$ using the inverse relationship $\mathbf{z}' = \boldsymbol{\psi}^{-1}(\mathbf{y}')$.

In Sect. 2.4, we consider some transformations of this class that often turn out to be useful. Before that, we highlight the importance of this simple relationship by recalling two examples that have been broadly used in the literature.

*Example 2.1* Given a random vector $(U_1, U_2)$ uniformly distributed on $(0, 1] \times (0, 1]$, the *Box-Muller transformation* [4],

$$\begin{cases} X_1 = \psi_1(U_1, U_2) = \sqrt{-2 \log U_1} \cos(2\pi U_2), \\ X_2 = \psi_2(U_1, U_2) = \sqrt{-2 \log U_1} \sin(2\pi U_2), \end{cases} \tag{2.3}$$

yields two independent standard Gaussian r.v.'s, $X_i \sim N(0, 1)$, $i = 1, 2$. The Box-Muller transformation can be derived in different ways, see, for instance, Example 2.12 in Sect. 2.4.2.

*Example 2.2* The inversion method (see Sect. 2.4.1) connects a uniform r.v. $U$ and the variable of interest $X$ using the inverse function of the target cdf $F_X(x)$. Specifically, we have $X = F_X^{-1}(U)$ with $U \sim \mathcal{U}([0, 1])$.

### Non-invertible Transformations

In this section we assume that the relationship between two univariate random variables $Y$ and $X$ (with densities $q(y)$ and $p_o(x)$, respectively) can be expressed through a non-monotonic transformation $Y = \psi(X)$ [22]. Consider a generalization of the inverse function defined as the family of sets

$$\psi^{-1}(y) = \{x \in \mathbb{R} : \psi(x) = y\}, \quad y \in \mathbb{R}.$$

Since $\psi$ is not monotonic, for a generic value $y$ the set $\psi^{-1}(y)$ contains more than one solution, i.e., in general $\psi^{-1}(y) = \{x_1, \ldots, x_n\}$, so that the inverse function is not uniquely defined. Therefore, if we are able to draw a sample $y'$ from $q(y)$, we still have to choose adequately one solution $x'$ out of the $n$ elements in $\psi^{-1}(y')$ in order to ensure that $x'$ is distributed according to $p_o(x)$.

For simplicity, let us consider the case $n = 2$, i.e., $\psi^{-1}(y) = \{x_1, x_2\}$ for all possible values of $y$. This means that the function $\psi(x)$ can be decomposed into two monotonic (and invertible) parts. Namely, for each value of $y$ it is possible to find a value $x_c \in \mathbb{R}$ such that $\psi_1(x) \triangleq \psi(x)$ is a monotonic function in the domain $x \in (-\infty, x_c]$ and $\psi_2(x) \triangleq \psi(x)$ is another monotonic function for $x \in [x_c, +\infty)$. Using the notation

$$x_1 = \psi_1^{-1}(y), \ \ x_2 = \psi_2^{-1}(y)$$

to indicate the two solutions of the equation $y = \psi(x)$, the pdf of the r.v. $Y = \psi(X)$ can be expressed as

$$q(y) = p_o(\psi_1^{-1}(y)) \left| \frac{d\psi_1^{-1}(y)}{dy} \right| + p_o(\psi_2^{-1}(y)) \left| \frac{d\psi_2^{-1}(y)}{dy} \right|, \tag{2.4}$$

where $p_o(x)$ is the pdf of $X$. Then, given a sample $y'$ from $q(y)$, we can obtain a sample $x'$ from $p_o(x)$ selecting $x_1' = \psi_1^{-1}(y')$ with probability

$$\begin{aligned} w_1 &= \frac{p_o(\psi_1^{-1}(y')) \left| \frac{d\psi_1^{-1}(y')}{dy} \right|}{q(y')} \\ &= \frac{p_o(\psi_1^{-1}(y')) \left| \frac{d\psi_1^{-1}(y')}{dy} \right|}{p_o(\psi_1^{-1}(y')) \left| \frac{d\psi_1^{-1}(y')}{dy} \right| + p_o(\psi_2^{-1}(y')) \left| \frac{d\psi_2^{-1}(y')}{dy} \right|}, \end{aligned} \tag{2.5}$$

or choosing $x_2' = \psi_2^{-1}(y')$ otherwise with probability $w_2 = 1 - w_1$. Moreover, since

$$\begin{aligned} \left| \frac{d\psi_i^{-1}(y')}{dy} \right| &= \left| \frac{1}{\frac{d\psi(\psi_i^{-1}(y'))}{dx}} \right| \\ &= \left| \frac{1}{\frac{d\psi(x_i')}{dx}} \right| \\ &= \left| \frac{d\psi(x_i')}{dx} \right|^{-1}, \quad i = 1, 2, \end{aligned}$$

we can rewrite the weight $w_1$ in Eq. (2.5) (that represents the probability of accepting $x_1'$) as

$$\begin{aligned} w_1 &= \frac{p_o(x_1') \left| \frac{d\psi(x_1')}{dx} \right|^{-1}}{p_o(x_1') \left| \frac{d\psi(x_1')}{dx} \right|^{-1} + p_o(x_2') \left| \frac{d\psi(x_2')}{dx} \right|^{-1}} \\ &= \frac{p_o(x_1') \left| \frac{d\psi(x_2')}{dx} \right|}{p_o(x_1') \left| \frac{d\psi(x_2')}{dx} \right| + p_o(x_2') \left| \frac{d\psi(x_1')}{dx} \right|}. \end{aligned} \tag{2.6}$$

In summary, if we have a non-monotonic transformation $Y = \psi(X)$, and the equation $x = \psi^{-1}(y)$ has $n = 2$ possible solutions, we can use the algorithm in Table 2.1 to draw from $p_o(x)$. Furthermore, considering $n$ generic possible solutions, it is straightforward to show that the probability of choosing the $k$th solution

**Table 2.1** Drawing samples using a non-monotonic transformation ($n = 2$)

| |
|---|
| 1. Set $i = 1$. Let $N$ be the number of desired samples from $p_o(x)$ |
| 2. Draw a sample $y' \sim q(y)$ in Eq. (2.4) |
| 3. Set $x_1' = \psi_1^{-1}(y')$ and $x_2' = \psi_2^{-1}(y')$ |
| 4. Draw $u' \sim \mathcal{U}([0, 1])$ |
| 5. If $u' \leq \frac{p_o(x_1')\left\|\frac{d\psi(x_2')}{dx}\right\|}{p_o(x_1')\left\|\frac{d\psi(x_2')}{dx}\right\|+p_o(x_2')\left\|\frac{d\psi(x_1')}{dx}\right\|}$ then set $x^{(i)} = x_1'$. Otherwise, set $x^{(i)} = x_2'$ |
| 6. Update $i = i + 1$. If $i > N$ then stop, else go back to step 2 |

$(1 \leq k \leq n)$ is

$$w_k = \frac{p_o(x_k') \left| \frac{d\psi(x_k')}{dx} \right|^{-1}}{\sum_{i=1}^{n} p_o(x_i') \left| \frac{d\psi(x_i')}{dx} \right|^{-1}}.$$

*Example 2.3* When the non-monotonic transformation is $Y = p_o(X)$, i.e., the transformation $\psi(x) = p_o(x)$ is exactly the pdf of $X$, the density $q(y)$ of $Y$ is called the *vertical density* [19, 26, 27]. See Sect. 2.4.2 for a detailed description of the vertical density approach.

### 2.3.2 *Many-to-One Transformations*

Given $Z_1, Z_2, \ldots, Z_m$ r.v.'s, such that $Z_i \in \mathcal{C}_i$, $i = 1, \ldots, m$, and with joint pdf $q(z_1, z_2, \ldots, z_m)$, one possibility to design a random sampler based on a transformation is to find a function $\phi : \cup_{i=1}^{m} \mathcal{C}_i \subseteq \mathbb{R}^m \rightarrow \mathbb{R}$ such that

$$X = \phi(Z_1, \ldots, Z_m) \tag{2.7}$$

is distributed according to the target pdf $p_o(x)$. Hence, if we are able to draw $(z_1', \ldots, z_m')$ from $q(z_1, \ldots, z_m)$, the sample $x' = \phi(z_1', \ldots, z_m')$ is distributed according to $p_o(x)$.

In order to determine the relationship between $q(z_1, \ldots, z_m)$ and $p_o(x)$, we have to study the following system of equations [8, 10],

$$\begin{cases} X = \phi(Z_1, \ldots, Z_m), \\ Y_2 = Z_2, \\ \quad \vdots \\ Y_m = Z_m, \end{cases} \tag{2.8}$$

where the last $m-1$ random variables involved in the equations,

$$Y_2 = Z_2, \ldots, Y_m = Z_m,$$

are chosen arbitrarily. We also assume that $\phi$ can be inverted w.r.t. $z_1$, i.e., $\frac{\partial \phi}{\partial z_1} \neq 0$, and the inverse transformation is given by

$$\begin{cases} Z_1 = \phi^{-1}(X, Y_2, \ldots, Y_m), \\ Z_2 = Y_2, \\ \vdots \\ Z_m = Y_m, \end{cases} \tag{2.9}$$

where $\phi^{-1}$ represents the solution of the equation $x = \phi(z_1, z_2, \ldots, z_m)$ w.r.t. the variable $z_1$, i.e., $z_1 = \phi^{-1}(x, z_2, \ldots, z_m)$. The Jacobian matrix of the transformation in Eq. (2.9) is

$$\mathbf{J}^{-1} = \begin{bmatrix} \frac{\partial \phi^{-1}}{\partial x} & \frac{\partial \phi^{-1}}{\partial y_2} & \cdots & \frac{\partial \phi^{-1}}{\partial y_m} \\ 0 & 1 & \cdots & 0 \\ \vdots & \vdots & \ddots & \vdots \\ 0 & 0 & \cdots & 1 \end{bmatrix},$$

implying that $|\det \mathbf{J}^{-1}| = \left|\frac{\partial \phi^{-1}}{\partial x}\right|$, and Eq. (2.2) reduces to

$$p(x, y_2, \ldots y_m) = q\left(\phi^{-1}(x, y_2, \ldots, y_m), y_2, \ldots, y_m\right) \left|\frac{\partial \phi^{-1}}{\partial x}\right|, \tag{2.10}$$

where $p(x, y_2, \ldots y_m)$ is the joint density of the vector $(X, Y_2, \ldots, Y_m)$. Finally, we have to marginalize Eq. (2.10) in order to obtain the target pdf $p_o(x)$, i.e.,

$$\begin{aligned} p_o(x) &= \underbrace{\int_{\mathcal{C}_2} \cdots \int_{\mathcal{C}_{m-1}} \int_{\mathcal{C}_m}}_{m-1} p(x, y_2, \ldots y_m) dy_2 \ldots dy_m \\ &= \int_{\mathcal{C}_2} \cdots \int_{\mathcal{C}_m} q\left(\phi^{-1}(x, y_2, \ldots, y_m), y_2, \ldots, y_m\right) \left|\frac{\partial \phi^{-1}}{\partial x}\right| dy_2 \ldots dy_m. \end{aligned} \tag{2.11}$$

*Example 2.4* The first equation of the Box-Muller transformation (see Example 2.1),

$$X = \phi(U_1, U_2) = \sqrt{-2 \log U_1} \cos(2\pi U_2),$$

converts two r.v.'s $U_1$ and $U_2$ uniformly distributed on $(0, 1]$, into a standard Gaussian r.v. $X \sim \mathcal{N}(0, 1)$.

*Example 2.5* Other relevant transformations are:

- The sum of random variables. For instance, for $m = 2$ we have

$$X = Z_1 + Z_2,$$

where $Z_1, Z_2 \in \mathcal{C}$ have joint pdf $q(z_1, z_2)$ and

$$p_o(x) = \int_{\mathcal{C}} q(x - y, y)\, dy.$$

If $Z_1 \sim q_1(z_1)$ and $Z_2 \sim q_2(z_2)$ are independent, then

$$p_o(x) = \int_{\mathcal{C}} q_1(x - y) q_2(y) dy,$$

and this technique is also known as the *convolution* method.

- The product or ratio of two random variables:

$$X = Z_1 Z_2,$$

or

$$X = \frac{Z_1}{Z_2}.$$

In the first case, we have

$$p_o(x) = \int_{\mathcal{C}} \frac{1}{|y|} q\left(\frac{x}{y}, y\right) dy, \tag{2.12}$$

and in the second,

$$p_o(x) = \int_{\mathcal{C}} |y| q(xy, y)\, dy.$$

See Example 2.7 and Sects. 2.4.4–5.2 for some specific applications.

*Example 2.6* A convolution method can be used for sampling the so-called Erlang distribution, that is a special class of Gamma distribution. Consider the target pdf

$$p_o(x) \propto p(x) = x^{\alpha-1} e^{-\beta x}, \quad x \geq 0, \ \alpha, \beta > 0,$$

that is a Gamma pdf. We also indicate as $X \sim \mathcal{G}(\alpha, \beta)$ a r.v. $X$ that has a Gamma density with parameters $\alpha$ and $\beta$. When $\alpha$ is an integer, the Gamma pdf becomes an Erlang pdf. Given a value $\alpha \in \mathbb{N}^+$, it is possible to write

$$X = \frac{1}{\beta} \sum_{i=1}^{\alpha} E_i,$$

where $X \sim \mathcal{G}(\alpha, \beta)$ and $E_i$, $i = 1, .., \alpha$, are exponential r.v.'s with scale parameter 1. Moreover, since it is possible to sample from exponential pdfs applying a log-transformation to a uniform random variable, i.e., $E_i = -\ln U_i$ (see Sect. 2.4.1), then we also have

$$X = -\frac{1}{\beta} \sum_{i=1}^{\alpha} \ln U_i = -\frac{1}{\beta} \ln \left[ \prod_{i=1}^{\alpha} U_i \right].$$

### Scale Transformation

A particular case of the many-to-one transformations is the product of two r.v.'s, where a r.v. $X$ can be expressed as the product

$$X = SZ. \tag{2.13}$$

We can interpret that the r.v. $S$ controls the *scale* of $Z$, i.e., given $S = s'$, then $s'Z$ is a scaled version of $Z$, and Eq. (2.13) is often termed a *scale mixture* [10, 15]. This kind of transformation has a wide use for generating samples from several target distributions (see, for instance, Sects. 2.4.2 and 2.4.4).

*Example 2.7* Consider the following pdf

$$p_o(x) \propto \exp(-|x|^\rho), \quad x \in \mathbb{R}, \quad \rho > 0,$$

as target distribution, and let $X \sim p_o(x)$. This distribution is often called *exponential power* or *generalized Gaussian* [10]. The parameter $\rho$ allows us to control the kurtosis of the pdf. This family includes the standard Gaussian pdf for $\rho = 2$, the Laplacian pdf for $\rho = 1$, and the uniform pdf for $\rho \to +\infty$. It is possible to write

$$X = SY^{1/\rho},$$

where $S \sim \mathcal{U}([-1, 1])$ and $Y \sim \mathcal{G}(1 + 1/\rho, 1)$, namely $Y$ has a Gamma pdf

$$q(y) \propto y^{1/\rho} \exp(-y), \quad y \geq 0.$$

Then, the r.v. $Z = Y^{1/\rho}$ has pdf

$$h(z) \propto z \exp(-z^{\rho}) z^{\rho-1} = z^{\rho} \exp(-z^{\rho}), \quad z \geq 0.$$

An alternative scale mixture for $p_o(x)$ can be obtained choosing $S$ as a discrete random variable taking two possible values $\{-1, 1\}$ with equal probability, and $Y \sim \mathcal{G}(1/\rho, 1)$.

**One-to-Many Transformations** This is the case of a system of equations converting one random variable into $M$ other random variables. This kind of transformation generates distributions called *singular*, i.e., pdfs defined over a curve (variety of dimension 1) embedded into an $M$-dimensional space. See Chap. 6, specifically Sect. 6.7, for further details.

### *2.3.3 Deconvolution Method*

Let us consider two r.v.'s $X$ and $Z$, with known joint pdf $f(x, z)$, such that $X$ has a density $p_o(x)$, $x \in \mathcal{D}$. Furthermore, let us assume that there is some known relationship among them,

$$Y = \varphi(X, Z), \tag{2.14}$$

where the r.v. $Y$ is distributed according to $q(y)$, $y \in \mathcal{C}$, and assume that we are able to evaluate $q(y)$ and to draw from it. Our goal is to generate samples from $p_o(x)$. If $\varphi$ is invertible w.r.t. the variable $z$, Eq. (2.10) can be rewritten as

$$p(x, y) = f(x, \varphi^{-1}(x, y)) \left| \frac{\partial \varphi^{-1}(x, y)}{\partial y} \right|,$$

where we have substituted $z = \varphi^{-1}(x, y)$. Moreover, $q(y)$ is a marginal density of $p(x, y)$, i.e., $q(y) = \int_{\mathcal{D}} p(x, y) dx$ and, obviously, $p_o(x)$ is the other marginal pdf, i.e., $p_o(x) = \int_{\mathcal{C}} p(x, y) dy$. Therefore, since we can express the joint pdf as $p(x, y) = h(x|y) q(y)$, we can draw from $p_o(x)$ following the procedure:

1. Generate $y' \sim q(y)$.
2. Draw $x' \sim h(x|y')$, where

$$h(x|y) = \frac{p(x, y)}{q(y)} = \frac{1}{q(y)} f(x, \varphi^{-1}(x, y)) \left| \frac{\partial \varphi^{-1}(x, y)}{\partial y} \right|. \tag{2.15}$$

Clearly, we need to be able to draw from $h(x|y)$. This technique is usually known as the "deconvolution" method, because it was first developed for the specific transformation $\varphi(X, Z) = X + Z$ [10].

### 2.3.4 Discrete Mixtures

Let us assume that the target pdf can be expressed as

$$p_o(x) = \omega_1 h_1(x) + \omega_2 h_2(x) + \ldots + \omega_n h_n(x) = \sum_{i=1}^{n} \omega_i h_i(x), \tag{2.16}$$

where $\omega_i \geq 0$, for $i = 1, \ldots, n$, $\sum_{i=1}^{n} \omega_i = 1$ and the functions $h_i(x)$ are densities from which we can easily draw. The sum in Eq. (2.16) is usually referred to as a discrete mixture of densities [24]. In order to draw from $p_o(x)$ we can follow these two simple steps:

1. Draw an index $j' \in \{1, \ldots, n\}$, according to the weights $\omega_i$, $i, \ldots, n$. Namely, generate a random index $j'$ with probability mass function (pmf) $\text{Prob}\{j' = i\} = \omega_i$.
2. Draw $x' \sim h_{j'}(x)$.

Two particular cases of discrete mixtures are described in the next two subsections.

#### Partition into Intervals

It is often useful to divide the domain of the target distribution, in order to split the sampling problem into different (maybe easier) independent subproblems. In this case, given a partition of the domain $\mathcal{D} = \cup_{i=1}^{n} \mathcal{D}_i$, such that $\mathcal{D}_i \cap \mathcal{D}_j = \emptyset$ for $i \neq j$, each density of the mixture is defined as $h_i(x) = p(x)$ for all $x \in \mathcal{D}_i$, $i = 1, .., n$, and the problem reduces to being able to draw from $p_o(x)$ restricted to different subsets of its domain, using possibly a different sampling method in each subset. This strategy is also known as *composition* or *decomposition* method [15, Chap. 2], [12, Chap. 4] and it is often used jointly with the rejection sampling (RS) technique (see Chaps. 3 and 4). For instance, examples of such approach are the so-called *patchwork algorithms* [18, 25, 29] or the *adaptive* RS (ARS) techniques [13, 14].

#### Pdf Expressed as an Infinite Series

If the target density can be expressed as an infinite series of densities, i.e.,

$$p_o(x) = \sum_{i=1}^{+\infty} \omega_i h_i(x), \tag{2.17}$$

then it is also possible to draw from $p_o(x)$, as long as we are able to sample from the discrete pmf $\text{Prob}\{i\} = \omega_i$, $i = 1, \ldots, +\infty$, as well as from each of the densities $h_i(x)$, for $i = 1, \ldots, +\infty$ [12, Chap. 2], [15, Chap. 3]. Note that it is always possible to draw from the pmf $\omega_i$ using the method described in Sect. 2.4.1, inverting numerically the cumulative distribution.

When the target is expressed as in Eq. (2.17), it is also possible to apply the RS principle to draw from $p_o(x)$ (Sect. 3.8.3). Interesting applications of discrete mixtures are also given in Sects. 2.6.3 and 2.6.4, for drawing from polynomial densities.

### 2.3.5 Continuous Mixtures: Marginalization

In some cases, a joint distribution $p(x, y)$ is known such that the target pdf is one of its marginals, i.e.,

$$p_o(x) = \int_C p(x, y)dy, \tag{2.18}$$

where $C$ is the support of $y$. Moreover, since we can express the joint pdf as $p(x, y) = h(x|y)q(y)$, we can also write

$$p_o(x) = \int_C h(x|y)q(y)dy. \tag{2.19}$$

Therefore, if we are able to draw a sample $y' \sim q(y)$ and then $x' \sim h(x|y')$, the sample $x'$ has density $p_o(x)$. The variable $y$ plays the role of an auxiliary variable.

This method is also known as *continuous mixture* of densities [10, Chap. 1] since the marginal pdf $q(y)$ can be considered as a weight function and $h(x|y)$ as an uncountable collection of densities indexed by $y$. More specifically, the weight $q(y^*)$ is associated to the density $h(x|y^*)$. The methods in Sects. 2.3.2 and 2.3.3 can also be seen as applications of this idea.

*Example 2.8* Consider, for instance, $h(x|y) = y\exp(-yx)$ and $q(y) = \exp(-y)$ for $x \geq 0$ and $y \geq 0$ (i.e., two exponential pdfs). Then, we have

$$\begin{aligned} p_o(x) &= \int_{y\geq 0} h(x|y)q(y)dy, \\ &= \int_{y\geq 0} y\exp(-yx)\exp(-y)dy = \frac{1}{(1+x)^2}. \end{aligned}$$

### 2.3.6 Order Statistics

Let us consider $n$ i.i.d. r.v.'s, $X_1, X_2, \ldots .X_n$, with each r.v. $X_i$ having a pdf $q(x)$ and a cdf $F_X(x) = \int_{-\infty}^{x} q(z)dz$. The corresponding *order statistics* are denoted as $X_{s_1}, X_{s_2}, \ldots, X_{s_n}$, for $s_k \in \{1, \ldots, n\}$ and $k = 1, \ldots, n$, such that $X_{s_1} \leq X_{s_2} \leq \ldots. \leq X_{s_n}$. Note that order statistics have practical importance in many statistical applications [6, 8, 10].

Methods for simulating order statistics are usually divided into two different classes: (1) algorithms for generating independent replications of a single order statistic (e.g., the minimum, the maximum or the median of the entire population) and (2) algorithms for generating all the order statistics, namely generating $n$ ordered samples (Sect. 3.8.1 provides an example).

An important observation, especially for the first class of methods, is that each variable $X_{s_i}$ has density [10, 15]

$$g_{s_i}(x) \propto q(x) F_X(x)^{i-1} (1 - F_X(x))^{n-i}, \tag{2.20}$$

where $q(x)$ and $F_X(x)$, are respectively, the pdf and the cdf of the r.v. $X_i$, $i = 1, \ldots, n$.

*Example 2.9* For uniform (in [0, 1]) r.v.'s $U_{s_1}, U_{s_2}, \ldots, U_{s_n}$ we have

$$g_{s_i}(x) = \frac{\Gamma(n+1)}{\Gamma(i)\Gamma(n-i+1)} x^{i-1} (1-x)^{n-i}, \quad x \in [0, 1], \tag{2.21}$$

where $\Gamma(\cdot)$ is the Gamma function [8]. Namely, the uniform ordered r.v.'s have a *beta distribution* [24] with parameters $i$ and $n - i + 1$. For instance, the r.v.'s $U_{s_1}$, $U_{s_2}$ and $U_{s_3}$ are distributed as $3(1-x)^2$, $6x(1-x)$ and $3x^2$, respectively. Therefore, note that ordering uniform r.v.'s with support in [0, 1] we can draw samples from beta distributions.

We can easily derive an important special case of the problem tackled in the previous example: the distribution of $U_{s_n} = \max(U_1, \ldots, U_n)$, i.e., the maximum of $n$ uniform r.v.'s in [0, 1]. In this case, we can write

$$\begin{aligned} \text{Prob}\{\max(U_1, \ldots, U_n) \leq x\} &= \text{Prob}\{U_1 \leq x, \ldots, U_n \leq x\} \\ &= \prod_{i=1}^{n} \text{Prob}\{U_i \leq x\} = x^n, \quad x \in [0, 1]. \end{aligned} \tag{2.22}$$

Therefore, we have seen that the distribution is $x^n$, i.e., the pdf is $g_{s_n}(x) = nx^{n-1}$ for $x \in [0, 1]$. Moreover, since we can write

$$\text{Prob}\{U \leq x^n\} = \text{Prob}\{U^{1/n} \leq x\} = x^n,$$

where $U \sim \mathcal{U}([0, 1])$, then $U_{s_n} = \max(U_1, \ldots, U_n)$ is distributed as the r.v. $U^{1/n}$. By a similar argument, it is possible to show that the maximum of $n$ i.i.d. r.v.'s, $X_1, X_2, \ldots, X_n$, has cdf $[F_X(x)]^n = \prod_{i=1}^{n} F_X(x)$ where $F_X(x)$ is the cdf of each r.v. $X_i$, i.e.,

$$\text{Prob}\{\max(X_1, \ldots, X_n) \leq x\} = [F_X(x)]^n. \tag{2.23}$$

In a similar vein, one can show that

$$\text{Prob}\{\min(X_1, \ldots, X_n) \leq x\} = 1 - [1 - F_X(x)]^n \,. \tag{2.24}$$

Finally, let us remark that the order statistics are also related to the so-called *uniform spacings* $D_j$, that are obtained by taking the differences between uniform order statistics, i.e.,

$$D_j = U_{s_j} - U_{s_{j-1}}, \qquad \text{for } 2 \leq j \leq n.$$

Uniform spacings are useful to draw exactly in *simplices* and *polytopes*, as we describe in Chap. 6.

## 2.4 Universal Direct Methods

The techniques described in this section rely on the ability to evaluate, at least, the target pdf, but they often demand additional assumptions. They are termed “universal” because they can be applied (at least theoretically) to a very large class of target distributions.

### 2.4.1 *Inversion Method*

Let $X$ be a random variable with pdf $p_o(x)$ and related cdf

$$F_X(x) = \text{Prob}\{X \leq x\} = \int_{-\infty}^{x} p_o(v)dv, \tag{2.25}$$

which is always a non-decreasing function. We define its generalized inverse as

$$F_X^{-1}(y) \triangleq \inf\{x \in \mathcal{D} : F_X(x) \geq y\}. \tag{2.26}$$

The following theorem provides a very useful tool to generate samples distributed according to the target density $p_o(x)$ using a uniform random distribution $\mathcal{U}([0, 1])$ as the random source.

**Theorem 2.1 ([10, 24])** *If* $U \sim \mathcal{U}([0, 1])$, *then the r.v.* $Z = F_X^{-1}(U)$ *has density* $p_o(x)$.

*Proof* The generalized inverse function satisfies, by definition,

$$\{(u, x) \in \mathbb{R}^2 : F_X^{-1}(u) \leq x\} = \{(u, x) \in \mathbb{R}^2 : u \leq F_X(x)\}.$$

Therefore, we have

$$F_Z(z) = \text{Prob}\{Z = F_X^{-1}(U) \leq z\} = \text{Prob}\{U \leq F_X(z)\} = F_X(z). \quad (2.27)$$

Thus, $X$ and $Z$ have the same cdf, $X \overset{d}{=} Z$, and the same density $p_o(x)$. □

Theorem 2.1 implies that when we have an analytical expression for the inverse function $F_X^{-1}(\cdot)$, then we can first generate a sample $u' \sim \mathcal{U}([0, 1])$, transform it, obtaining $x' = F_X^{-1}(u')$, and the resulting sample $x'$ is distributed according to $p_o(x)$.

In general, it is straightforward to show that, given a r.v.'s $Y$ with cdf $F_Y(y)$, the r.v.'s defined through the monotonic transformation

$$Z = F_X^{-1}(F_Y(Y)) \quad (2.28)$$

is distributed according to $p_o(x)$, i.e., $X \overset{d}{=} Z$. Table 2.2 provides some examples of application of the inversion method for generating standard distributions. The last column of Table 2.2 provides a simplified form obtained noting that $1-u \sim \mathcal{U}([0, 1])$ (if $u \sim \mathcal{U}([0, 1])$) and making use of the periodicity of $F_X^{-1}$ in the last case.

The inversion method enables us to easily generate i.i.d. random numbers from a generic pdf $p_o(x)$, but we need to know the cdf $F_X$ and its inverse $F_X^{-1}$ analytically. In many practical cases both steps are intractable. In other cases, we are able to find $F_X(x)$, but it is impossible to invert it [10, Chaps. 3–7],[9, 21].

**Table 2.2** Examples of known $F_X$ and $F_X^{-1}$ for applying the inversion method

| pdf $p_o(x)$ | $F_X(x)$ | $F_X^{-1}(u)$ | Simplified form |
|---|---|---|---|
| Exponential<br>$\lambda e^{-\lambda x}$,<br>$x \geq 0$ | $1 - e^{-\lambda x}$ | $-\frac{1}{\lambda}\log(1-u)$ | $-\frac{1}{\lambda}\log(u)$ |
| Triangular<br>$\frac{2}{a}(1-\frac{x}{a})$,<br>$0 \leq x \leq a$ | $\frac{2}{a}(x - \frac{x^2}{2a})$ | $a(1-\sqrt{1-u})$ | $a(1-\sqrt{u})$ |
| Pareto<br>$\frac{\alpha b^\alpha}{x^{\alpha+1}}$,<br>$x \geq b > 0$ | $1 - \left(\frac{b}{x}\right)^\alpha$ | $\frac{b}{(1-u)^{1/\alpha}}$ | $\frac{b}{u^{1/\alpha}}$ |
| Weibull<br>$\rho\lambda^\rho x^{\rho-1}e^{-(\lambda x)^\rho}$,<br>$x \geq 0$ | $1 - e^{-(\lambda x)^\rho}$ | $\left[-\frac{1}{\lambda^\rho}\log(1-u)\right]^{1/\rho}$ | $\left[-\frac{1}{\lambda^\rho}\log(u)\right]^{1/\rho}$ |
| Cauchy<br>$\frac{1}{\pi}\frac{a}{x^2+a^2}$,<br>$x \in \mathbb{R}$ | $\frac{1}{2} + \frac{1}{\pi}\arctan\left(\frac{x}{a}\right)$ | $a\tan\left[\pi(u-\frac{1}{2})\right]$ | $a\tan(\pi u)$ |

### Numerical Inversion of $F_X(x) = u$

In many cases, the cdf $F_X(x)$ is known but it is impossible to invert it analytically, i.e., $F_X^{-1}$ is unknown. A straightforward way to tackle this difficulty is to solve numerically the equation $F_X(x) = u$. For instance, we could use the well-known *bisection* method, the *secant* method, or the *Newton-Raphson* method [1, 5]. A complete analysis of the different performances can be found in [10, Chap. 2].

For continuous r.v.'s, the inversion method is exact only when an explicit expression for $F_X^{-1}$ is available, since any numerical technique used will only provide an approximate solution. On the other hand, for discrete r.v.'s the time required to obtain an exact inversion is finite and the method can be exact. Alternative strategies, when $F_X$ is given but $F_X^{-1}$ is unknown, can be found in Sects. 3.6.2 and 3.7.

*Example 2.10* The cdf of a univariate standard Gaussian density cannot be computed analytically and, as a consequence, the inverse cdf is unknown. However, both functions (the cdf and its inverse) can be expressed as an infinite series [1, 23]. Therefore, they can be approximately calculated (with a known approximation degree) truncating the infinite sum to a finite number of terms. Hence, the numerical (i.e., approximate) inversion method can be applied to generate samples.

### Truncated Random Variables

Let $X \in \mathcal{D}$ be a r.v. with pdf $p_o(x)$ and cdf $F_X(x)$. Moreover, consider a r.v.

$$X_T \sim q(x) = \left[\frac{1}{F_X(b) - F_X(a)}\right] p_o(x),$$

taking values in a restricted interval $[a, b] \subset \mathcal{D}$. The easiest procedure for generating samples from $q(x)$ is to use the inversion method in the following way:

1. Compute $a' = F_X(a)$ and $b' = F_X(b)$.
2. Draw a sample $u'$ uniformly in $[a', b'] \subset [0, 1]$.
3. Calculate $x' = F_X^{-1}(u')$.

### Order Statistics

A sequence of order statistics $X_{s_1}, \ldots, X_{s_N}$ (see Sect. 2.3.6) of $N$ i.i.d. r.v.'s $X_1, \ldots, X_N$ with the same cdf $F_X(x)$ can be obtained as

$$X_{s_i} = F_X^{-1}(U_{s_i}),$$

where each $U_{s_i}$ is an order statistic of a sequence of uniform r.v.'s in $[0, 1]$. It is interesting to observe that this procedure can be more efficient than generating the $X_i$, $i = 1, \ldots, N$, (when possible) and then sorting them [10, 15].

### Maximum of $N$ i.i.d. Random Variates

Given $N$ i.i.d. random variables $X_1, \ldots, X_N$ with cdf $F_X(x)$, four alternative procedures to generate the maximum $Z = \max(X_1, \ldots, X_N)$ are the following [10, Chaps. 2 and 14]:

1. Generate all the $X_i$, $i = 1, \ldots, N$, using some appropriate algorithm and then take the maximum $Z = \max(X_1, \ldots, X_N)$.
2. Generate $N$ uniform i.i.d. random variables $U_i \sim \mathcal{U}([0, 1])$, take the maximum value $Y = \max(U_1, \ldots, U_N)$ and then set $Z = F_X^{-1}(Y)$.
3. Generate a uniform r.v. $U \sim \mathcal{U}([0, 1])$ and compute $Z = (F_X^N)^{-1}(U)$ where $(F_X^N)^{-1}$ represents the inverse function of
$$F_X^N(z) = [F_X(z)]^N.$$
We recall that $Z$ has cdf $F_X^N(z)$ (see Sect. 2.3.6).
4. Generate a uniform r.v. $U \sim \mathcal{U}([0, 1])$ and calculate $Z = F_X^{-1}(U^{1/N})$ (we recall that $U^{1/N}$ is distributed as the maximum of $N$ i.i.d. uniform random variables, see Sect. 2.3.6).

The last three procedures involve the inversion method.

### Dependent Random Variates

The inversion method can also be used to draw correlated random variates [12, 15]. Consider, for instance, two random variates $X$ and $Y$ with cdfs $F_X$ and $F_Y$, respectively. They can be generated using inversion by drawing two uniform r.v.'s, $U \sim \mathcal{U}([0, 1])$ and $V \sim \mathcal{U}([0, 1])$, and then taking $X = F_X^{-1}(U)$ and $Y = F_X^{-1}(V)$. Note that, if $U$ and $V$ are dependent, then $X$ and $Y$ will be dependent as well. Maximal positive or negative dependencies are obtained when $U = V$ or $U = 1-V$. For further details, see Chap. 6.

### Inversion for Multivariate Targets

The inversion method cannot be directly applied to multivariate distributions because the many-to-one cdf $F_X$ does not have a proper inverse in this case (see Chap. 6, specifically Sect. 6.3.2). However, since the target density can be factorized as

$$p_o(\mathbf{x}) \propto p(x_1)p(x_2|x_1)\cdots p(x_M|x_1, \ldots, x_{M-1}), \tag{2.29}$$

where $\mathbf{x} = [x_1, \ldots, x_M]^\top$, it is straightforward to apply the chain rule to draw a sample $\mathbf{x}'$ from $p_o(\mathbf{x})$:

1. draw $u_1' \sim \mathcal{U}([0,1])$ and set $x_1' \sim F_{X_1}^{-1}(u_1')$,
2. draw $u_2' \sim \mathcal{U}([0,1])$ and set $x_2' \sim F_{X_2|X_1=x_1'}^{-1}(u_2')$,

$\vdots$

M) draw $u_M' \sim \mathcal{U}([0,1])$, set $x_M' \sim F_{X_2|X_1=x_1',\ldots,X_M=x_M'}^{-1}(u_M')$ and $\mathbf{x}' = [x_1', \ldots, x_M']^\top$.

Here $F_{X_j|X_1,\ldots,X_{j-1}}(x_j)$ represents the cdf corresponding to the conditional density $p(x_j|x_1, \ldots, x_{j-1}), j = 1, \ldots, M$.

### 2.4.2 *Vertical Density Representation (VDR)*

In the previous section, we have studied the transformation of the r.v. of interest $X$ by way of its cdf $F_X$ and found that $U = F_X(X) \sim \mathcal{U}([0,1])$.

In this section, we consider the r.v. $Z = p(X)$, where $X$ is distributed as $p_o(x) = \frac{1}{c}p(x)$, $c = \int_{\mathcal{D}} p(x)dx$, i.e., the transformation is the pdf of $X$ [26–28]. First of all, for the sake of simplicity, we consider a monotonically decreasing pdf $p_o(x)$. In this case, the pdf of $Z$ is

$$q(z) = p_o(p^{-1}(z)) \left| \frac{dp^{-1}(z)}{dz} \right| \propto -z\frac{dp^{-1}(z)}{dz}, \quad \text{for } 0 < z \leq M, \tag{2.30}$$

where $p^{-1}(z)$ is the inverse function of $p(x)$ and $M = \max_{x \in \mathcal{D}} p(x) < +\infty$. The density $q(z)$ is called *vertical density* since it can be interpreted as the density of the ordinate of $p(x)$ [26–28]. More generally, for a non-monotonic multivariate pdf, $p_o(\mathbf{x}) \propto p(\mathbf{x})$, $\mathbf{x} \in \mathbb{R}^m$, the vertical density $q(z)$ is defined as

$$q(z) \propto -z\frac{dA(z)}{dz}, \tag{2.31}$$

where $A(z) = |\mathcal{O}(z)|$ is the Lebesgue measure of the set

$$\mathcal{O}(z) = \{\mathbf{x} \in \mathcal{D} : p(\mathbf{x}) \geq z\}. \tag{2.32}$$

Figure 2.1 depicts an example of the set $\mathcal{O}(z)$ for a generic univariate and unimodal pdf.

From a practical point of view, the VDR is really interesting when $q(z)$ is easy to draw from. In this case a simple random sampler can be designed as outlined in Table 2.3.

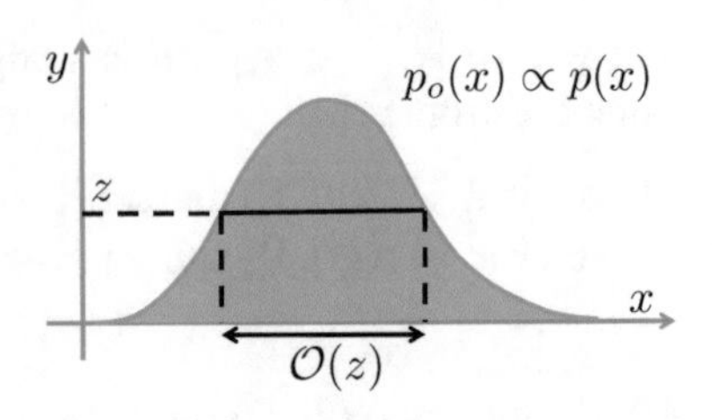


**Fig. 2.1** A generic pdf $p_o(x)$ and the set $\mathcal{O}(z)$ for a given value $z$

**Table 2.3** Vertical density representation (VDR)

| |
|---|
| 1. Set $i = 1$. Let $N$ be the number of desired samples from $p_o(\mathbf{x})$ |
| 2. Generate a sample $z'$ from the vertical density $q(z)$ in Eqs. (2.30) and (2.31) |
| 3. Generate a sample $\mathbf{x}^{(i)}$ *uniformly* on the set $\mathcal{C}(z') = \{\mathbf{x} \in \mathcal{D} : p(\mathbf{x}) = z'\}$ i.e., the boundary of $\mathcal{O}(z')$, and set $i = i + 1$ |
| 4. If $i > N$ then stop, else go back to step 2 |

The main advantages of the VDR approach are [28]:

- The vertical density $q(z)$ is always a univariate pdf, even if $p_o(\mathbf{x})$ is multivariate, i.e., $\mathbf{x} \in \mathbb{R}^m$.
- If $p_o(\mathbf{x})$ is bounded, the domain of the vertical pdf $q(z)$ is always bounded, i.e., $z \in (0, M]$, where $M = \max_{x \in \mathcal{D}} p(x)$.
- If $p_o(\mathbf{x})$ is unimodal, then $q(z)$ is a monotonic function, and there are specific methods to draw from monotonic pdfs (see, for instance, Sect. 3.6.1).

Due to these appealing features, the VDR strategy is really interesting to draw from multivariate target pdfs. However, its applicability is related to the ability of drawing efficiently from $q(z)$ and generating uniform samples in $\{\mathbf{x} \in \mathcal{D} : p(\mathbf{x}) = z'\}$. In the sequel, we provide some examples.

*Example 2.11* Consider an exponential target pdf $p_o(x) = \lambda e^{-\lambda x}$, $x \geq 0$, $\lambda > 0$. In this case, we have

$$q(z) = -z \frac{d(-\frac{1}{\lambda}\log(z))}{dz} = \frac{1}{\lambda}, \quad 0 < z \leq \lambda,$$

i.e., the vertical density of an exponential pdf is a uniform distribution in $(0, \lambda]$. Then, to draw from $p_o(x)$ we should

1. Draw $z' \sim \mathcal{U}((0, \lambda])$.
2. Take $x' = -\frac{1}{\lambda}\log(z')$.

Indeed, the set $\mathcal{C}(z') = \{x \in \mathcal{D} : p(x) = z'\}$ contains only one point $x' = -\frac{1}{\lambda}\log(z')$.

*Example 2.12* Consider now a multivariate standard Gaussian distribution

$$p_o(\mathbf{x}) = \frac{1}{(2\pi)^{m/2}} \exp\left(-\frac{1}{2}\mathbf{x}^\top \mathbf{x}\right), \quad \mathbf{x} \in \mathbb{R}^m. \tag{2.33}$$

The set $\mathcal{O}(z)$ is a hypersphere with volume[1]

$$A(z) = \frac{\pi^{m/2}}{\frac{m}{2}\Gamma\left(\frac{m}{2}\right)}\left(-2\log[(2\pi)^{m/2}z]\right)^{m/2}. \tag{2.34}$$

Hence, the vertical density is

$$q(z) = \frac{2\pi^{m/2}}{\Gamma\left(\frac{m}{2}\right)}\left(-2\log[(2\pi)^{m/2}z]\right)^{m/2-1}, \quad 0 < z \leq \frac{1}{(2\pi)^{m/2}}. \tag{2.35}$$

Note that $q(z)$ is a monotonic function for all $m \in \mathbb{N}$. When $m$ is even, there is a direct method to draw from it by multiplying i.i.d. uniform r.v.'s, as described in Sect. 2.6.1.

For $m = 1$, the vertical density is $q(z) = \frac{2}{\sqrt{-2\log(\sqrt{2\pi}z)}}$ for $0 < z \leq \frac{1}{\sqrt{2\pi}}$. In this case, the method would be to (a) draw a sample $z' \sim q(z)$ and (b) set $x' = +\sqrt{-2\log(\sqrt{2\pi}z')}$ or $x' = -\sqrt{-2\log(\sqrt{2\pi}z')}$ with probability $1/2$ in both cases. The most interesting case occurs for $m = 2$, since

$$q(z) = 2\pi, \quad 0 < z \leq \frac{1}{2\pi},$$

becomes a uniform distribution. Therefore, for $m = 2$, i.e. $\mathbf{x} = [x_1, x_2]^\top$, we can sample from $p_o(x)$ applying the following procedure:

1. Draw $z' \sim \mathcal{U}((0, \frac{1}{2\pi}])$.
2. Choose a point uniformly in a circumference with radius $r(z') = \sqrt{-2\log(2\pi z')}$, since

$$\mathcal{C}(z) = \{[x_1, x_2]^\top \in \mathbb{R}^2 : \quad x_1^2 + x_2^2 = [r(z')]^2\}.$$

   This can be easily done by drawing $\theta' \sim \mathcal{U}([0, 2\pi])$ and setting

$$x_1' = r(z')\sin(\theta'), \quad x_2' = r(z')\cos(\theta'),$$

   so that $\mathbf{x}' = [x_1', x_2']^\top \sim p_o(\mathbf{x})$.

Note that this case coincides exactly with the Box-Muller transformation (see Example 2.1).

*Example 2.13* Consider the class of multivariate exponential power distributions

$$p_o(\mathbf{x}) = c_m \exp\left(-(\mathbf{x}^\top\mathbf{x})^{m/2}\right), \quad \mathbf{x} \in \mathbb{R}^m, m \geq 1. \tag{2.36}$$

[1] Note that $\frac{\pi^{m/2}}{\frac{m}{2}\Gamma(\frac{m}{2})}$ is the volume of the unit sphere.

It is possible to show [20, 28] that the vertical density corresponding to $p_o(\mathbf{x})$ is a uniform distribution in $(0, c_m]$, i.e.,

$$q(z) = \frac{1}{c_m}, \quad 0 < z \leq c_m.$$

Hence sampling from $p_o(x)$ can be accomplished by drawing first $z' \sim \mathcal{U}((0, \frac{1}{c_m}])$ and then sampling uniformly on the set

$$\mathcal{C}(z) = \left\{ \mathbf{x} \in \mathbb{R}^m : \quad ||\mathbf{x}||^2 = -\left[\log\left(\frac{z'}{c_m}\right)\right]^{2/m} \right\}.$$

*Example 2.14* The relationship between the vertical density and the corresponding pdf can be used in both directions. For instance, consider as a target pdf the following density

$$q(z) \propto [-2\log(z)]^{m/2-1}, \quad 0 < z \leq 1,$$

shown in Eq. (2.35). To draw samples from it, we can (a) generate a sample $\mathbf{x}' = [x_1, \ldots, x_m] \in \mathbb{R}^m$ where $x_i \sim \mathcal{N}(0, 1)$ for $i = 1, \ldots, m$, and (b) compute $z' = p(\mathbf{x}') = \exp\left(-\frac{1}{2}\mathbf{x}'^{\top}\mathbf{x}'\right)$.

**An Alternative Interpretation of the VDR Approach**

The VDR method can be seen as the continuous mixture (see Sect. 2.3.5)

$$p_o(\mathbf{x}) = \int_0^M h(\mathbf{x}|z)q(z)dz, \tag{2.37}$$

where $q(z)$ is the vertical density corresponding to $p_o(\mathbf{x})$, $M = \max_{x \in \mathcal{D}} p_o(\mathbf{x})$, and the conditional pdf $h(\mathbf{x}|z)$ is a uniform distribution on $\mathcal{C}(z)$.

*Example 2.15* In Example 2.12 with $m = 1$, the vertical density is $q(z) = \frac{2}{\sqrt{-2\log(\sqrt{2\pi}z)}}$ and the conditional distribution $h(x|z)$ is formed by two delta functions, i.e.,

$$h(x|z) = \frac{1}{2}\delta(x - g(z)) + \frac{1}{2}\delta(x + g(z)),$$

with $g(z) = \sqrt{-2\log(\sqrt{2\pi}z)}$.

It is interesting to observe that the VDR version shown previously is only one possible solution to obtain a continuous mixture like Eq. (2.37) where $h(\mathbf{x}|z)$ is uniform and $q(z)$ is a univariate pdf with bounded domain. Indeed, considering

$$A(z) = |\mathcal{O}(z)|, \tag{2.38}$$

where $|\mathcal{O}(z)|$ is the Lebesgue measure of $\mathcal{O}(z) = \{\mathbf{x} \in \mathcal{D} : p(\mathbf{x}) \geq z\}$, another possible VDR-type method is given by the following steps:

1. Draw $z'$ according to $A(z) = |\mathcal{O}(z)|$,
2. Generate $\mathbf{x}'$ uniformly in $\mathcal{O}(z)$ (not just in the boundary $\mathcal{C}(z)$).

This method is called *VDR-type 2* in [11] and coincides exactly with the *inverse-of-density* method described in Sect. 2.4.4, where $A(z)$ is denoted as $p^{-1}(z)$. The inverse-of-density approach is also based on a basic result of simulation theory recalled below.

### 2.4.3 *The Fundamental Theorem of Simulation*

Many broadly applied Monte Carlo sampling techniques (inverse-of-density method, rejection sampling, slice sampling, etc.) rely on a simple result that we enunciate below.

**Theorem 2.2 ([24, Chap. 2])** *Drawing samples from a unidimensional r.v. X with density $p_o(x) \propto p(x)$ is equivalent to sampling uniformly on the bidimensional region defined by*

$$\mathcal{A}_0 = \{(x, y) \in \mathbb{R}^2 : 0 \leq y \leq p(x)\}. \tag{2.39}$$

*Namely, if $(x', y')$ is uniformly distributed on $\mathcal{A}_0$, then $x'$ is a sample from $p_o(x)$.*

*Proof* Let us consider a random pair $(X, Y)$ uniformly distributed on the region $\mathcal{A}_0$ in Eq. (2.39), and let $q(x, y)$ be its joint pdf, i.e.,

$$q(x, y) = \frac{1}{|\mathcal{A}_0|}\mathbb{I}_{\mathcal{A}_0}(x, y), \tag{2.40}$$

where $\mathbb{I}_{\mathcal{A}_0}(x, y)$ is the indicator function on $\mathcal{A}_0$ and $|\mathcal{A}_0|$ is the Lebesgue measure of the set $\mathcal{A}_0$. Clearly, we can also write $q(x, y) = q(y|x)q(x)$. The theorem is proved if the marginal density $q(x)$ is exactly $p_o(x)$.

Since $(X, Y)$ is uniformly distributed on the set $\mathcal{A}_0$ defined in Eq. (2.39), we have $q(y|x) = 1/p(x)$ for $0 \leq y \leq p(x)$, or, in a more compact form,

$$q(y|x) = \frac{1}{p(x)}\mathbb{I}_{\mathcal{A}_0}(x, y). \tag{2.41}$$

Therefore, we can express the joint pdf as

$$q(x, y) = q(y|x)q(x) = \frac{q(x)}{p(x)}\mathbb{I}_{\mathcal{A}_0}(x, y). \tag{2.42}$$

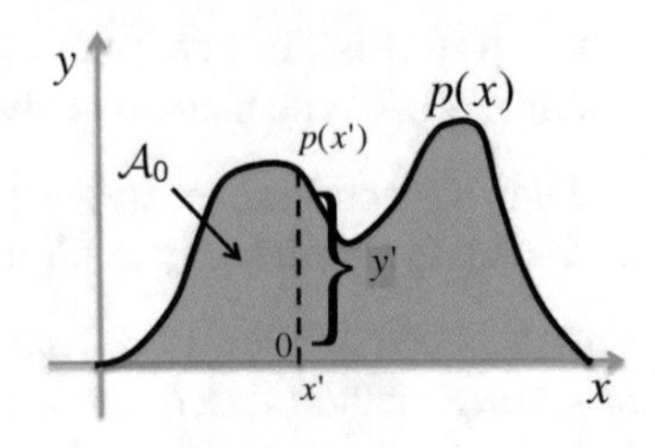


**Fig. 2.2** The area $\mathcal{A}_0$ below the target function $p(x) \propto p_o(x)$

Taking Eqs. (2.40) and (2.42) together and solving for $q(x)$ yields

$$q(x) = \frac{1}{|\mathcal{A}_0|} p(x) = p_o(x). \tag{2.43}$$

□

Theorem 2.2 states that, if we are able to draw a pair $(x', y')$ uniformly distributed on the region $\mathcal{A}_0$, the coordinate $x'$ is marginally distributed according to $p_o(x)$. Many Monte Carlo techniques simulate jointly the random variables $(X, Y)$ and then consider only the first sample $x'$. The variable $Y$ plays the role of an *auxiliary* variable.

Figure 2.2 depicts a target function $p_o(x) \propto p(x)$ and the shaded area corresponding to the set $\mathcal{A}_0$.

This result can be extended to more general *multivariate uniform distributions with non-uniform marginals*, where at least one of these marginals is the target pdf. The same approach is also valid if the random variable $\mathbf{X}$ is a vector. This is also an example of construction of a continuous mixture described in Sect. 2.3.5.

The inverse-of-density method, described below, and the rejection sampling principle (Chap. 3) are clear examples of how this simple idea can be used to design Monte Carlo sampling algorithms.

### 2.4.4 *Inverse-of-Density Method*

In this section, we present the inverse-of-density method (IoD) [10, Chap. 4], [16], also known as *Khintchine's method* [17]. First we consider monotonic target densities, but it can be easily extended to more generic pdfs.

#### IoD for Monotonic Univariate Target pdfs

Given a monotonic target pdf $p_o(x)$ and defining $y = p_o(x)$, we indicate as $p_o^{-1}(y)$ the corresponding inverse function of the target density. Note that $p_o^{-1}(y)$ is also a normalized density, since it describes the same region $\mathcal{A}_0$ defined by $p_o(x)$, as shown

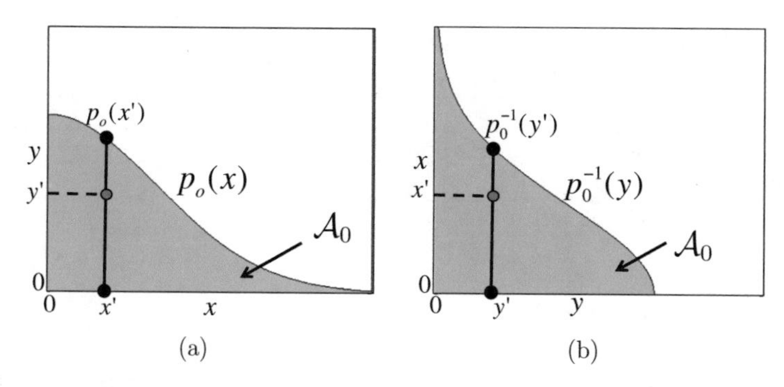


**Fig. 2.3** Two alternative ways to draw a random point $(x', y')$ uniformly in the area $\mathcal{A}_0$. **(a)** We can first draw $x' \sim p_o(x)$ and then $y' \sim \mathcal{U}([0, p_o(x')])$. **(b)** Alternatively, we can first draw $y' \sim p_o^{-1}(y)$ and then $x' \sim \mathcal{U}([0, p_o^{-1}(y')])$

in Fig. 2.3. Therefore, we can write

$$\mathcal{A}_0 = \{(x, y) \in \mathbb{R}^2 : 0 \leq y \leq p_o(x)\} = \{(y, x) \in \mathbb{R}^2 : 0 \leq x \leq p_o^{-1}(y)\}.$$

Thus in order to generate samples $(x', y')$ uniformly in $\mathcal{A}_0$ we can proceed in two alternative ways:

1. Draw $x'$ from $p_o(x)$ and then $y'$ uniformly in the interval $[0, p_o(x')]$, i.e, $y' \sim \mathcal{U}([0, p_o(x')])$ (see Fig. 2.3a),
2. Draw $y'$ from $p_o^{-1}(y)$ and then $x'$ uniformly in the interval $[0, p_o^{-1}(y')]$, i.e., $x' \sim \mathcal{U}([0, p_o^{-1}(y')])$ (see Fig. 2.3b).

Both procedures generate points $(x', y')$ uniformly distributed on the region $\mathcal{A}_0$. Moreover, from the fundamental theorem of simulation, the first coordinate $x'$ is distributed according to the target pdf $p_o(x)$, while the second coordinate $y'$ is distributed according to the inverse pdf $p_o^{-1}(y)$. Hence, if we are able to draw samples $y'$ from $p_o^{-1}(y)$, we can use the second procedure to generate samples $x'$ from $p_o(x)$.

Note that generating a sample $x'$ uniformly in the interval $[0, a]$, i.e., $x' \sim \mathcal{U}([0, a])$, is equivalent to drawing a sample $z'$ uniformly in $[0, 1]$ and then multiplying it by $a$, i.e. $x' = z'a$. Similarly, given a known value $y'$, drawing a sample $x'$ uniformly in the interval $[0, p_o^{-1}(y')]$, i.e. $x' \sim \mathcal{U}([0, p_o^{-1}(y')])$, is equivalent to generating a sample $z'$ uniformly in $[0, 1]$ and then taking $x' = z'p_o^{-1}(y')$. The algorithm described in Table 2.4 uses exactly the latter procedure. Obviously, to use this technique we need the ability to draw from the inverse pdf $p_o^{-1}(y)$.

The IoD method can be summarized by the following relationship

$$X = Up_o^{-1}(Y), \tag{2.44}$$

where $X$ has density $p_o(x)$, $U \sim \mathcal{U}([0, 1])$ and $Y$ is distributed according to $p_o^{-1}(y)$.

**Table 2.4** IoD algorithm for monotonic target densities (version 1)

| |
|---|
| 1. Set $i = 1$. Let $N$ be the number of desired samples from $p_o(x)$ |
| 2. Draw a sample $y' \sim p_o^{-1}(y)$ |
| 3. Draw $u'$ uniformly in $[0, 1]$, i.e., $z' \sim \mathcal{U}([0, 1])$ |
| 4. Set $x^{(i)} = u' p_o^{-1}(y')$ and $i = i + 1$ |
| 5. If $i > N$ then stop, else go back to step 2 |

### IoD for Generic Target pdfs

More generally, for a non-monotonic multivariate pdf $p_o(\mathbf{x}) \propto p(\mathbf{x})$ with $\mathbf{x} \in \mathcal{D} \subseteq \mathbb{R}^m$, the inverse pdf $p_o^{-1}(y)$ can be defined as

$$p_o^{-1}(y) = A(y) = |\mathcal{O}(y)|, \tag{2.45}$$

where $A(y) = |\mathcal{O}(y)|$ is the Lebesgue measure of the set

$$\mathcal{O}(y) = \{\mathbf{x} \in \mathcal{D} : p(\mathbf{x}) \geq y\}.$$

In this general case, if we can draw samples from $p_o^{-1}(y)$, we can use the following algorithm to generate samples from $p_o(\mathbf{x})$:

1. Draw $y'$ from $p_o^{-1}(y)$,
2. Draw $\mathbf{x}'$ uniformly on $\mathcal{O}(y')$.

This technique coincides exactly with the VDR type 2 method [11, 16] that we briefly described in the final part of Sect. 2.4.2.

*Example 2.16* Consider a standard Gaussian distribution,

$$p_o(x) = \frac{1}{\sqrt{2\pi}} \exp\left(-\frac{x^2}{2}\right).$$

The corresponding inverse pdf is

$$p_o^{-1}(y) = 2\sqrt{-2\log(2\pi y)}, \quad 0 < y \leq \frac{1}{2\pi}.$$

Therefore, if $y'$ is drawn from $p_o^{-1}(y)$, then

$$x' \sim \mathcal{U}([-\sqrt{-2\log(2\pi y')}, \sqrt{-2\log(2\pi y')}]),$$

is marginally distributed according to $p_o(x)$.

**Table 2.5** Khintchine's algorithm for a monotonic target pdf (IoD version 2)

| |
|---|
| 1. Set $i = 1$. Let $N$ be the number of desired samples from $p_o(x)$ |
| 2. Draw a sample $w'$ from $q(w)$ in Eq. (2.46) |
| 3. Draw $u' \sim \mathcal{U}([0, 1])$ |
| 4. Set $x^{(i)} = u'w'$ and $i = i + 1$ |
| 5. If $i > N$ then stop, else go back to step 2 |

**Khintchine's Method for Monotonic Target pdfs**

In the literature, it is possible to find the IoD method in an alternative form. For the sake of simplicity, we consider a monotonic target density $p_o(x)$ but the results can be extended to more general pdfs [16]. Consider the transformed variable $W = p_o^{-1}(Y)$, where $Y$ has a pdf $p_o^{-1}(y)$. As a consequence, the density of $W$ is given by

$$q(w) = p_o^{-1}(p_o(w))\left|\frac{dp_o}{dw}\right| = w\left|\frac{dp_o}{dw}\right|. \tag{2.46}$$

The function $q(w)$ in Eq. (2.46) is the vertical density associated to the inverse pdf $p_o^{-1}(y)$ (see Sect. 2.4.2). This is yet another link between the IoD and the VDR approaches [16]. Using the r.v. $W$, we can express Eq. (2.44) as

$$X = UW, \tag{2.47}$$

where $U \sim \mathcal{U}([0, 1])$ and $W$ is distributed according to $q(w)$ in Eq. (2.46). Table 2.5 outlines this alternative form of the IoD method.

## 2.5 Tailored Techniques

In this section, we describe some algorithms designed to generate samples from specific classes of distributions. Therefore, they are not universal tools that one can always rely upon (as some of the methods described earlier are). However, when applicable, they can be fast and efficient, and may be preferred over more general methods.

### 2.5.1 *Recursive Methods*

In this section, we provide an example of recursive techniques taken from [10, Chap. 4]. Consider three random variables $Y \sim h(x)$, $Z \sim q(x)$ and $X \sim p_o(x)$, such that their densities are related by way of a mixture,

$$h(x) = \alpha q(x) + (1 - \alpha)p_o(x), \tag{2.48}$$

where $0 < \alpha < 1$. We assume that $q(x)$ is easy to draw from and assume that there is a known transformation

$$X = \phi(Y, W), \tag{2.49}$$

where $W$ is a r.v. with pdf $g(w)$ easy to draw from and recall $Y \sim h(x)$. Our goal is to generate samples according to $p_o(x)$. We can observe that:

- In order to use the relationship in Eq. (2.49), we need realizations of $Y$ and $W$, i.e., we need to draw from $h(x)$ and, clearly, also from $g(w)$.
- To draw from $h(x)$, we need to generate a sample from $q(x)$ with probability $\alpha$ or a sample from $p_o(x)$ with probability $1 - \alpha$.

Therefore, we obtain that, with probability $\alpha$, $X \stackrel{d}{=} \phi(Z, W)$ and, with probability $1 - \alpha$,

$$X \stackrel{d}{=} \phi(X', W) \stackrel{d}{=} \phi(\phi(Y', W'), W), \tag{2.50}$$

where $(Y', W')$ is a random vector with the same distribution as the pair $(Y, W)$. These considerations lead to the algorithm shown in Table 2.6.

Note that we never use, or draw from, the pdf $h(x)$, and we do not evaluate $p_o(x)$ either. The expected number of iterations in the loop from step 4 and 6 is $1/\alpha$, since the number of uniform r.v.'s $u'$ to be sampled is geometrically distributed with parameter $\alpha$.

*Example 2.17* Consider a Gamma target density

$$p_o(x) = \frac{1}{\Gamma(\alpha)} x^{\alpha-1} \exp(-x), \quad x > 0, \tag{2.51}$$

with $0 < \alpha < 1$. Given this $p_o(x)$ in Eq. (2.51), we can write

$$\underbrace{x\frac{dp_o}{dx}}_{h(x)} = \alpha \underbrace{\frac{1}{\Gamma(\alpha+1)} x^{\alpha} \exp(-x)}_{q(x)} + (1-\alpha) p_o(x),$$

**Table 2.6** Recursive method

| |
|---|
| 1. Set $i = 1$. Let $N$ be the number of desired samples from $p_o(x)$, given the relationships in Eqs. (2.48) and (2.49) |
| 2. Draw $w' \sim g(w)$ and $z' \sim q(x)$ |
| 3. Set $x' = \phi(z', w')$ |
| 4. Draw $u' \sim \mathcal{U}([0, 1])$ |
| 5. If $u' \leq \alpha$, then set $x^{(i)} = x'$ and $i = i + 1$ |
| 6. If $u' > \alpha$, then set $x' = \phi(x', w')$ and repeat from step 4 |
| 7. If $i > N$ then stop, else go back to step 2 |

where $q(x)$ is another Gamma pdf with parameter $\alpha + 1$. There are efficient generators for the Gamma pdf with parameter greater than 1 [6, 7]. Moreover, using the Khintchine's method [see Eq. (2.47)], we know that

$$X = \phi(Y, U) = UY,$$

where $X$ has pdf $p_o(x)$, whereas $U \sim \mathcal{U}([0, 1])$ and $Y$ has density $h(x) = x\frac{dp_o}{dx}$. Therefore the recursive algorithm in Table 2.6 can be applied and, in this case, it can be elegantly summarized as

$$X = Z\prod_{i=1}^{L} U_i, \tag{2.52}$$

where $Z \sim q(x)$ is a Gamma r.v. with parameter $\alpha + 1$, $U_i \sim \mathcal{U}([0, 1])$, $i = 1, \ldots L$, and $L$ has a geometric pmf with parameter $\alpha$.

### 2.5.2 Convex Densities

Consider a decreasing target $p_o(x)$ with support $x \in [0, +\infty)$ and satisfying

$$\frac{d^2 p_o(x)}{dx^2} \geq 0,$$

i.e., $p_o(x)$ is a convex function. Different methods tailored to this class of pdfs are available [10]. The algorithm described here is based on the scale transformation

$$X = VY, \tag{2.53}$$

where $V$ has a triangular pdf, $q(v) = 2(1 - v)$, $v \in [0, 1]$, and $Y$ is distributed as

$$g(y) = \frac{y^2}{2}\frac{d^2 p_o(y)}{dy^2}, \quad y \geq 0, \tag{2.54}$$

with $\frac{d^2 p_o(y)}{dy^2} \geq 0$.[2] In other words, using Eq. (2.12), the target $p_o(x)$ has the following integral representation

$$p_o(x) = \int_x^{+\infty} \frac{1}{y}2\left(1 - \frac{x}{y}\right) g(y)dy, \tag{2.55}$$

[2]Note that $g(y)$ is always a proper normalized pdf. This can be easily proved integrating by parts twice.

**Table 2.7** Generator for convex densities

| |
|---|
| 1. Set $i = 1$. Let $N$ be the number of desired samples from a convex $p_o(x) \propto p(x)$ |
| 2. Generate $u_1, u_2 \sim \mathcal{U}([0, 1])$ and set $v' = \min(u_1, u_2)$ |
| 3. Draw $y'$ from $g(y) = \frac{y^2}{2}\frac{d^2 p}{dy^2}$ |
| 4. Set $x^{(i)} = v'y'$ and $i = i + 1$ |
| 5. If $i > N$ then stop, else go back to step 2 |

where $\frac{1}{y}$ is the determinant of the Jacobian matrix of the transformation (2.53). To prove that Eq. (2.55) holds, we replace Eq. (2.54) in (2.55) obtaining

$$\begin{aligned}\int_x^{+\infty} \frac{2}{y}\left(1 - \frac{x}{y}\right)\frac{y^2}{2}\frac{d^2 p_o}{dy^2}dy &= \int_x^{+\infty}\left(y\frac{d^2 p_o}{dy^2} - x\frac{d^2 p_o}{dy^2}\right)dy \\ &= \int_x^{+\infty} y\frac{d^2 p_o}{dy^2}dy - x\int_x^{+\infty}\frac{d^2 p_o}{dy^2}dy \\ &= \int_x^{+\infty} y\frac{d^2 p_o}{dy^2}dy - x\frac{dp_o}{dy}\end{aligned}$$

and integrating by parts

$$\int_x^{+\infty} y\frac{d^2 p_o}{dy^2}dy - x\frac{dp_o}{dy} = x\frac{dp_o}{dy} - \int_x^{+\infty}\frac{dp_o}{dy}dy - x\frac{dp_o}{dy} = p_o(x).$$

The algorithm based on the relationship given by Eq. (2.53) relies on the ability to draw from $g(y)$. It is summarized in Table 2.7. Note that, given $U_1, U_2 \sim \mathcal{U}([0, 1])$, the r.v. $V = \min(U_1, U_2)$ is distributed as a triangular distribution $q(v) = 2(1 - v)$, $v \in [0, 1]$.

## 2.6 Examples

We now present a collection of examples that serve both to illustrate the methods in Sects. 2.3–2.5 and to introduce a few derivations and "tricks" that will be useful in subsequent chapters.

### 2.6.1 *Multiplication of Independent Uniform Random Variates*

The multiplication and the ratio of i.i.d. uniform r.v.'s play a relevant role in many random sampling schemes. First, let us consider two independent r.v.'s $U_i \sim$

$\mathcal{U}([0,1])$, for $i = 1, 2$, and the product r.v.

$$X_2 = U_1 U_2. \tag{2.56}$$

As we have seen in Sect. 2.3.2, to find the distribution of $X_2$ we can consider the transformation

$$\begin{cases} X_2 = U_1 U_2, \\ Y = U_2, \end{cases}$$

and the corresponding inverse transformation

$$\begin{cases} U_1 = X_2/Y, \\ U_2 = Y, \end{cases}$$

Since the determinant of the Jacobian matrix is $1/Y$, following the procedure in Sect. 2.3.2, the joint pdf of the vector $(X_2, Y)$ is

$$p(x_2, y) = q\left(\frac{x_2}{y}, y\right)\left|\frac{1}{y}\right|, \quad 0 \leq x_2 \leq 1, x \leq y \leq 1,$$

where $q(u_1, u_2) = 1$ for $(u_1, u_2) \in [0,1] \times [0,1]$, is the joint pdf of the vector $(U_1, U_2)$. Through marginalization we obtain

$$\begin{aligned} p_o(x_2) &= \int_{x_2}^{1} q\left(\frac{x_2}{y}, y\right)\left|\frac{1}{y}\right| dy \\ &= \int_{x_2}^{1} \frac{1}{y} dy = -\log(x_2), \end{aligned}$$

for all $x_2 \in [0,1]$, i.e., $X_2 = U_1 U_2$ is distributed as $p_o(x_2) = -\log(x_2)$ for $0 \leq x_2 \leq 1$. Now, let us consider

$$X_3 = U_1 U_2 U_3, \tag{2.57}$$

which can be rewritten as

$$X_3 = X_2 U_3,$$

where $X_2$ has a pdf $p_o(x_2) = -\log(x_2)$ independent from $U_3$. As suggested in Sect. 2.3.2, to obtain the pdf of $X_3$ we can consider again the system of equations

$$\begin{cases} X_3 = X_2 U_3, \\ Y = U_3, \end{cases}$$

so that

$$\begin{cases} X_2 = X_3/Y, \\ U_3 = Y, \end{cases}$$

The determinant of the Jacobian of this inverse transformation is again $1/Y$. Therefore,

$$p(x_3, y) = q\left(\frac{x_3}{y}, y\right)\left|\frac{1}{y}\right|,$$

where $q(x_2, u_3) = -\log(x_2)$ for $(x_2, u_3) \in [0, 1] \times [0, 1]$ is the joint pdf of $(X_2, U_3)$. Then through marginalization

$$p_o(x_3) = \int_{x_3}^{1} -\log\left(\frac{x_3}{y}\right)\frac{1}{y}dy = \frac{[\log(x_3)]^2}{2}.$$

Proceeding in the same way it is possible to obtain the pdf of the r.v. obtained multiplying $n$ i.i.d. uniform r.v.'s,

$$X_n = U_1 U_2 \cdots U_n = \prod_{i=1}^{n} U_i,$$

by an induction argument, leading to the expression

$$p(x_n) = \frac{(-1)^{n-1}}{(n-1)!}[\log(x_n)]^{n-1}, \quad \text{for } 0 \leq x_n \leq 1, n \in \mathbb{N}. \tag{2.58}$$

Densities of the type of Eq. (2.58) are needed, e.g., for generating multivariate Gaussian distributions, as we have shown in Example 2.12.

### 2.6.2 *Sum of Independent Uniform Random Variates*

Another relevant case is the sum of $n$ i.i.d. uniform r.v.'s, i.e.,

$$X_n = \sum_{i=1}^{n} U_i,$$

where $U_i \sim \mathcal{U}([0,1])$. It is possible to prove that the cdf of $X_n$ is [10]

$$
\begin{aligned}
F_{X_n}(x) &= \frac{1}{n!}\sum_{i=0}^{n}(-1)^i\binom{n}{i}(x-i)_+^n \\
&= \frac{1}{n!}\left(x_+^n - \binom{n}{1}(x-1)_+^n + \binom{n}{2}(x-2)_+^n + \ldots + (x-n)_+^n\right),
\end{aligned}
$$

where $(\cdot)_+$ denotes the positive part of the argument, i.e., $(g(x))_+ = \max\{g(x), 0\}$. For instance, with $n = 2$ we have

$$
F_X(x) = \frac{1}{2}\left(x_+^2 - 2(x-1)_+^2 + (x-2)_+^2\right),
$$

so that the pdf of $X_2$ is

$$
p_o(x_2) = x_+ - 2(x-1)_+ + (x-2)_+ = \begin{cases} 0 & \text{for } x \leq 0, \\ x & \text{for } 0 < x \leq 1, \\ 2-x & \text{for } 1 < x \leq 2, \\ 0 & \text{for } x > 2. \end{cases}
$$

Namely, the shape of $p_o(x_2)$ is an isosceles triangle. More generally, $p_o(x_n)$ is formed by polynomial pieces of degree $n-1$ where each piece is defined in the interval $\mathcal{I}_i = (i, i+1]$ for $i = 0, 1, 2, \ldots, n-1$.

### 2.6.3 Polynomial Densities with Non-negative Coefficients

Consider a target distribution of the form

$$
p_o(x) \propto p(x) = \sum_{i=0}^{n} c_i x^i, \quad 0 \leq x \leq 1,
$$

where $c_i \geq 0$. Since the $c_i$'s are all non-negative, we can interpret $p_o(x)$ as a discrete mixture of pdfs $x^i$, $i = 0, \ldots, n$. Hence, we can apply the following procedure to sample from $p_o(x)$:

1. Draw an index $j' \in \{0, \ldots, n\}$ according to the pmf $w_j = \frac{c_j}{\sum_{i=0}^{n} c_i}$, $j = 0, \ldots, n$.
2. Generate a sample from $h_{j'}(x) = x^{j'}$, for instance, by way of the inversion method: draw $u \sim \mathcal{U}([0,1])$ and set $x' = u^{1/(j'+1)}$. Note that this approach is always feasible because the support of $x$ is bounded, i.e., $x \in [0, 1]$.

### 2.6.4 *Polynomial Densities with One or More Negative Constants*

Consider, again, a target pdf of the form

$$p_o(x) \propto p(x) = \sum_{i=0}^{n} c_i x^i, \quad 0 \le x \le 1, \tag{2.59}$$

where the $c_i$'s are all real constants. Let us define the sets

$$\mathcal{A} = \{k \in \{0, \ldots, n\} : c_k \ge 0\}, \quad \mathcal{B} = \{j \in \{0, \ldots, n\} : c_j < 0\},$$

i.e., $\mathcal{A}$ is the collection of indices $k$ for which $c_k \ge 0$ whereas $\mathcal{B}$ is the collection of indices $j$ for which $c_j < 0$. If $\mathcal{B}$ is not empty, the target in Eq. (2.59) cannot be interpreted as a discrete mixture. Then, drawing from $p_o(x)$ in Eq. (2.59) is not as straightforward as in the previous section. A possible solution, given by [2, 3, 6, 10], can be applied whenever

$$a_0 = c_0 + \sum_{j \in \mathcal{B}} c_j \ge 0.$$

In this case, we rewrite the target as

$$p_o(x) \propto p(x) = a_0 + \sum_{k \in \mathcal{A}} c_k x^k - \sum_{j \in \mathcal{B}} c_j (1 - x^j), \quad 0 \le x \le 1, \tag{2.60}$$

where $a_0 \ge 0$ by assumption. If we denote $a_k = c_k \ge 0$ for $k \in \mathcal{A}$ and $a_j = -c_j > 0$ for $j \in \mathcal{B}$, then it is apparent that the pdf in Eq. (2.60) can be interpreted as a discrete mixture, with coefficients (weights) $w_\ell = \frac{a_\ell}{\sum_{i=0}^{n} a_i}$, $\ell = 0, \ldots, n$.

Table 2.8 describes the resulting algorithm, where we have used the fact that the r.v. $Z = U_1^{1/\beta} U_2$ with $\beta > 1$ and $U_1, U_2 \sim \mathcal{U}([0, 1])$ is distributed as

$$q(z) = \frac{\beta}{\beta - 1}(1 - x^{\beta - 1}), \quad 0 \le x \le 1.$$

## 2.7 Summary

In this chapter we have described several basic random sampling methods which are often used as building blocks for more sophisticated schemes. We have grouped these techniques into three categories:

- Transformation methods.
- Universal methods.
- Tailored methods.

**Table 2.8** Sampling from polynomial densities with some negative coefficients

| |
|---|
| 1. Set $i = 1$. Let $N$ be the number of desired samples from a polynomial density $p_o(x)$ expressed as in Eq. (2.60) |
| 2. Draw a discrete r.v. $L = \ell' \in \{0, \ldots, n\}$ according to the pmf $w_\ell = \frac{a_\ell}{\sum_{m=0}^{n} a_m}$, $\ell = 0, \ldots, n$ |
| 3. If $\ell' \in \mathcal{A}$, then draw $u \sim \mathcal{U}([0, 1])$, set $x^{(i)} = u^{1/(\ell'+1)}$ and $i = i + 1$ |
| 4. If $\ell' \in \mathcal{B}$, then draw $u_1, u_2 \sim \mathcal{U}([0, 1])$, set $x^{(i)} = u_1^{1/(\ell'+1)} u_2$ and $i = i + 1$ |
| 5. If $i > N$ then stop, else go back to step 2 |

For the first group we have established fundamental results relating sets of r.v.'s by way of deterministic transformations, invertible or not. We have also covered the use of order statistics and mixtures.

The latter are "raw" techniques. This means that, while sometimes it is possible to use them directly in application problems, they appear more often as tools to devise more practical schemes. For instance, they are employed to derive and justify the so-called universal methods of Sect. 2.3, including the inversion method, the vertical density representation (VDR), and the inverse-of-density (IoD) method. These algorithms are termed universal because they are theoretically applicable to virtually any target pdf. In practice, however, they are subject to significant constraints, e.g., the need to invert the cdf analytically for the inversion method, or to invert the pdf (and then draw from the resulting distribution) for the VDR and IoD methods. We have made a special effort in highlighting the close connections among the various methods.

In the third group we have included some methods that can only be applied when the target density satisfies stronger assumptions, e.g., convexity or a certain recursive structure.

Finally, we have looked into a number of examples that show how some of the previous methods can be put to work in practice, but also allow us to introduce a few specific derivations and relationships that will be useful in subsequent chapters. This includes handling frequently-appearing operations of r.v.'s, as sums and products, or drawing from pdfs with polynomial form.

To conclude, we summarize some of the main relationships obtained in this chapter in tabular form, for quick reference when applied in subsequent chapters. In particular, Table 2.9 displays the main transformations of r.v.'s (inversion, VDR, IoD and scale transformation for convex pdfs), while Table 2.10 collects several transformations that work specifically with uniform random variables. The latter are important because often we need to devise suitable samplers based on the availability of a uniform source.

**Table 2.9** Summary of the main techniques described in this chapter

| Method | Transformation | Auxiliary variables |
|---|---|---|
| Inversion | $X = F^{-1}(U)$ | $U \sim \mathcal{U}([0,1])$ |
| Vertical density representation (VDR) | $X = p_o^{-1}(Z)$ | $Z \sim -z\frac{dp_o^{-1}}{dz}$ |
| Inverse-of-density (IoD) (VDR type 2) | $X = Up_o^{-1}(Y)$ | $Y \sim p_o^{-1}(y)$<br>$U \sim \mathcal{U}([0,1])$ |
| Khintchine's method (IoD, version 2) | $X = UZ$<br>$Z = p_o^{-1}(Y)$ | $Z \sim -z\frac{dp_o}{dz}$<br>$U \sim \mathcal{U}([0,1])$<br>$Y \sim p_o^{-1}(y)$ |
| For convex pdfs | $X = VY$ | $Y \sim \frac{y^2}{2}\frac{d^2p_o}{dy^2}$<br>$V \sim \min(U_1, U_2)$<br>$U_i \sim \mathcal{U}([0,1]), i = 1, 2$ |

**Table 2.10** Main transformations of uniform random variables in [0, 1]

| Transformation | Density |
|---|---|
| $X = \max(U_1, \ldots, U_n)$ | $q(x) = nx^{n-1}$,<br>$0 \le x \le 1$ |
| $X = U^{1/n}$ | $q(x) = nx^{n-1}$,<br>$0 \le x \le 1$ |
| $X = \min(U_1, \ldots, U_n)$ | $q(x) = n(1-x)^{n-1}$,<br>$0 \le x \le 1$ |
| $X = \prod_{i=1}^n U_i$ | $q(x) = \frac{(-1)^{n-1}}{(n-1)!}[\log(x_n)]^{n-1}$,<br>$0 \le x \le 1$ |
| $X = \sum_{i=1}^n U_i$ | $q(x) = \frac{1}{(n-1)!}\sum_{i=0}^n(-1)^i\binom{n}{i}(x-i)_+^{n-1}$,<br>$0 \le x \le n$ |

## References


1. F.S. Acton, *Numerical Methods That Work* (The Mathematical Association of America, Washington, DC, 1990)
2. J.H. Ahrens, U. Dieter, Computer methods for sampling from the exponential and normal distributions. Commun. ACM **15**, 873–882 (1972)
3. J.H. Ahrens, U. Dieter, Computer methods for sampling from gamma, beta, Poisson and binomial distributions. Computing **12**, 223–246 (1974)
4. G.E.P. Box, M.E. Muller, A note on the generation of random normal deviates. Ann. Math. Stat. **29**, 610–611 (1958)
5. R.L. Burden, J.D. Faires, *Numerical Analysis* (Brooks Cole, Belmont, 2000)
6. J. Dagpunar, *Principles of Random Variate Generation* (Clarendon Press, Oxford/New York, 1988)
7. J. Dagpunar, *Simulation and Monte Carlo: With Applications in Finance and MCMC* (Wiley, Chichester, 2007)
8. M.H. DeGroot, M.J. Schervish, *Probability and Statistics*, 3rd edn. (Addison-Wesley, New York, 2002)

9. L. Devroye, Random variate generation for unimodal and monotone densities. Computing **32**, 43–68 (1984)
10. L. Devroye, *Non-uniform Random Variate Generation* (Springer, New York, 1986)
11. K.T. Fang, Z.H. Yang, S. Kotz, Generation of multivariate distributions by vertical density representation. Statistics **35**(3), 281–293 (2001)
12. J.E. Gentle, *Random Number Generation and Monte Carlo Methods* (Springer, New York, 2004)
13. W.R. Gilks, Derivative-free adaptive rejection sampling for Gibbs sampling. Bayesian Stat. **4**, 641–649 (1992)
14. W.R. Gilks, P. Wild, Adaptive rejection sampling for Gibbs sampling. Appl. Stat. **41**(2), 337–348 (1992)
15. W. Hörmann, J. Leydold, G. Derflinger, *Automatic Nonuniform Random Variate Generation* (Springer, New York, 2003)
16. M.C. Jones, On Khintchine's theorem and its place in random variate generation. Am. Stat. **56**(4), 304–307 (2002)
17. A.Y. Khintchine, On unimodal distributions. Izvestiya NauchnoIssledovatel'skogo Instituta Matematiki i Mekhaniki **2**, 1–7 (1938)
18. A.J. Kinderman, J.G. Ramage, Computer generation of normal random variables. J. Am. Stat. Assoc. **71**(356), 893–898 (1976)
19. S. Kotz, M.D. Troutt, On vertical density representation and ordering of distributions. Statistics **28**, 241–247 (1996)s
20. T.J. Kozubowski, On the vertical density of the multivariate exponential power distribution. Statistics **36**, 219–221 (2002)
21. G. Marsaglia, The exact-approximation method for generating random variables in a computer. Am. Stat. Assoc. **79**(385), 218–221 (1984)
22. J.R. Michael, W.R. Schucany, R.W. Haas, Generating random variates using transformations with multiple roots. Am. Stat. **30**(2), 88–90 (1976)
23. J.G. Proakis, *Digital Communications*, 4th edn. (McGraw-Hill, Singapore, 2000)
24. C.P. Robert, G. Casella, *Monte Carlo Statistical Methods* (Springer, New York, 2004)
25. E. Stadlober, H. Zechner, The patchwork rejection technique for sampling from unimodal distributions. ACM Trans. Model. Comput. Simul. **9**(1), 59–80 (1999)
26. M.D. Troutt, A theorem on the density ordinate and an alternative interpretation of the Box-Muller method. Statistics **22**, 463–466 (1991)
27. M.D. Troutt, Vertical density representation and a further remark on the Box-Muller method. Statistics **24**, 81–83 (1993)
28. M.D. Troutt, W.K. Pang, S.H. Hou, *Vertical Density Representation and Its Applications* (World Scientific, Singapore, 2004)
29. A.W. Van Kemp, Patchwork rejection algorithms. J. Comput. Appl. Math. **31**(1), 127–131 (1990)

# Chapter 3
# Accept–Reject Methods

**Abstract** The accept/reject method, also known as *rejection sampling* (RS), was suggested by John von Neumann in 1951. It is a classical Monte Carlo technique for *universal sampling* that can be used to generate samples virtually from any target density $p_o(x)$ by drawing from a simpler proposal density $\pi(x)$. The sample is either accepted or rejected by an adequate test of the ratio of the two pdfs, and it can be proved that accepted samples are actually distributed according to the target distribution. Specifically, the RS algorithm can be viewed as choosing a subsequence of i.i.d. realizations from the proposal density $\pi(x)$ in such a way that the elements of the subsequence have density $p_o(x)$.

In this chapter, we present the basic theory of RS as well as different variants found in the literature. Computational cost issues and the range of applications are analyzed in depth. Several combinations with other Monte Carlo techniques are also described.

## 3.1 Introduction

This chapter is devoted to introduce the *rejection sampling* (RS) technique, suggested first by John von Neumann in 1951 [43]. The RS method is probably the most general technique to produce independent samples from a given distribution. This universality (theoretically, it can be applied to draw from any kind of pdf) makes it very appealing.

The RS algorithm requires the ability to evaluate the density of interest $p_o(x)$ up to a multiplicative constant (which is most often the case in practical applications). However, an important limitation of RS methods is the need to analytically establish a bound for the ratio of the target and proposal densities, since there is a lack of general procedures for the computation of tight bounds.

The rest of the chapter is organized as follows. We first describe the basic theory and the RS algorithm in Sect. 3.2. Different variants and generalizations are also presented in this section, and computational cost issues are analyzed in Sect. 3.3.

Section 3.4 introduces a little-known improvement of the standard RS algorithm called *band rejection* [11, 35]. First, we describe a slightly more general version and



L. Martino et al., *Independent Random Sampling Methods*, Statistics and Computing, https://doi.org/10.1007/978-3-319-72634-2_3

then present the band rejection method introduced in [11, 35] as a special case. The main limitation of this technique is that it can be applied only when the target has a bounded domain. However, when it can be applied, better performance than using the standard RS approach can be obtained, in general.

Section 3.5 presents another variant of the standard RS method, the *acceptance-complement* method [20], that needs additional assumptions on the target pdf $p_o(x)$. Specifically, if the target can be expressed as a discrete mixture of two simpler pdfs, in certain cases the acceptance-complement technique can be applied, turning out to be more efficient than a standard RS scheme.

In Sect. 3.6, a particular class of rejection samplers (known as *strip* samplers) is described, in which samples are drawn from a finite mixture of piecewise uniform densities with disjoint support. Namely, the proposal pdf consists of disjoint rectangular pieces. Proposals of this type are simple and intuitive to design yet flexible enough to attain good acceptance rates. Unfortunately, they can only be used when the target has a bounded domain. However, in Sect. 3.6.2 a combination of the inversion and rejection methods is described for the case in which the cdf $F_X(x)$ is computable, but not invertible. This technique, called *inversion-rejection* [12, 13], extends the strip samplers to handle target densities which are not bounded or have an infinite support.

Section 3.7 describes how to combine efficiently a suitable transformation of a random variable and the RS algorithm. The resulting technique is called in different ways: *transformed rejection* method in [18, 44], *almost exact inversion* method in [13, Chap. 3], and also *exact-approximation method* in [28]. Finally, some relevant applications of the RS technique are presented in Sect. 3.8 and the use of samples generated by an RS scheme for building an efficient Monte Carlo estimator is discussed in Sect. 3.9.

## 3.2 Rejection Sampling

Consider a function $p(x) \propto p_o(x)$ and a proposal density $\pi(x)$ which is easy to simulate. Additionally, choose a constant $L$ such that $L\pi(x)$ is an envelope function for $p(x)$, i.e.,

$$L\pi(x) \geq p(x),$$

for all $x \in \mathcal{D}$. The constant $L$ is an upper bound for the ratio $p(x)/\pi(x)$, i.e.,

$$L \geq \frac{p(x)}{\pi(x)} \quad \forall x \in \mathcal{D}.$$

In the standard rejection sampling algorithm [30, 43], we first draw a sample from the proposal pdf, $x' \sim \pi(x)$, and then accept it with probability

$$p_A(x') = \frac{p(x')}{L\pi(x')} \leq 1.$$

Otherwise, the proposed sample $x'$ is discarded. In Fig. 3.1, we can see a graphical representation of the rejection sampling technique. The RS procedure can be outlined as:

1. Draw $x'$ from $\pi(x)$.
2. Generate $v'$ uniformly in the interval $[0, L\pi(x')]$, i.e., $v' \sim \mathcal{U}([0, L\pi(x')])$.
3. If the point $(x', v')$ belongs to the region

$$\mathcal{A}_0 = \{(x, y) \in \mathbb{R}^2 : 0 \leq y \leq p(x)\}$$

below the target function $p(x)$, the sample $x'$ is accepted.
4. Otherwise, when the point $(x', v')$ falls into the region between the functions $L\pi(x)$ and $p(x)$, the sample $x'$ is rejected.

We can also summarize this procedure in an equivalent way: first draw a sample $x'$ from $\pi(x)$ and $u' \sim \mathcal{U}([0, 1])$. If $u'L\pi(x') \leq p(x')$, we accept $x'$. Table 3.1 describes how we can generate $N$ samples from the target pdf $p_o(x) \propto p(x)$ according to the standard rejection sampling algorithm. The RS technique is based on the following theorem.

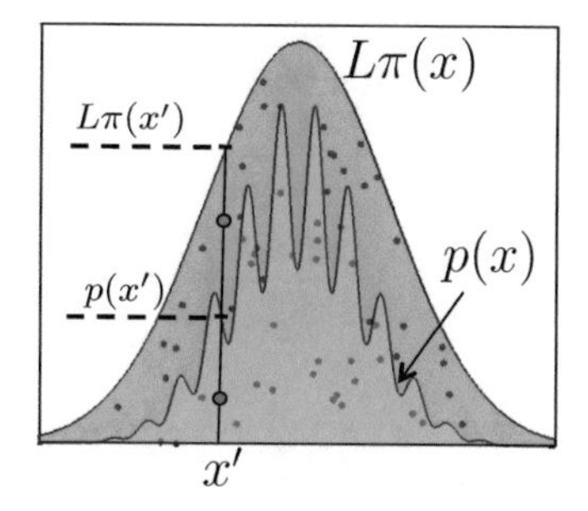


**Fig. 3.1** Graphical description of the RS procedure. The green area corresponds to the region $\mathcal{A}_0$ defined by $p(x)$. The red region indicates the area between the functions $L\pi(x)$ and $p(x)$. First, a sample $x'$ is generated from the proposal pdf $\pi(x)$ and another coordinate $v' \sim \mathcal{U}([0, L\pi(x')])$. If the point $(x', v')$ belongs to the region $\mathcal{A}_0$ (green area) the sample $x'$ is accepted. Otherwise, it is discarded

**Table 3.1** Rejection sampling algorithm

| |
|---|
| 1. Set $i = 1$. Let $N$ be the number of desired samples from $p_o(x)$ |
| 2. Draw a sample $x' \sim \pi(x)$ and $u' \sim \mathcal{U}([0, 1])$ |
| 3. If $u' \leq \frac{p(x')}{L\pi(x')}$, then set $x^{(i)} = x'$ and $i = i + 1$ |
| 4. Else, if $u' > \frac{p(x')}{L\pi(x')}$, then discard $x'$ and go back to step 2 |
| 5. If $i > N$, then stop. Else go back to step 2 |

**Theorem 3.1 ([13, 37])** *Let the r.v.'s $X_1$ and $X_2$ have pdfs $\pi(x)$ and $p_o(x) \propto p(x)$, respectively, and let $U$ have a uniform distribution $\mathcal{U}([0,1])$. If there exists a bound $L \geq p(x)/\pi(x)$ $\forall x \in \mathcal{D}$, then*

$$Prob\left\{ X_1 \leq y \middle| U \leq \frac{p(X_1)}{L\pi(X_1)} \right\} = Prob\{X_2 \leq y\}. \tag{3.1}$$

*Proof* Assuming, without lack of generality, that $\mathcal{D} = \mathbb{R}$ and recalling that $U \sim \mathcal{U}([0,1])$ and $X_1$ is distributed according to $\pi(x)$, we can write

$$\text{Prob}\left\{ X_1 \leq y \middle| U \leq \frac{p(X_1)}{L\pi(X_1)} \right\} = \frac{\text{Prob}\left\{ X_1 \leq y, U \leq \frac{p(X_1)}{L\pi(X_1)} \right\}}{\text{Prob}\left\{ U \leq \frac{p(X_1)}{L\pi(X_1)} \right\}}$$

$$= \frac{\int_{-\infty}^{y} \int_{0}^{\frac{p(x)}{L\pi(x)}} \pi(x) du dx}{\int_{-\infty}^{+\infty} \int_{0}^{\frac{p(x)}{L\pi(x)}} \pi(x) du dx}.$$

Then, integrating first w.r.t. $u$ and after some trivial calculations, we arrive at the expression

$$\text{Prob}\left\{ X_1 \leq y \middle| U \leq \frac{p(X_1)}{L\pi(X_1)} \right\} = \frac{\int_{-\infty}^{y} p(x) dx}{\int_{-\infty}^{+\infty} p(x) dx}.$$

Furthermore, since $p_o(x) \propto p(x)$, i.e., $p_o(x) = \frac{1}{c} p(x)$ with $c = \int_{-\infty}^{+\infty} p(x) dx$, we can rewrite the expression above as

$$\text{Prob}\left\{ X_1 \leq y \middle| U \leq \frac{p(X_1)}{L\pi(X_1)} \right\} = \frac{\int_{-\infty}^{y} c p_o(x) dx}{c} = \int_{-\infty}^{y} p_o(x) dx.$$

Finally, since the r.v. $X_2$ has density $p_o(x)$, we can write

$$\text{Prob}\left\{ X_1 \leq y \middle| U \leq \frac{p(X_1)}{L\pi(X_1)} \right\} = \text{Prob}\{X_2 \leq y\} = \int_{-\infty}^{y} p_o(x) dx,$$

so that the expression in Eq. (3.1) is verified. □

### 3.2.1 Acceptance Rate

The average probability of accepting one proposed sample $x'$ is usually called *acceptance rate*. This is the main figure of merit of a rejection sampler. In the sequel, we denote the acceptance rate as $\hat{a}$. In Sect. 3.3, we provide an exhaustive description of the computational cost of a rejection sampler and show that

$$\hat{a} = \frac{c}{L},$$

where $c = \int_{\mathcal{D}} p(x)dx$, hence the importance of finding a bound $L$ that is as small as possible.

*Example 3.1* Consider a Gamma density,

$$p_o(x) = \frac{1}{\Gamma(\alpha)} x^{\alpha-1} e^{-x}, \quad x \geq 0, \ \alpha > 0.$$

where $\Gamma(\alpha)$ is the Gamma function. Several RS schemes, using different proposals easy to draw from, have been described in the literature [11]. For instance, in [10], a log-logistic proposal pdf is used,

$$\pi(x) = \lambda\mu \frac{x^{\lambda-1}}{(\mu + x^\lambda)^2}, \quad x \geq 0,$$

with $\mu = \alpha^\lambda$, $\lambda = \sqrt{2\alpha - 1}$ for $\alpha \geq 1$, and $\lambda = \alpha$ for $\alpha < 1$. The bound $L$ such that $L\pi(x) \geq p_o(x)$ is

$$L = \frac{4\alpha^\alpha e^{-\alpha}}{\Gamma(\alpha)\sqrt{2\alpha - 1}}.$$

### 3.2.2 Distribution of the Rejected Samples

Theorem 3.1 clarifies that the distribution of the samples accepted in an RS algorithm is the target pdf, when the bound $L$ is such that $L\pi(x) \geq p(x)$, $\forall x \in \mathcal{D}$. Namely, an accepted sample $x'_a$ is distributed as

$$q_a(x) = p_o(x).$$

On the other hand, it is straightforward to show that a sample $x'_r$ rejected in the RS test is distributed as

$$q_r(x) \propto L\pi(x) - p(x). \tag{3.2}$$

The validity of the RS scheme depends on a suitable choice of $L$. Indeed, if the constant $L$ does not guarantee that $L\pi(x) \geq p(x)$, $\forall x \in \mathcal{D}$, both densities, $q_a(x)$ and $q_r(x)$, are different, as shown in Sect. 3.2.3.

### 3.2.3 Distribution of the Accepted and Rejected Samples with Generic L > 0

In certain cases, it may be necessary to know the distribution of the samples accepted ($x'_a$) or rejected ($x'_r$) in an RS-type step when the inequality $L\pi(x) \geq p_o(x)$ is not guaranteed to hold, i.e., $L\pi(x) < p_o(x)$ for all $x \in \mathcal{D}_1$ whereas $L\pi(x) \geq p_o(x)$ for all $x \in \mathcal{D}_2$ with $\mathcal{D} = \mathcal{D}_1 \cup \mathcal{D}_2$ and $\mathcal{D}_1 \cap \mathcal{D}_2 = \emptyset$ (namely, the RS condition is satisfied only in certain parts of the domain $\mathcal{D}$). In this case, the accepted samples are distributed as

$$q_a(x) \propto \min\{p(x), L\pi(x)\}. \tag{3.3}$$

Indeed, given $x'$ drawn from $\pi(x)$, if we have $L\pi(x') < p(x')$, i.e., $x' \in \mathcal{D}_1$, the sample $x'$ is accepted with probability 1. Hence, the samples that belong to the region $\mathcal{D}_1$ are distributed according to $\pi(x)$. Otherwise, if the sample $x'$ belongs to a region where $L\pi(x') \geq p(x')$, i.e., $x' \in \mathcal{D}_2$, then $x'$ is distributed according to $p_o(x) \propto p(x)$ (as proved in Theorem 3.1). Similarly, we can state that the rejected samples are distributed as

$$q_r(x) \propto \big(L\pi(x) - p(x)\big)_+ \tag{3.4}$$

where $(\cdot)_+$ denotes the positive part of $(\cdot)$.

### 3.2.4 Different Application Scenarios

The most favorable scenario to use the RS algorithm occurs when $p(x)$ is bounded with bounded domain. Indeed, in this case the proposal pdf $\pi(x)$ can be chosen as a uniform density (i.e., the easiest possible proposal). Moreover, the calculation of the bound $L$ for the ratio $p(x)/\pi(x)$ is converted into the problem of finding an upper bound for the target function $p(x)$, which is a simpler issue in general (see Sect. 3.6.1).

Otherwise, when $p(x)$ is unbounded or its domain is infinite, the proposal $\pi(x)$ cannot be uniform and more elaborate schemes have to be sought. Section 3.7 and Chap. 5 are devoted to describe methods that address this issue by transforming $p(x)$ so that it is embedded within a finite region.

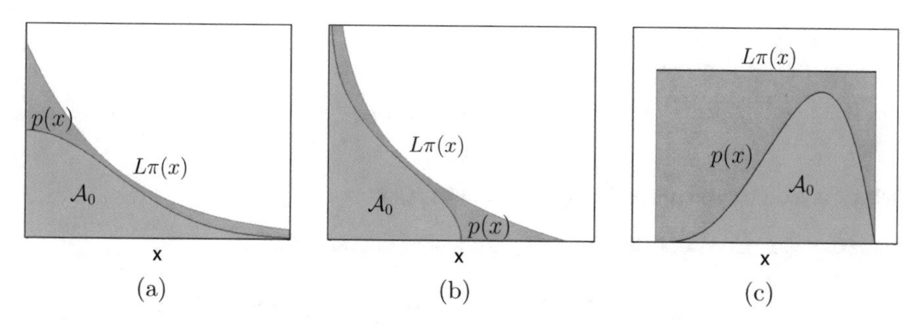


**Fig. 3.2** Three possible cases for the density $p_o(x) \propto p(x)$ with three possible envelope functions $L\pi(x)$: **(a)** bounded with an infinite domain, **(b)** unbounded in a finite domain, and **(c)** bounded with a finite domain. A uniform proposal distribution can only be used in the last case

Figure 3.2 illustrates these three possible cases described above. There exists a fourth possible scenario, when the function $p(x)$ is unbounded with infinite support. However, we can consider it as a combination of the cases in Fig. 3.2a, b.

### 3.2.5 *Butcher's Version of the Rejection Sampler*

Consider a target density that can be expressed as the product of two functions,

$$p_o(x) \propto p(x) = \pi(x)g(x), \quad x \in \mathcal{D}, \tag{3.5}$$

where $\pi(x)$ is a pdf easy to draw from, and $0 \leq g(x) \leq 1$ (not necessarily a cdf). In this case, the RS method for generating samples from $p_o(x)$ can take the interesting form described below [6]:

1. Draw $x' \sim \pi(x)$ and $u' \sim \mathcal{U}([0, 1])$.
2. If $u' \leq g(x')$, accept $x'$. Otherwise, if $u' > g(x')$, reject $x'$.

Indeed, in this case the ratio between target and proposal used in the RS test is exactly

$$\frac{p_o(x)}{\pi(x)} = \frac{\pi(x)g(x)}{\pi(x)} = g(x),$$

and, since we have assumed $0 \leq g(x) \leq 1$, we do not need any additional bound $L$. The acceptance rate, i.e., the average probability of accepting one proposed sample $x'$, is

$$\hat{a} = \int_{\mathcal{D}} \pi(x) g(x) dx,$$

which is exactly the normalizing constant of $p_o(x)$.

*Example 3.2* Consider the following truncated Gamma density,

$$p_o(x) \propto x^{\alpha-1} e^{-x}, \quad \text{with } 0 < x \leq 1, \ \alpha > 0.$$

We can set $\pi(x) = \alpha x^{\alpha-1}$ and $g(x) = e^{-x}$. Random samples from $\pi(x)$ can be easily obtained by the inversion method, so that a rejection sampler to draw from $p_o(x)$ is given by the following steps:

1. Draw $u_1 \sim \mathcal{U}([0, 1])$ and set $x' = u_1^{1/\alpha}$.
2. Draw $u_2 \sim \mathcal{U}([0, 1])$.
3. If $u_2 \leq e^{-x'}$, then accept $x'$.

### 3.2.6 *Vaduva's Modification of the Butcher's Method*

In Butcher's version the function $g(x)$ is not necessarily a cdf. If $g(x)$ can be expressed as $g(x) = 1 - F_Q(x)$ or $g(x) = F_Q(x)$, where $F_Q(x)$ is the cdf of a r.v. $Q$ with pdf $q(x)$, a modification of the previous RS scheme can be designed. Indeed, Vaduva [42] proposed the following variant of the RS method for drawing from a target $p_o(x) \propto \pi(x)[1 - F_Q(x)]$:

1. Draw $x' \sim \pi(x)$ and $z' \sim q(x)$.
2. If $x' \leq z'$, accept $x'$. Otherwise, reject $x'$.

If $g(x) = F_Q(x)$, we only need to replace $x' \leq z'$ with $x' \geq z'$ in the acceptance step. Vaduva's algorithm is entirely equivalent to Butcher's version of the RS method. Indeed, considering the inequality $x' \leq z'$ and applying to both sides the (decreasing) transformation $1 - F_Q(x)$, we obtain

$$1 - F_Q(x') \geq 1 - F_Q(z').$$

Since $z' \sim q(x)$, the sample

$$v' = F_Q(z') \sim \mathcal{U}([0, 1]),$$

is uniformly distributed in [0, 1] (see the inversion method in Sect. 2.4.1). Moreover, since $u' = 1 - v' \sim \mathcal{U}([0, 1])$ and $g(x') = 1 - F_Q(x')$ we can write

$$g(x') \geq u',$$

which is exactly the acceptance test in Butcher's version of the RS algorithm. Considerations in a similar fashion can be done for the case $g(x) = F_Q(x)$.

When both variants can be applied, the choice between drawing directly $u'$ and computing $g(x) = 1 - F_Q(x)$ on the one hand (Butcher's method) or generating $z'$ from $q(x)$ on the other hand (Vaduva's method) depends mainly upon the relative speeds of calculating $F_Q(x)$ and generating $z'$ with pdf $q(x)$.

*Example 3.3* Consider again the truncated Gamma density of Example 3.2, i.e.,

$$p_o(x) \propto x^{\alpha-1}e^{-x}, \quad \text{with } 0 < x \leq 1, \ \alpha > 0.$$

Vaduva's method consists in drawing $x' \sim \pi(x)$ and $z' \sim q(x) = e^{-x}$ ($g(x) = 1 - F_Q(x)$ with $F_Q(x) = 1 - e^{-x}$). The sample is accepted if $x' \leq z'$.

### 3.2.7 Lux's Extension

Assume that the target pdf, defined for all $x \in \mathbb{R}^+$, can be expressed as

$$p_o(x) = \pi(x) \int_{-\infty}^{r(x)} \left( \frac{1}{\int_0^{r^{-1}(y)} \pi(z)dz} \right) q(y)dy, \quad x > 0, \tag{3.6}$$

where $r(x) : \mathbb{R}^+ \to \mathbb{R}^+$ is a strictly *decreasing* function or

$$p_o(x) = \pi(x) \int_{-\infty}^{r(x)} \left( \frac{1}{\int_{r^{-1}(y)}^{+\infty} \pi(z)dz} \right) q(y)dy, \quad x > 0, \tag{3.7}$$

if $r(x)$ is a strictly *increasing* function. In these cases, Lux [13, 27] suggested the following algorithm:

1. Draw $x' \sim \pi(x)$ and $y \sim q(y)$.
2. If $y' \leq r(x')$ accept $x'$. Otherwise reject $x'$.

It is not difficult to show that the acceptance rate in this case is given by

$$\hat{a} = \int_0^{+\infty} F_Y(r(x))\pi(x)dx,$$

where $F_Y$ is the cdf of $Y \sim q(y)$. Usually, a convenient choice of $q(y)$ is $q(y) \propto \int_0^{r^{-1}(y)} \pi(z)dz$ or $q(y) \propto \int_{r^{-1}(y)}^{+\infty} \pi(z)dz$, depending on whether $r(x)$ is decreasing or increasing.

*Example 3.4* Consider the Gamma pdf as target density,

$$p_o(x) = \frac{1}{\Gamma(\alpha)} x^{\alpha-1} e^{-x}, \quad x \geq 0, \alpha > 0,$$

where $\Gamma(\alpha)$ is the Gamma function. We can choose $r(x) = x^{\alpha-1}$ (hence, $r^{-1}(y) = y^{1/(\alpha-1)}$), $\pi(x) = e^{-x}$ and

$$q(y) = \frac{1}{\Gamma(\alpha)} e^{-y^{1/(\alpha-1)}}, \quad y \geq 0, \tag{3.8}$$

such that

$$\begin{aligned} p_o(x) &= \pi(x) \int_{-\infty}^{r(x)} \left( \frac{1}{\int_{r^{-1}(y)}^{+\infty} \pi(z) dz} \right) q(y) dy, \\ &= e^{-x} \int_{-\infty}^{r(x)} \frac{1}{e^{-y^{1/(\alpha-1)}}} \frac{1}{\Gamma(\alpha)} e^{-y^{1/(\alpha-1)}} dy, \\ &= \frac{1}{\Gamma(\alpha)} e^{-x} x^{\alpha-1}, \quad x \geq 0. \end{aligned}$$

Drawing from $\pi(x) = e^{-x}$ is straightforward using the inversion method. Therefore, if we can draw from $q(y)$, then Lux's method becomes readily applicable. If $\alpha > 1$, note that the r.v. $Y \sim q(y)$ can be obtained as $Y = Z^{\alpha-1}$, where $Z$ is distributed as

$$h(z) = \frac{1}{\Gamma(\alpha-1)} z^{\alpha-2} e^{-z}, \quad z \geq 0,$$

i.e., another Gamma pdf with parameter $\alpha - 1$ (that, in some cases, we can draw from with other sampling methods). Therefore, we have found that a Gamma pdf with parameter $\alpha > 1$ can be generated by drawing from another Gamma pdf with parameter $\alpha - 1$ and then sampling (repeatedly, until accepting) from an exponential density. The acceptance rate of this method is $\hat{a} = 2^{1-\alpha}$.

## 3.3 Computational Cost

The computational efficiency of the RS algorithm depends on three factors:

1. The computational cost of generating samples from the proposal pdf.
2. The computational cost of evaluating the ratio of the target density over the proposal density, which is needed for the rejection test. In turn, the cost of this operation depends essentially on the difficulty in evaluating the target (which can be costly in many problems). The evaluation of the proposal is normally not an issue.

3. The acceptance rate. Note that this rate depends strictly on the discrepancy between the shapes of the proposal and the target pdfs.

Different proposal pdfs are often compared according to the first and last factors. Namely, two proposal densities to be used within the same RS algorithm (i.e., for the same target pdf) should be compared in terms of the expected cost of generating samples from the proposal and the probability of accepting these samples (or, equivalently, the discrepancy between the target and the proposal densities).

In this section, we analyze the acceptance rate (Sect. 3.3.1) and discuss how to speed up the computation of the ratio in the RS test (Sect. 3.3.2). Finally, we also describe a procedure to reduce the mean number of uniform random variates needed in the RS algorithm (Sect. 3.3.3).

## 3.3.1 Further Considerations About the Acceptance Rate

The fundamental figure of merit of a rejection sampler is the mean acceptance rate, i.e., the expected number of accepted samples over the total number of proposed candidates. In practice, finding a tight upper bound $L$ and, in general, a "good" envelope function, such that, $L\pi(x) \geq p(x)$, is crucial for the performance of a rejection sampling algorithm. We can formally define the acceptance rate as

$$\hat{a} = \text{Prob}\left\{U \leq \frac{p(X)}{L\pi(X)}\right\}, \tag{3.9}$$

where $X \sim \pi(x)$ and $U \sim \mathcal{U}([0,1])$. The probability in Eq. (3.9) represents the acceptance condition in an RS scheme and can be calculated as

$$\begin{aligned}\hat{a} &= \int_{\mathcal{D}} \left[\int_0^{\frac{p(x)}{L\pi(x)}} du\right] \pi(x)dx \\ &= \int_{\mathcal{D}} \frac{p(x)}{L\pi(x)} \pi(x)dx = \frac{\int_{\mathcal{D}} p(x)dx}{L} = \frac{c}{L},\end{aligned} \tag{3.10}$$

where $1/c$ is the normalization constant of $p(x)$ and the proposal pdf $\pi(x)$ is assumed to be normalized.[1] Note that in cases where $p(x) = p_o(x)$, hence $c = 1$, the constant $L$ is necessarily larger or equal than 1, with equality if, and only if, $\pi(x) = p_o(x)$. Moreover, for the ratio $p(x)/\pi(x)$ to remain bounded, it is necessary that $\pi(x)$ has tails that decay to zero at the same rate or slower than those of $p(x)$.

[1] Otherwise, if $c_\pi = \int_{\mathcal{D}} \pi(x)dx \neq 1$, the acceptance rate is given by the more general expression $\hat{a} = \frac{c}{Lc_\pi}$.

Unfortunately, the RS method has an important structural limitation: even if we are able to find the optimal bound

$$L^* = \sup_{x \in \mathcal{D}} \frac{p(x)}{\pi(x)},$$

the acceptance rate can be far away from 1, depending on the difference in shape between $p(x)$ and $\pi(x)$. More specifically, the efficiency of the RS algorithm is a function of the discrepancy between the shapes of $\pi(x)$ and $p(x)$. Let us assume an unnormalized proposal $\pi(x)$ and let us denote as $c_\pi = \int_\mathcal{D} \pi(x)dx$ the normalizing constant of $\pi(x)$. Indeed, observe that the probability of rejecting a sample is $\hat{r} = 1 - \hat{a}$, that is

$$\begin{aligned} \hat{r} &= 1 - \frac{c}{Lc_\pi} = \frac{Lc_\pi - c}{Lc_\pi}, \\ &= \frac{1}{Lc_\pi}\left(\int_D L\pi(x)dx - \int_D p(x)dx\right), \\ &= \frac{1}{Lc_\pi}\left(\int_D |L\pi(x) - p(x)|dx\right) = \frac{1}{Lc_\pi} d(L\pi, p), \end{aligned} \tag{3.11}$$

where we have assumed unnormalized $\pi(x) \geq 0$ and $p(x) \geq 0$, and $d(L\pi, p)$ denotes the $L_1$ distance between $L\pi(x)$ and $p(x)$. This drawback is an important issue, since accept–reject algorithms can generate many "useless" samples when rejecting. For this reason, different schemes [7, 8, 24, 25] have been proposed to improve the efficiency of the standard RS method in estimation problems (see Sect. 3.9). Furthermore, many adaptive accept–reject schemes have been designed. The basic idea of these methods is to iteratively build up a sequence of "good" proposal densities $\{\pi_t(x)\}_{t=0}^{+\infty}$ which become closer and closer to the target density $p_o(x)$ and, as a consequence, improve the acceptance rate. Adaptive schemes are the topic of Chap. 4.

*Example 3.5* The Nakagami distribution is widely used in radio communication applications to model the wireless fading channel, due to its good agreement with empirical channel measurements for some urban multi-path environments [36]. The Nakagami target density is $p_o(x) = \frac{1}{c}p(x)$, where $\frac{1}{c} = 2m^m/(\Omega\Gamma(m))$ and

$$p(x) = x^{2m-1} \exp\left(-\frac{m}{\Omega}x^2\right), \qquad x \geq 0, \tag{3.12}$$

where $m \geq 0.5$ is the fading parameter, which indicates the fading depth, and $\Omega > 0$ is the average received power. When $m$ is an integer or half-integer (i.e., $m = \frac{n}{2}$ with $n \in \mathbb{N}$), independent samples can be generated through the square root of a sum of squares of $n$ zero-mean i.i.d. Gaussian r.v.'s [3, 36]. For $m \neq \frac{n}{2}$, several RS schemes for drawing i.i.d. samples from a Nakagami-$m$ pdf have been proposed [33]. Here, we first consider the RS method given in [26], where another Nakagami density is used as a proposal pdf, i.e.,

$$\pi_1(x) \propto g_1(x) = \alpha_p \, x^{2m_p - 1} \exp\left(-\frac{m_p}{\Omega_p} x^2\right), \qquad x \geq 0, \tag{3.13}$$

with $m_p = n/2$, $n = \lfloor 2m \rfloor$ (and $\lfloor x \rfloor$ denoting the integer part of $x \in \mathbb{R}$), and the remaining parameters ($\alpha_p$ and $\Omega_p$) adjusted to obtain the same location and value of the maximum in the proposal as in the target,

$$\Omega_p = \frac{2m_p}{2m_p - 1} x_{\max}^2 = \Omega \frac{m_p(2m-1)}{m(2m_p-1)},$$

where $x_{\max}$ is the location of the maximum of the Nakagami pdf, obtained solving $\frac{dp(x)}{dx} = 0$, which results in

$$x_{\max} = \sqrt{\frac{(2m-1)\Omega}{2m}}. \tag{3.14}$$

We also set

$$\alpha_p = \frac{p(x_{\max})}{g_1(x_{\max})} = \exp(m_p - m) \left(\frac{\Omega(2m-1)}{2m}\right)^{m-m_p}.$$

It is possible to prove that $g_1(x) \geq p(x)$, $x \geq 0$, $\forall m \geq 0.5$ and $\forall \Omega > 0$. Alternatively, in [45] a truncated Gaussian density is suggested as a proposal pdf. Namely,

$$\pi_2(x) \propto g_2(x) = b \exp\big(-a(x - x_{\max})^2\big), \quad x \geq 0, \tag{3.15}$$

with $a = \frac{m}{\Omega}$ and $b = p(x_{\max})$. Once more, we have $g_2(x) \geq p(x)$, $x \geq 0$, $\forall m \geq 0.5$ and $\forall \Omega > 0$.

We analyze the performance of these two RS techniques in terms of the acceptance rate. The acceptance rate of the first scheme can be obtained analytically,

$$\hat{a}_1 = (2e)^{m-m_p} \frac{\Gamma(m)(2m_p-1)^{m_p}}{\Gamma(m_p)(2m-1)^m}, \tag{3.16}$$

with $\Gamma(m)$ denoting the Gamma function, whereas the acceptance rate of the second scheme can be approximated for $m \geq 4$ as

$$\hat{a}_2 \approx \frac{e^{m-\frac{1}{2}} \Gamma(m)(2m-1)^{\frac{1}{2}-m}}{\sqrt{\pi} 2^{m+\frac{1}{2}}}. \tag{3.17}$$

Note that in both cases the acceptance rate is independent from the average received power, $\Omega$. Figure 3.3 shows the acceptance rate, obtained empirically after drawing

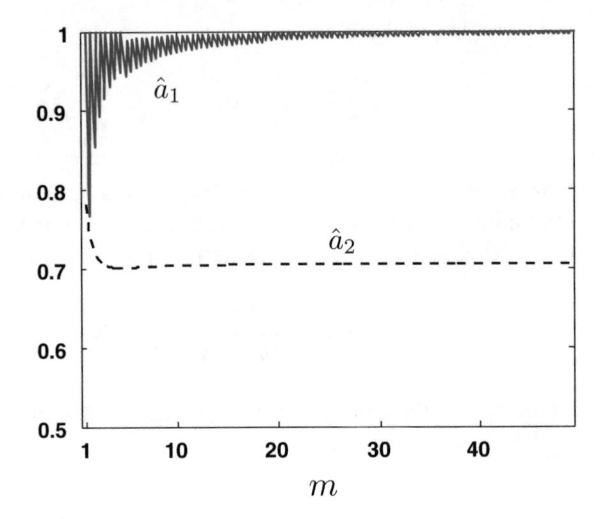


**Fig. 3.3** The acceptance rates of Example 3.5 using the proposal $\pi_1(x)$ (continuous line) and $\pi_2(x)$ (dashed line) for $1 \leq m \leq 50$. The acceptance rate is independent of $\Omega$ for both of the proposed schemes

$N = 6 \cdot 10^5$ independent samples, for both approaches and several values of the fading depth, $m$. It can be seen that the first technique, using the proposal $\pi_1(x)$, is extremely efficient. Indeed, in this numerical example it provides exact sampling (i.e., $a_{R1} = 1$) when $m$ is integer or half-integer, since the proposal is equal to the target in these cases.

### 3.3.2 *Squeezing*

If we observe carefully the RS algorithm, described in Table 3.1, it is apparent that testing the acceptance condition is a fundamental but, in some cases, expensive step. In order to speed up the RS technique, we may seek a simple function $\varphi(x)$ that is easy to evaluate and satisfies

$$\varphi(x) \leq p(x). \tag{3.18}$$

Such function is often termed a *squeeze function*. The basic idea of squeezing is to add a previous test involving the function $\varphi(x)$ in order to avoid the evaluation of $p(x)$. The method can be summarized in the following way:

1. Draw $x' \sim \pi(x)$ and $u' \sim \mathcal{U}([0, 1])$.
2. If $u' L\pi(x') \leq \varphi(x')$, then accept $x'$ without evaluating the target function $p(x)$.
3. Otherwise, if $u' L\pi(x') \leq p(x')$, then accept $x'$. Else, reject $x'$.

Figure 3.4 illustrates the squeeze principle. If the point $(x', u' L\pi(x'))$ belongs to the area below $\varphi(x)$ (darker green area), the sample $x'$ is accepted at step 2. Otherwise, we check whether the point stays in the lighter green area and in this case we also

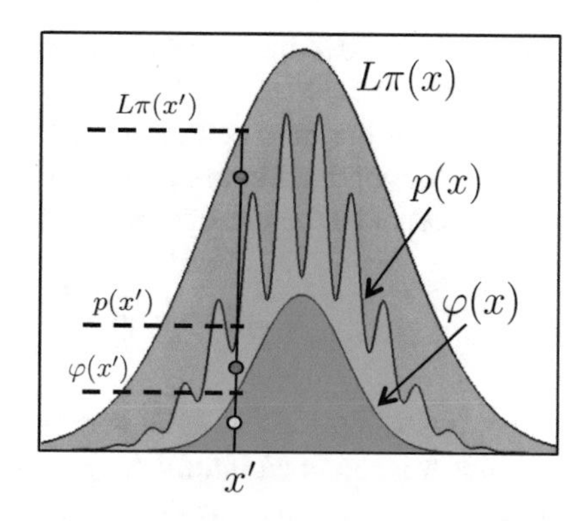


**Fig. 3.4** Squeeze principle: we first check whether the point $(x', u'L\pi(x'))$ falls within the darker green region, and in such case the sample $x'$ is already accepted. Note that in order to check this we do not need to evaluate the function $p(x)$, but only the (simpler) squeeze function $\varphi(x) \leq p(x)$

**Table 3.2** Sibuya's modified RS algorithm

| |
|---|
| 1. Set $i = 1$. Let $N$ be the number of desired samples from $p_o(x)$ |
| 2. Draw $u' \sim \mathcal{U}([0, 1])$ |
| 3. Draw $x' \sim \pi(x)$ |
| 4. If $u' \leq \frac{p(x')}{L^*\pi(x')}$, then set $x^{(i)} = x'$, $i = i + 1$ and go to step 6 |
| 5. Else, if $u' > \frac{p(x')}{L^*\pi(x')}$, then discard $x'$ and repeat from step 3 |
| 6. If $i > N$, then stop. Else go back to step 2 |

accept $x'$ at step 3. The sample $x'$ is discarded if the point $(x', u'L\pi(x'))$ falls within the red region.

### 3.3.3 *Sibuya's Modified Rejection Method*

In this subsection, we present an alternative procedure to implement the RS scheme that turns out of theoretical interest, although its applicability is actually limited. In a standard RS scheme, we must generate a uniform sample $u'$ each time we perform an RS test to accept or discard a proposed sample $x'$. Therefore, one way in which we can reduce the computational complexity of the rejection sampler is by minimizing the number of draws from the uniform distribution. Here, we present an alternative procedure that reduces considerably the number of uniform variates necessary in an RS method. However, the practical application of this approach is restricted, since the optimal bound $L^*$ is needed and performance turns out worse than a standard RS scheme [13, Chap. 2], [15] in terms of the acceptance rate.

If the optimal value $L^* = \sup \frac{p(x)}{\pi(x)}$ is known, it is possible to apply a variant of RS outlined in Table 3.2. This modified rejection method was proposed by M. Sibuya in [38] and analyzed later in [15]. Sibuya observed that we can fix a uniform sample $u'$ in the RS test and change only the proposed sample $x'$ until it is accepted.

Note that the ratio $\sup \frac{p(x)}{L^*\pi(x)}$ can reach the value 1 only if $L^* = \sup_{x\in\mathcal{D}} \frac{p(x)}{\pi(x)}$, i.e., it is the optimal bound. Sibuya's method can be applied only in this situation. Indeed, consider the use of a constant $L > L^*$. In this case, we have

$$0 < M = \sup \frac{p(x)}{L\pi(x)} < 1, \qquad \text{for every } x \in \mathcal{D}.$$

Then, if a sample $u' > M$ is generated in the step 2 of Table 3.2, the algorithm remains trapped in an infinite loop, since we can never obtain a sample $x'$ such that $\frac{p(x')}{L\pi(x')} \geq u'$ (hence we never accept $x'$).

## 3.4 Band Rejection Method

A little-known but important improvement of the standard RS algorithm was proposed by W. Payne [35] and later on extended using non-uniform proposal pdfs in [11, Chap. 3]. The technique can be applied only when the target has a bounded domain. In this section, we describe a generalized version and present the so-called band RS method introduced in [11, 35] as a special case.

### *3.4.1 Preliminaries*

The band rejection (BR) scheme is better described if we consider first an alternative graphical representation of the RS method. Consider a target pdf $p_o(x) \propto p(x)$ and a proposal $\pi(x)$, with $x \in \mathcal{D} = [a, b]$ (i.e., a bounded domain). Given a bound $L \geq \frac{p(x)}{\pi(x)}$, Fig. 3.5a depicts the function $\phi(x; L) = \frac{p(x)}{L\pi(x)}$, $x \in \mathcal{D}$, and the acceptance and rejection sets denoted as

$$\mathcal{A} = \{(x, u) \in \mathcal{D} \times [0, 1] : \quad u \leq \phi(x; L)\}$$

and

$$\mathcal{R} = \{(x, u) \in \mathcal{D} \times [0, 1] : \quad u > \phi(x; L)\}$$

respectively. Note that

$$0 \leq \phi(x; L) \leq \frac{1}{L} \sup_{x\in\mathcal{D}} \frac{p(x)}{\pi(x)} \leq 1.$$

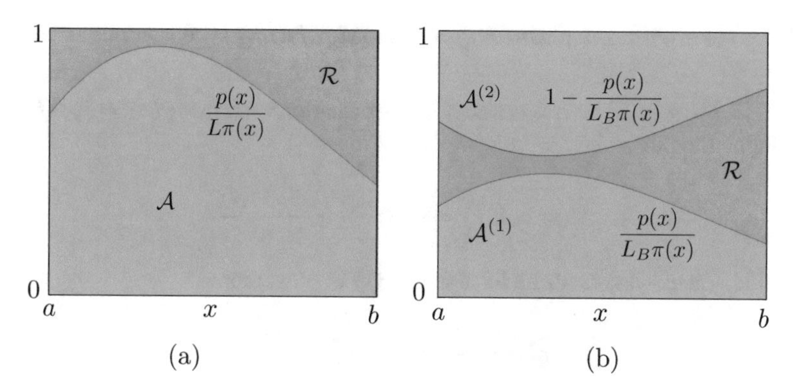


**Fig. 3.5** (**a**) Yet another graphical interpretation of the RS method: $\mathcal{A}$ and $\mathcal{R}$ represent the acceptance and rejection sets. (**b**) An alternative RS scheme with an extended acceptance region

If the optimal bound $L = L^* = \sup_{x\in\mathcal{D}} \frac{p(x)}{\pi(x)}$ is selected, then the function $\phi(x; L^*) = \frac{p(x)}{L^*\pi(x)}$ reaches the value 1 for some $x \in \mathcal{D}$. Obviously, in order to obtain good RS schemes we would like the set $\mathcal{A}$ to be as large as possible, ideally including the whole rectangle $[a, b] \times [0, 1]$ (see Fig. 3.5a).

Let us assume that we have chosen and fixed a proposal pdf $\pi(x)$ and a bound $L$. In this case, the only remaining possibility to improve the acceptance rate of the RS algorithm is to post-process and reuse the rejected samples, converting them into samples distributed according to the target $p_o(x)$. Hence, we need to relate in some way the points $(x, u)$ in the region $\mathcal{R}$ to the target pdf $p_o(x)$.

In order to clarify how this can be done, we first introduce a (non-practical) alternative to the standard RS scheme:

- Set $L_B = 2L$.
- Draw $x' \sim \pi(x)$ and $u' \sim \mathcal{U}([0, 1])$.
- If $u' \leq \phi(x; L_B) = \frac{p(x)}{L_B\pi(x)}$, then accept $x'$.
- Else, if $u' \geq 1 - \phi(x; L_B)$, then accept $x'$. Otherwise, reject $x'$.

In this alternative RS technique we have a "second chance" for reusing an initially rejected sample. However, although the method is valid (as shown in the sequel), the acceptance rate is exactly the same as in the standard RS method. Figure 3.5b illustrates the algorithm by depicting the two acceptance regions, $\mathcal{A}^{(1)}$ and $\mathcal{A}^{(2)}$, corresponding to the two acceptance conditions. The question now is whether it is possible to modify the previous procedure in a suitable way to increase the acceptance rate of the standard RS scheme. The answer is positive, as described below.

### 3.4.2 Generalized Band Rejection Algorithm

Let $\rho : [a, b] \to [a, b]$ be a decreasing transformation such that $\rho(a) = b$, $\rho(b) = a$, and define the pdf

$$q(x) \propto h(x) = -p(\rho^{-1}(x))\frac{d\rho^{-1}(x)}{dx}. \tag{3.19}$$

Note that, since $p_o(x)$ is normalized and $p_o(x) \propto p(x)$, then

$$q(x) = -p_o(\rho^{-1}(x))\frac{d\rho^{-1}(x)}{dx}.$$

Moreover, consider the following upper bound

$$L_B \geq \frac{p(x) + h(x)}{\pi(x)}. \tag{3.20}$$

We outline the generalized band rejection (GBR) algorithm in Table 3.3 using the notation just introduced. The GBR algorithm contains two possible acceptance tests, the first one with probability $\frac{p(x')}{L_B\pi(x')}$ and the second one with probability $\frac{h(x')}{L_B\pi(x')}$. Hence, note that the function $h(x)$ is also involved in the second acceptance condition (at step 3.3 of Table 3.3).

For $\rho^{-1}(x) = x$ the GBR scheme yields the "non practical" method of Sect. 3.4.1, which has exactly the same acceptance probability as the standard rejection sampler. However, other choices of $\rho$ may yield improvements over the standard technique. A suitable and simple example is $\rho^{-1}(x) = a + b - x$, so that $h(x) = p(a + b - x)$. Figure 3.6 illustrates the acceptance and rejection regions of the GBR sampler for this choice of $\rho$. Comparing Figs. 3.5b and 3.6, it can be observed that the area of rejection region $\mathcal{R}$ (red area) is smaller in Fig. 3.6 than in Fig. 3.5b. This means that the acceptance rate for the case in Fig. 3.6 is greater.

The validity of the GBR algorithm is granted by the following theorem (a simpler version can be found [11, Chap. 3]).

**Table 3.3** Generalized band rejection algorithm

| |
|---|
| 1. Set $i = 1$. Let $N$ be the number of desired samples from $p_o(x)$ |
| 2. Find a suitable upper bound $L_B \geq \frac{p(x)+h(x)}{\pi(x)}$, with $h(x)$ given in Eq. (3.19) |
| 3. Draw samples $x' \sim \pi(x)$ and $u' \sim \mathcal{U}([0, 1])$ |
| 4. If $u' \leq \frac{p(x')}{L_B\pi(x')}$, then accept $x^{(i)} = x'$, set $i = i + 1$ and jump to step 6 |
| 5. If $u' > 1 - \frac{h(x')}{L_B\pi(x')}$, set $x^{(i)} = \rho^{-1}(x')$ and set $i = i + 1$. Otherwise, reject $x'$ |
| 6. If $i > N$, then stop. Else, go back to step 3 |

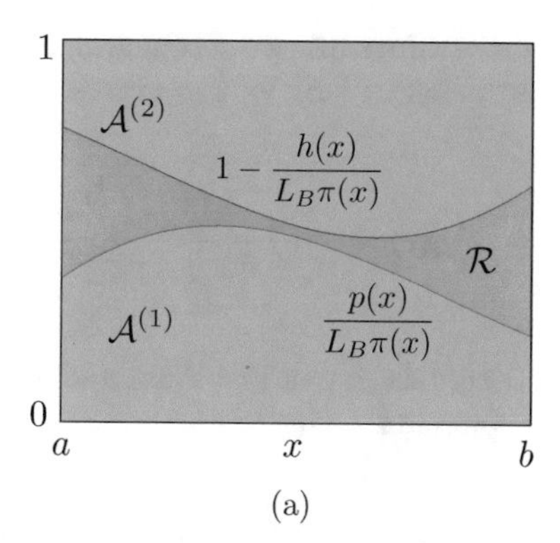


**Fig. 3.6** Band rejection with $h(x) = p(a+b-x)$, i.e., $\rho^{-1}(x) = a+b-x$

**Theorem 3.2** *Let the r.v.'s $X_1$ and $X_2$ have pdfs $\pi(x)$ and $p_o(x) \propto p(x)$, respectively, and let $U$ have a uniform distribution $\mathcal{U}([0,1])$. If there exists a bound $L_B \geq \frac{p(x)+h(x)}{\pi(x)}$, $\forall x \in \mathcal{D} = [a,b]$, where $h(x) = -p(\rho^{-1}(x))\frac{d\rho^{-1}}{dx}$ (with $\rho$ a decreasing function such that $\rho(a) = b$ and $\rho(b) = a$), then*

$$
\begin{aligned}
P_{tot}(y) :=& Prob\{X_2 \leq y\} \\
=& \frac{c_1}{c_1+c_2} Prob\left\{ X_1 \leq y \middle| U \leq \frac{p(X_1)}{L_B\pi(X_1)} \right\} \\
&+ \frac{c_2}{c_1+c_2} Prob\left\{ X_1 \geq \rho^{-1}(y) \middle| U \leq 1 - \frac{h(X_1)}{L_B\pi(X_1)} \right\},
\end{aligned}
\tag{3.21}
$$

*where $c_1 = \int_a^b p(x)dx$ and $c_2 = \int_a^b h(x)dx$ are the normalizing constants of $p_o(x) = \frac{1}{c_1}p(x)$ and $q(x) = \frac{1}{c_2}h(x)$, respectively.*

*Remark 3.1* Indeed, the probability corresponding to the first acceptance test is

$$
P_1(y) = \text{Prob}\left\{ X_1 \leq y \middle| U \leq \frac{p(X_1)}{L_B\pi(X_1)} \right\},
$$

while the probability of the second acceptance condition can be written as

$$
\begin{aligned}
P_2(y) &= \text{Prob}\left\{ \rho(X_1) \leq y \middle| 1 - U \geq \frac{h(X_1)}{L_B\pi(X_1)} \right\} \\
&= \text{Prob}\left\{ X_1 \geq \rho^{-1}(y) \middle| U \leq 1 - \frac{h(X_1)}{L_B\pi(X_1)} \right\}.
\end{aligned}
$$

so that the probability describing the whole algorithm is a convex combination of the previous ones, i.e., $P_{tot}(y) = \frac{c_1}{c_1+c_2}P_1(y) + \frac{c_2}{c_1+c_2}P_2(y)$.

*Proof* Since the two acceptance tests are disjoint events (corresponding to the green areas in Fig. 3.6), we can express $P_{tot}(y) = \frac{c_1}{c_1+c_2}P_1 + \frac{c_2}{c_1+c_2}P_2$ as

$$P_{tot}(y) = \frac{c_1}{c_1+c_2}\frac{\int_a^y \int_0^{\frac{p(x)}{L_B\pi(x)}} \pi(x)dudx}{\int_a^b \int_0^{\frac{p(x)}{L_B\pi(x)}} \pi(x)dudx} + \frac{c_2}{c_1+c_2}\frac{\int_{\rho^{-1}(y)}^b \int_{1-\frac{h(x)}{L_B\pi(x)}}^1 \pi(x)dudx}{\int_a^b \int_{1-\frac{h(x)}{L_B\pi(x)}}^1 \pi(x)dudx}.$$

Then, integrating first w.r.t. $u$ and after some trivial calculations, we arrive at the expression

$$\begin{aligned}
P_{tot}(y) &= \frac{c_1}{c_1+c_2}\frac{\int_a^y \frac{p(x)}{L_B}dx}{\int_a^b \frac{p(x)}{L_B}dx} + \frac{c_2}{c_1+c_2}\frac{\int_{\rho^{-1}(y)}^b \frac{h(x)}{L_B}dx}{\int_a^b \frac{h(x)}{L_B}dx} \\
&= \frac{c_1}{c_1+c_2}\frac{\int_a^y p(x)dx}{\int_a^b p(x)dx} + \frac{c_2}{c_1+c_2}\frac{\int_{\rho^{-1}(y)}^b h(x)dx}{\int_a^b h(x)dx} \\
&= \frac{c_1}{c_1+c_2}\frac{c_1\int_a^y p_o(x)dx}{c_1} + \frac{c_2}{c_1+c_2}\frac{c_2\int_{\rho^{-1}(y)}^b q(x)dx}{c_2}, \\
&= \frac{c_1}{c_1+c_2}\int_a^y p_o(x)dx + \frac{c_2}{c_1+c_2}\int_{\rho^{-1}(y)}^b q(x)dx.
\end{aligned}$$

Furthermore, recalling that $q(x) = -p_o(\rho^{-1}(x))\frac{d\rho^{-1}}{dx}$ and performing the change of variables $x = \rho(z)$ (recall that $\rho^{-1}(a) = b$ by assumption), we can rewrite the expression above as

$$\begin{aligned}
P_{tot}(y) &= \frac{c_1}{c_1+c_2}\int_a^y p_o(x)dx - \frac{c_2}{c_1+c_2}\int_y^a p_o(z)dz, \\
&= \frac{c_1}{c_1+c_2}\int_a^y p_o(x)dx + \frac{c_2}{c_1+c_2}\int_a^y p_o(z)dz, \\
&= \frac{c_1+c_2}{c_1+c_2}\int_a^y p_o(x)dx, \\
&= \int_a^y p_o(x)dx = \text{Prob}\{X_2 \leq y\}.
\end{aligned}$$

so that Eq. (3.21) is verified. □

Given the previous theorem, it is straightforward to verify that the acceptance rate of the GBR algorithm is

$$\hat{a}_{GBR} = \frac{c_1+c_2}{L_B}, \tag{3.22}$$

where $c_1 = \int_{\mathcal{D}} p(x)dx$ and $c_2 = \int_{\mathcal{D}} h(x)dx$. Given a proposal $\pi(x)$, the GBR algorithm can outperform the standard RS method for a proper choice of the upper bound $L_B$ and the transformation $\rho$. The Payne-Dagpunar's scheme, described in the following section, provides an example of how to choose $\rho$ in the GBR setup.

### 3.4.3 Payne-Dagpunar's Band Rejection

The band rejection (BR) method was introduced in [11, 35] using

$$\rho^{-1}(x) = a + b - x, \qquad x \in [a, b],$$

which is the case shown in Fig. 3.6. Observe that in this case we have

$$h(x) = p(a + b - x),$$

and $c_1 = c_2$. If the best $L_B$ is available, i.e.,

$$L_B^* = \sup_{x \in \mathcal{D}} \frac{p(x) + p(a + b - x)}{\pi(x)},$$

then one can show that the BR scheme always provides better performance than a standard RS algorithm with the same proposal and using the best bound,

$$L^* = \sup_{x \in \mathcal{D}} \frac{p(x)}{\pi(x)}.$$

Indeed, we have

$$\hat{a}_{RS} = \frac{c_1}{L^*} \leq \hat{a}_{BR} = 2\frac{c_1}{L_B^*}, \tag{3.23}$$

where $\hat{a}_{RS}$ is the acceptance rate of the standard RS technique and $\hat{a}_{BR}$ is the acceptance rate of the BR method, since

$$\sup_{x \in \mathcal{D}} \frac{p(x) + p(a + b - x)}{\pi(x)} \leq 2 \sup_{x \in \mathcal{D}} \frac{p(x)}{\pi(x)},$$

i.e., $L_B^* \leq 2L^*$. Therefore, using the optimal upper bound the BR scheme always outperforms or, at least, obtains the same performance as the standard RS algorithm.

Moreover, since $L^* \leq L_B^* \leq 2L^*$, the best scenario for a BR scheme occurs when $L_B^* = L^*$, obtaining $\hat{a}_{BR} = 2\hat{a}_{RS}$ from Eq. (3.23), i.e., twice the acceptance rate w.r.t. the corresponding standard RS method.This is exactly the case in the example below.

*Example 3.6* Consider a triangular density

$$p_o(x) \propto p(x) = 1 - x, \quad 0 \le x \le 1.$$

Hence, we have $x \in \mathcal{D} = [a = 0, b = 1]$, and $c_1 = \int_{\mathcal{D}} p(x)dx = \frac{1}{2}$. Moreover, we choose $\rho^{-1}(x) = 1 - x$, so that

$$q(x) \propto h(x) = p(1 - x) = x, \quad 0 \le x \le 1,$$

and $c_2 = \int_{\mathcal{D}} h(x)dx = \frac{1}{2}$. We also use a uniform proposal pdf, $\pi(x) = 1, 0 \le x \le 1$, and compute the (optimal) upper bound

$$L_B^* = \sup_{x \in [0,1]} \frac{p(x) + h(x)}{\pi(x)} = \sup_{x \in [0,1]} (1 - x + x) = 1.$$

Therefore, in this example, the band rejection algorithm provides an exact direct sampler (i.e., the acceptance rate is $\frac{c_1 + c_2}{L_B^*} = 1$). The algorithm to generate the $i$th sample is:

- Draw $x' \sim \pi(x) = \mathcal{U}([0, 1])$ and another sample $u' \sim \mathcal{U}([0, 1])$.
- If $u' \le p(x') = 1 - x'$, then set $x^{(i)} = x'$. Otherwise, set $x^{(i)} = \rho^{-1}(x') = 1 - x'$.

Indeed, note that the second acceptance condition $u' > 1 - h(x') = 1 - x'$ always holds if the previous inequality, $u' \le p(x') = 1 - x'$, is not satisfied. A careful look at of the algorithm above shows that it can be summarized as (a) draw $u_1, u_2 \sim \mathcal{U}([0, 1])$ and (b) take $x' = \min(u_1, u_2)$. The acceptance rate of a standard RS algorithm, using the same proposal pdf $\pi(x) = \mathcal{U}([0, 1])$, is $\hat{a}_{RS} = 1/2$.

## 3.5 Acceptance-Complement Method

If we impose additional assumptions on the target $p_o(x)$, then we can construct variations of the rejection sampling algorithm that turn out to be more efficient for some problems. Let us assume that we are able to decompose our target pdf as the sum of two terms

$$p_o(x) = g_1(x) + g_2(x), \tag{3.24}$$

and define $\omega_1 = \int_{\mathcal{D}} g_1(x)dx$, $\omega_2 = \int_{\mathcal{D}} g_2(x)dx$. Note that, since $\omega_1 + \omega_2 = 1$, we can rewrite Eq. (3.24) as the discrete mixture

$$p_o(x) = \omega_1 h_1(x) + \omega_2 h_2(x) = \omega_1 h_1(x) + (1 - \omega_1)h_2(x), \tag{3.25}$$

where $h_1(x) \triangleq g_1(x)/\omega_1$ and $h_2(x) \triangleq g_2(x)/\omega_2$ are proper pdfs. Hence, if we are able to draw from $h_1(x)$ and $h_2(x)$, we can use the procedure for discrete mixtures described in Sect. 2.3.4 to generate samples from the target $p_o(x)$.

Let us consider the case in which we are able to draw directly from $h_2(x)$ but not from $h_1(x)$. However, we assume that there is a known pdf $\pi(x)$ such that

$$\pi(x) \geq g_1(x) = \omega_1 h_1(x), \tag{3.26}$$

and, therefore, it is possible to draw from $h_1$ by rejection sampling. Note that $\pi(x)$ is a normalized density (i.e., $\int_{\mathcal{D}} \pi(x)dx = 1$), unlike $g_1(x)$, as $\int_{\mathcal{D}} g_1(x)dx = \omega_1 < 1$. For this reason, we can set $L = 1$ in the inequality $L\pi(x) \geq g_1(x)$, recalling that we assume Eq. (3.26). Indeed, we will show that the acceptance-complement can only be applied with $L = 1$.

In this scenario, the standard procedure to generate a sample $x'$ from $p_o(x)$, as seen in Sect. 2.3.4, is:

1. Draw an index $j' \in \{1, 2\}$ with probabilities given by the weights $\omega_1, \omega_2 = 1-\omega_1$.
2. If $j' = 2$, then generate $x' \sim h_2(x)$.
3. If $j' = 1$, then:
   (a) Draw $x' \sim \pi(x)$ and $u' \sim \mathcal{U}([0, 1])$.
   (b) If $u' \leq \frac{\omega_1 h_1(x')}{\pi(x')}$, then return $x'$.
   (c) Otherwise, if $u' > \frac{\omega_1 h_1(x')}{\pi(x')}$, repeat from step (a).

An interesting variant of this approach, that avoids rejection steps, is the so-called acceptance-complement method [20, 21]. In order to draw a sample $x'$ from $p_o(x)$, the acceptance-complement algorithm performs the following steps.

1. Draw $x' \sim \pi(x)$ and $u' \sim \mathcal{U}([0, 1])$.
2. If $u' \leq \frac{\omega_1 h_1(x')}{\pi(x')}$, then accept $x'$.
3. Otherwise, if $u' > \frac{\omega_1 h_1(x')}{\pi(x')}$, then draw $x'' \sim h_2(x)$ and accept it.

Table 3.4 enumerates all the steps required by the acceptance-complement algorithm. Since we avoid any rejection, this approach is computationally more efficient and faster than the original procedure.

Furthermore, the acceptance-complement technique still yields exact, i.i.d. samples from the target distribution. This is guaranteed by the following theorem.

**Theorem 3.3 ([13, 20])** *Let us consider a target pdf* $p_o(x) = g_1(x) + g_2(x)$, *where* $g_j(x) = \omega_1 h_j(x)$, *with* $j = 1, 2$. *Moreover, consider three independent r.v.'s* $X_1$, $X_2$, *and* $U$ *such that* $X_1$ *has pdf* $\pi(x)$, $X_2$ *has pdf* $h_2(x)$, $U \sim \mathcal{U}([0, 1])$, *and* $\pi(x) \geq g_1(x) = \omega_1 h_1(x)$, $\forall x \in \mathcal{D}$ *with* $\mathcal{D} \equiv \mathbb{R}$. *In this case, we can state*

$$\begin{aligned} P(y) = Prob\left\{X_1 \leq y \middle| U \leq \frac{\omega_1 h_1(X_1)}{\pi(X_1)}\right\} + Prob\left\{X_2 \leq y \middle| U > \frac{\omega_1 h_1(X_1)}{\pi(X_1)}\right\} = \\ = Prob\{X_3 \leq y\}, \end{aligned} \tag{3.27}$$

*where* $X_3$ *has density* $p_o(x) \propto p(x)$.

**Table 3.4** Acceptance-complement algorithm

| |
|---|
| 1. Set $i = 1$. Let $N$ be the number of desired samples from $p_o(x)$ |
| 2. Find a suitable decomposition of the form of Eq. (3.25), i.e., $p_o(x) = g_1(x) + g_2(x)$, where $g_j(x) = \omega_1 h_j(x)$, with $j = 1, 2$ |
| 3. Draw samples $x'$ from $\pi(x)$ and $u'$ from $\mathcal{U}([0, 1])$ |
| 4. If $u' \leq \frac{g_1(x')}{\pi(x')}$ then accept $x^{(i)} = x'$ and set $i = i + 1$ |
| 5. If $u' > \frac{g_1(x')}{\pi(x')}$ generate $x''$ from $h_2(x) = \frac{g_2(x)}{\omega_2}$, set $x^{(i)} = x''$ and set $i = i + 1$ |
| 6. If $i > N$ then stop, else go back to step 3 |

*Proof* From Theorem 3.1, we can write

$$\text{Prob}\left\{X_1 \leq y \middle| U \leq \frac{\omega_1 h_1(X_1)}{\pi(X_1)}\right\} = \omega_1 \int_{-\infty}^{y} h_1(x)dx. \tag{3.28}$$

Moreover, the second term in (3.27) can be expressed as

$$\text{Prob}\left\{X_2 \leq y \middle| U > \frac{\omega_1 h_1(X_1)}{\pi(X_1)}\right\} = \text{Prob}\left\{X_2 \leq y\right\} \text{Prob}\left\{U > \frac{\omega_1 h_1(X_1)}{\pi(X_1)}\right\}, \tag{3.29}$$

due to the independence of the two events, since the r.v.'s $X_1$, $X_2$, and $U$ are independent. Thus we can also write

$$\begin{aligned}
\text{Prob}\left\{X_2 \leq y \middle| U > \frac{\omega_1 h_1(X_1)}{\pi(X_1)}\right\} &= \\
&= \left(\int_{-\infty}^{y} h_2(x)dx\right)\left(1 - \text{Prob}\left\{U \leq \frac{\omega_1 h_1(X_1)}{\pi(X_1)}\right\}\right) \\
&= \left(\int_{-\infty}^{y} h_2(x)dx\right)\left(1 - \int_{-\infty}^{+\infty} \frac{\omega_1 h_1(x)}{\pi(x)}\pi(x)dx\right) \\
&= \left(\int_{-\infty}^{y} h_2(x)dx\right)\left(1 - \omega_1 \int_{-\infty}^{+\infty} h_1(x)dx\right) \\
&= \left(\int_{-\infty}^{y} h_2(x)dx\right)(1 - \omega_1),
\end{aligned} \tag{3.30}$$

where Eq. (3.30) results from Prob$\{U \leq Y\} = E[Y]$ with $Y \leq 1$, since $U \sim \mathcal{U}([0, 1])$. Furthermore, since $\omega_2 = 1 - \omega_1$, we have

$$\text{Prob}\left\{X_2 \leq y \middle| U > \frac{\omega_1 h_1(X_1)}{\pi(X_1)}\right\} = \omega_2 \int_{-\infty}^{y} h_2(x)dx. \tag{3.31}$$

Therefore, given Eqs. (3.28) and (3.31) we can write

$$\begin{aligned} P(y) &= \omega_1 \int_{-\infty}^{y} h_1(x)dx + \omega_2 \int_{-\infty}^{y} h_2(x)dx \\ &= \int_{-\infty}^{y} p_o(x)dx = \text{Prob}\{X_3 \leq y\}, \end{aligned} \tag{3.32}$$

since $p_o(x) = \omega_1 h_1(x) + \omega_2 h_2(x)$. □

Finally let us remark that, if we use the acceptance-complement procedure of Table 3.4 with a proposal $\pi(x)$ such that

$$L\pi(x) \geq g_1(x) = \omega_1 h_1(x),$$

where $L \neq 1$, then the generated samples $x'$ are distributed according to the density

$$q(x) = \frac{\omega_1}{L} h_1(x) + \left(1 - \frac{\omega_1}{L}\right) h_2(x), \tag{3.33}$$

that is different from the target pdf $p_o(x)$.

## 3.6 RS with Stepwise Proposals

One natural way to classify rejection samplers is according to the type of proposal density they use to generate candidate samples. In this section, we focus on one such class. In particular, we study strip methods, i.e., RS algorithms in which samples are drawn from a finite mixture of piecewise uniform densities with disjoint support. Proposals of this type are simple and intuitive to design, yet flexible enough to attain good acceptance rates.

We start with a description of standard strip samplers in Sect. 3.6.1, followed by a generalization (*the inversion-rejection* method) in Sect. 3.6.2 that extends the technique to handle target densities which are not bounded or have an infinite support.

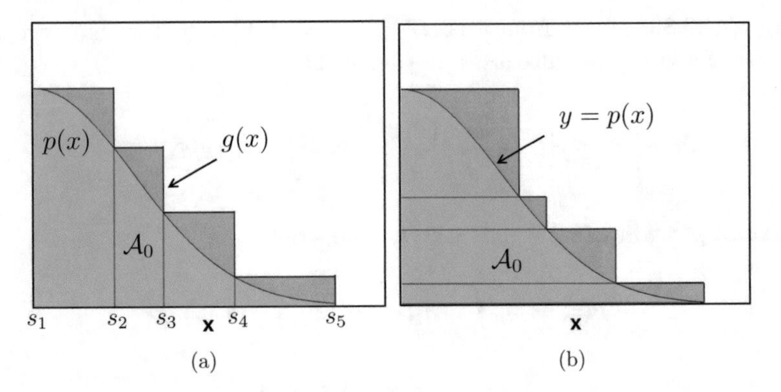


**Fig. 3.7** Examples of construction of a region $\mathcal{R}$ formed by four rectangles that covers the region $\mathcal{A}_0$ ($\mathcal{A}_0 \subseteq \mathcal{R}$) below $p(x)$. The bounded function $p(x) \propto p_o(x)$ is defined in a bounded domain. **(a)** The region $\mathcal{R}$ is composed of vertical bars. **(b)** The region $\mathcal{R}$ is composed of horizontal bars

### *3.6.1 Strip Methods*

The goal of this class of algorithms is to make the design of the sampling method as simple and fast as possible [19, Chap. 5]. Indeed, the central idea is straightforward: we simply cover the region $\mathcal{A}_0$ in Eq. (2.39) below the target pdf $p_o(x)$ with a union of rectangles forming an envelope region for $\mathcal{A}_0$, as depicted in Fig. 3.7. Clearly when the number of rectangles increases, this "strip" approximation of $\mathcal{A}_0$ becomes tighter and tighter.

The resulting algorithms are quite simple and fast, but they only work with bounded pdfs with finite support in their basic formulation.[2] Moreover, it is necessary to determine where $p_o(x)$ is increasing and where it is decreasing. For this reason, and in order to simplify the treatment, in this section we consider a monotonically decreasing target pdf $p_o(x) \propto p(x)$.

A strip method relies on building a stepwise envelope function

$$g(x) \geq p(x), \quad x \in \mathcal{D}.$$

There are two possibilities to construct it:

1. The first approach uses *vertical strips* (see Fig. 3.7a) and is known as *Ahrens method*, after J. Ahrens [1, 2, 17].
2. The second way, that uses *horizontal strips* (see Fig. 3.7b), is also called the *ziggurat method* [29].

[2] However, there exist some variations in the literature [34], [41, Chap. 4] that may extend the applicability of this technique.

In the sequel, we focus our attention on the first approach as the scheme with horizontal strips is equivalent. Consider, for the moment, a bounded pdf $p_o(x)$ with bounded domain $\mathcal{D} = [a, b]$. We choose a set of support points

$$\mathcal{S} = \{s_1 = a, s_2, \ldots, s_n, s_{n+1} = b\}$$

where $s_i \in \mathcal{D}$, $\forall i$ with $s_1 < s_2 < \ldots. < s_{n+1}$. We can define the $n$ intervals, $\mathcal{D}_i = [s_i, s_{i+1}]$, $1 \leq i \leq n$, that form a partition of the full support, i.e., $\mathcal{D} = \cup_{i=1}^n \mathcal{D}_i$. Moreover, since we assume that the function $p(x) \propto p_o(x)$ is decreasing (i.e., $p(s_i) \geq p(s_{i+1})$ for $i = 1, \ldots, n$), the rectangular set

$$\mathcal{R}_i \triangleq [s_i, s_{i+1}] \times [0, p(s_i)],$$

embeds the area below $p(x)$ for all $x \in \mathcal{D}_i$. Hence, the region composed by the union of these rectangular pieces, $\mathcal{R} = \cup_{i=1}^n \mathcal{R}_i$, covers the area below $p(x)$ for $x \in \mathcal{D} = [a, b]$. Furthermore, the stepwise function

$$g(x) \triangleq \begin{cases} p(s_1), & x \in \mathcal{D}_1, \\ \vdots & \\ p(s_i), & x \in \mathcal{D}_i, \\ \vdots & \\ p(s_n), & x \in \mathcal{D}_n. \end{cases} \tag{3.34}$$

yields an upper bound for $p(x)$, i.e., $p(x) \leq g(x)$ $\forall x$. Therefore, we can easily define a proposal density $\pi(x) \propto g(x)$ as a mixture of uniform pdfs

$$\pi(x) \triangleq \sum_{i=1}^{n} \omega_i \mathbb{I}_{\mathcal{D}_i}(x), \tag{3.35}$$

where $\mathbb{I}_{\mathcal{D}_i}(x)$ is the indicator function for the set $\mathcal{D}_i$, and the weights are defined as

$$\omega_i = \frac{|\mathcal{R}_i|}{\sum_{i=1}^n |\mathcal{R}_i|}, \tag{3.36}$$

where $|\mathcal{R}_i| = p(s_i)(s_{i+1} - s_i)$, for $i = 1, \ldots, n$. Hence, if we first draw an index $j'$ with $\text{Prob}\{j = j'\} = \omega_{j'}$ and a uniform sample $x'$ from $\mathcal{U}([s_{j'}, s_{j'+1}])$, then $x'$ is distributed as the proposal pdf $\pi(x)$ in Eq. (3.35). Thus, recalling also the envelope function $g(x) \propto \pi(x)$ in Eq. (3.34), the algorithm for drawing a sample $x'$ can be summarized as follows.

1. Draw $x' \sim \pi(x)$ in Eq. (3.35) and $u' \sim \mathcal{U}([0, 1])$.
2. If $u' \leq \frac{p(x')}{g(x')}$, then accept $x'$. Otherwise, discard it.

**Table 3.5** Vertical strip algorithm for a decreasing pdf $p_o(x)$

| |
|---|
| 1. Set $i = 1$. Let $N$ be the number of desired samples from $p_o(x)$ |
| 2. Draw an index $j' \in \{1, \dots, n\}$ with $\text{Prob}\{j = j'\} = \omega_{j'}$, as defined in Eq. (3.36) |
| 3. Generate a point $(x', u_2')$ uniformly within the rectangle $\mathcal{R}_{j'}$, i.e., $x' \sim \mathcal{U}([s_{j'}, s_{j'+1}])$ and $u_2' \sim \mathcal{U}([0, p(s_{j'})])$ |
| 4. If $u_2' \leq p(x')$, then accept $x^{(i)} = x'$ and set $i = i + 1$ |
| 5. Otherwise, if $u_2' > p(x')$, discard $x'$ and go back to step 2 |
| 6. If $i > N$, then stop. Else, go back to step 2 |

Table 3.5 provides a detailed description of the vertical strip algorithm. Recall that $g(x) = p(s_i)\ \forall x \in \mathcal{D}_i$.

Different strategies have been studied to choose the positions of the support points in order to decrease the overall computational cost of the rejection sampler [13, Chap. 8], [19, Chap. 5], [23]. Moreover, strip methods can easily improve the proposal pdf by adding new support points adaptively, following the schemes described in Chap. 4.

The next technique, called inversion-rejection method, can be seen as an extension of the strip methods to deal with unbounded pdfs or densities with unbounded domain.

### 3.6.2 Inversion-Rejection Method

In many cases, we may be able to calculate analytically the cdf $F_X(x)$ but not to invert it. Therefore, the inversion method in Sect. 2.4.1 cannot be applied. To overcome this problem, numerical inversion methods have been proposed in the literature [19, Chap. 7]. However, this approach can only be considered approximate, since the generated samples are not drawn *exactly* from $p_o(x)$. Furthermore, it can be computationally demanding.

Other approaches that start from the inversion principle and guarantee exact sampling have been studied. An example is the inversion-rejection method [12, 13], which is a combination of the inversion and rejection algorithms for the case when the cdf $F_X(x)$ is computable but not invertible.

For simplicity, we describe this technique for a bounded and decreasing target pdf, $p_o(x)$, with infinite support $\mathcal{D} = [a, +\infty)$. However, the algorithm can be easily extended to more general pdfs. Consider an infinite sequence of support points, sorted in ascending order,

$$\mathcal{S} = \{s_1 = a, s_2, \dots, s_i, s_{i+1} \dots\},$$

with $s_i \in \mathcal{D}$, $s_i < s_{i+1}$ for $i = 1, \dots, +\infty$.

Observe that the sequence is fixed but does not need to be stored: we can compute $s_i$ from the index $i$ or from the previous point $s_{i-1}$. For instance, equi-spaced intervals can be chosen such that $s_{i+1} - s_i = \delta$. This is really a crucial issue in order to design efficient samplers, as illustrated in [13, Chap. 7]. Moreover, we assume that the cdf $F_X(x)$ can be evaluated and a global upper bound $M$ is also available, i.e.,

$$M \geq p(x), \tag{3.37}$$

where $p(x) \propto p_o(x)$. The algorithm consists of the following steps.

1. Generate $v' \sim \mathcal{U}([0, 1])$.
2. Find the index $j'$ such that

$$F_X(s_j') \leq v' \leq F_X(s_{j'+1}). \tag{3.38}$$

   Thus, the interval $\mathcal{D}_{j'} = [s_{j'}, s_{j'+1}] \subset \mathcal{D}$ is chosen with probability Prob$\{j = j'\}$ $= F_X(s_{j'+1}) - F_X(s_{j'})$. This is always possible, since we can use, e.g., a *sequential search* to find $s_{j'}$ [11, 13, 19].
3. Draw a sample $x'$ uniformly from $\mathcal{D}_{j'} = [s_{j'}, s_{j'+1}]$, i.e., $x' \sim \mathcal{U}([s_{j'}, s_{j'+1}])$.
4. Generate $u'$ from $\mathcal{U}([0, 1])$.
5. If $Mu' \leq p(x')$, then accept $x'$.
6. Otherwise, if $Mu' > p(x')$, then discard $x'$.

Hence, we first select an interval $\mathcal{D}_i = [s_i, s_{i+1}]$ with probability $F_X(s_{i+1}) - F_X(s_i)$ by inversion. Since the interval $\mathcal{D}_i$ is closed and $p_o(x)$ is bounded, we can use a uniform pdf in $\mathcal{D}_i$ as proposal density and draw a sample $x'$ from $p_o(x)$ by rejection. Since we have assumed $p(x)$ to be a decreasing function, this procedure can be easily improved if we define a stepwise envelope function, $g(x) = p(s_i)$, $\forall x \in \mathcal{D}_i$, so that $g(x) \geq p(x)$ for all $x \in \mathcal{D}$ to be used in the rejection sampler. The inversion-rejection algorithm is described in Table 3.6.

**Table 3.6** Inversion-rejection algorithm for a decreasing pdf $p_o(x)$

| |
|---|
| 1. Set $i = 1$. Let $N$ be the number of desired samples from $p_o(x)$ |
| 2. Draw a index $j'$ with Prob$\{j = i\} = F_X(s_{i+1}) - F_X(s_i)$, $i = 1, \ldots, +\infty$, by inversion |
| 3. Generate a pair $x' \sim \mathcal{U}([s_{j'}, s_{j'+1}])$ and $u' \sim \mathcal{U}([0, g(x')])$, where $g(x) = p(s_{j'})$ for all $x \in [s_{j'}, s_{j'+1}]$ |
| 4. If $u' \leq p(x')$, then accept $x^{(i)} = x'$ and set $i = i + 1$ |
| 5. Otherwise, if $u' > p(x')$, discard $x'$ and go back to step 2 |
| 6. If $i > N$, then stop. Else, go back to step 2 |

Obviously, the algorithm in Table 3.6 is also a *strip* technique. However, unlike the strip algorithms in Sect. 3.6.1, this technique can be applied to unbounded pdfs and densities with infinite support as long as the cdf $F_X(x)$ can be evaluated. Hence, in this sense the inversion-rejection algorithm can be considered as an extension of the strip methods. On the other hand, it can also be viewed as a numerical inversion method with the addition of a rejection step.

The next technique is also related to the inversion algorithm. It tries to replace a non-invertible cdf, $F_X(x)$, with another invertible function which is as close as possible to $F_X(x)$.

## 3.7 Transformed Rejection Method

The transformed rejection method, due to [18, 44], is also called *almost exact inversion method* in [13, Chap. 3] and *exact-approximation method* in [28]. It is strongly related to the *ratio-of-uniforms* described in Chap. 5. Before applying the RS method, this technique transforms a generic target pdf $p_o(x)$ into another density bounded and with a bounded domain.[3] This is achieved by applying a suitable transformation $f$ to the target random variable $X$. This idea was suggested by several authors [5, 13, 16, 18, 28, 44] and is summarized in Table 3.7.

In this section we study the conditions required to obtain suitable transformations that modify target pdfs of the type displayed in Fig. 3.2a, b (bounded with unbounded support and unbounded with bounded support, respectively), converting them into a pdf of the type depicted in Fig. 3.2c (bounded with bounded support). We provide a detailed description of the different possibilities.

### *Bounded Target* $p_o(x)$ *with Unbounded Support*

Let $p_o(x)$ be a bounded density with unbounded support $\mathbb{R}$ (see Fig. 3.8a), and let us consider a monotonic differentiable function $f : \mathbb{R} \to (0, 1)$. If $X \sim p_o(x)$, then the transformed random variable $Z = f(X)$ has density

$$\rho(z) = p_o(f^{-1}(z)) \left| \frac{df^{-1}(z)}{dz} \right| = p(f^{-1}(z)) |\dot{f}^{-1}(z)|, \quad \text{for} \quad z \in (0, 1), \tag{3.39}$$

[3] Indeed, as discussed in Sect. 3.2.4, the simplest scenario to use the RS algorithm occurs when the target density is bounded with bounded support. In this case, the proposal $\pi(x)$ can be a uniform pdf and other RS schemes such as the band rejection method (Sect. 3.4) can be applied.

**Table 3.7** Transformed rejection sampling algorithm

| |
|---|
| 1. Set $i = 1$. Let $N$ be the number of desired samples from $p_o(x)$ |
| 2. Find a suitable transformation $Z = f(X)$ such that the r.v. $Z$ has a density $\rho(z) = p_o(f^{-1}(z)) \left\lvert \frac{df^{-1}(z)}{dz} \right\rvert$ which is bounded with a bounded domain |
| 3. Draw samples $z' \sim \pi(z)$ and $u' \sim \mathcal{U}([0, 1])$ |
| 4. If $u' \leq \frac{\rho(z')}{L\pi(z')}$, then set $x^{(i)} = f^{-1}(z')$ and $i = i + 1$ |
| 5. Otherwise, if $u' > \frac{\rho(z')}{L\pi(z')}$, discard $z'$ and go back to step 3 |
| 6. If $i > N$, then stop. Else, go back to step 3 |

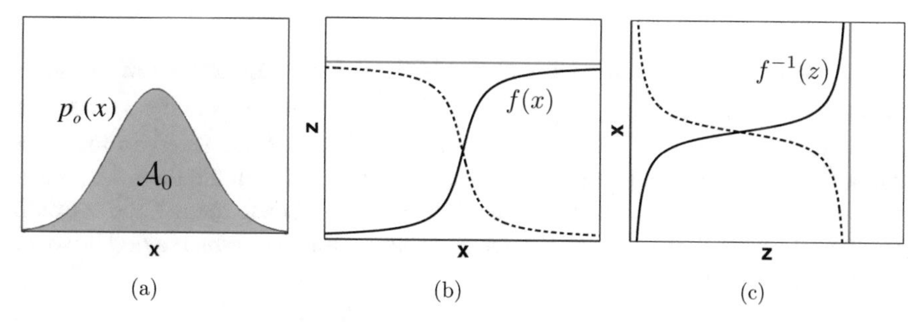


**Fig. 3.8** **(a)** A bounded target pdf $p_o(x)$ with an unbounded domain. **(b)** Two possible examples, monotonically increasing (solid line) and monotonically decreasing (dashed line), of the transformation $f(x)$ with horizontal asymptotes at $x = 0$ and $x = 1$. **(c)** The corresponding inverse transformations $f^{-1}(z)$ with vertical asymptotes at $z = 0$ and $z = 1$

where $f^{-1}(z)$ is the inverse of $f(x)$ and $\dot{f}^{-1}(z)$ denotes the first derivative of $f^{-1}(z)$. Obviously, the domain of $\rho(z)$, $\mathcal{D}_Z = (0, 1)$, is bounded. However, the density $\rho(z)$ can still be unbounded (i.e., it may have vertical asymptotes) depending on the choice of the transformation $f(x)$. Indeed, taking a closer look at (3.39), we notice that, although the first term $p(f^{-1}(z))$ is bounded (since $p_o(x)$ is assumed to be bounded), the second term, $|\dot{f}^{-1}(z)|$, is unbounded in general, since

$$\lim_{z\to 0} \left\lvert \frac{df^{-1}(z)}{dz} \right\rvert = \lim_{z\to 1} \left\lvert \frac{df^{-1}(z)}{dz} \right\rvert = \infty. \tag{3.40}$$

This is due to the fact that $f(x)$ must have horizontal asymptotes, since it is a monotonic continuous function that converts the infinite support into a finite domain, $\mathcal{D}_Z = (0, 1)$. Figure 3.8b, c provides an example.

Hence, it is clear from (3.39) and (3.40) that the density $\rho(z)$ only remains bounded when the tails of $p_o(x)$ decay to zero quickly enough, namely, faster than the derivative $\frac{df^{-1}(z)}{dz} = \left(\frac{df(x)}{dx}\right)^{-1}$ diverges when $z \to z^* \in \{0, 1\}$ (where $z^*$ denotes a vertical asymptote). More formally, let us note that the limit

$$\begin{aligned} L_1 &= \lim_{z\to z^*} \rho(z) \\ &= \lim_{z\to z^*} p\big(f^{-1}(z)\big)\left|\dot{f}^{-1}(z)\right| \\ &= \lim_{z\to z^*} \frac{p\big(f^{-1}(z)\big)}{\left|\dot{f}(x)\right|_{x=f^{-1}(z)}} \\ &= \lim_{x\to f^{-1}(z^*)} \frac{p(x)}{\left|\dot{f}(x)\right|} < \infty, \end{aligned}$$

with $z^* \in \{0, 1\}$, is finite if, and only if, $p\big(f^{-1}(z)\big)$ is an infinitesimal of the same or higher order than $\left|\dot{f}(x)\right|_{x=f^{-1}(z)}$ at $z = z^*$. Observe also that better acceptance rates can be obtained if $f(x)$ is similar to the cdf $F_X(x)$ of $X$. Indeed, in this case the pdf $\rho(z)$ becomes flatter and closer to a uniform density. In the limit, i.e., when $f(x) = F_X(x)$, then $\rho(z)$ is the uniform pdf in $\mathcal{D}_Z = (0, 1)$ and the method is exact. For this reason, this technique is also termed *almost exact inversion method* by some authors (see, e.g., [13]).

### Unbounded Target pdf $p_o(x)$ with Bounded Support

A similar methodology can also be applied when the target pdf, $p_o(x)$, is unbounded but has a bounded support, $\mathcal{D} = (a, b)$. Using again a monotonic, continuous function, $f : (a, b) \to (0, 1)$, with continuous derivative, we can also transform $p_o(x)$ into a bounded density with bounded domain, $\mathcal{D}_Z = (0, 1)$. Without loss of generality, let us assume that $p_o(x)$ has only one vertical asymptote at $x = x^*$, i.e., $\lim_{x\to x^*} p(x) = \infty$. Now, let us consider $Z = f(X)$ with $X \sim p_o(x)$. We already know that the density of $Z$ is given by

$$\rho(z) = p_o\big(f^{-1}(z)\big)\left|\dot{f}^{-1}(z)\right|, \qquad \text{for} \quad z \in (0, 1). \tag{3.41}$$

Unfortunately, although $\left|\dot{f}^{-1}(z)\right|$ is bounded (since we assume $\dot{f}^{-1}(z)$ to be continuous), $\rho(z)$ is unbounded in general, as the first term diverges, i.e.,

$$\lim_{z\to f(x^*)} p\big(f^{-1}(z)\big) = \lim_{x\to x^*} p(x) = \infty.$$

However, we notice that now the limit of interest,

$$\begin{aligned} L_2 &= \lim_{z \to f(x^*)} \rho(z) \\ &= \lim_{z \to f(x^*)} p(f^{-1}(z)) \left| \dot{f}^{-1}(z) \right| \\ &= \lim_{x \to x^*} p(x) \left| \dot{f}^{-1}(z) \right|_{z=f(x)} \\ &= \lim_{x \to x^*} \frac{p(x)}{\left| \dot{f}(x) \right|} < \infty, \end{aligned} \tag{3.42}$$

can be finite if $\left|\dot{f}^{-1}(z)\right|_{z=f(x)} \to 0$, or equivalently $\left|\dot{f}(x)\right| \to \infty$, when $x \to x^*$ (since $p_o(x) \to \infty$ when $x \to x^*$). Moreover, the limit $L_2$ is finite if, and only if, $1/\left|\dot{f}(x)\right|$ is an infinitesimal of equal or higher order than $1/p(x)$ at $x = x^*$.

### *3.7.1 Transformed Rejection and IoD Method*

The idea introduced above consists in converting the unbounded area below the target $p_o(x) \propto p(x)$ (e.g, see Fig. 2.2), defined by the set $\mathcal{A}_0$ in Eq. (2.39), into a bounded region. As we have seen in Sects. 2.4.3 and 2.4.4, given a monotonic target pdf, $p_o(x)$, we have another density, $p_o^{-1}(y)$, associated to the set $\mathcal{A}_0$. In this section, we show a complementary study, analyzing suitable transformations $h(y)$ applied to the r.v. $Y \sim p_o^{-1}(y)$, such that the resulting pdf, $q(\theta)$, of the r.v. $\Theta = h(Y) \sim q(\theta)$, has a bounded pdf with bounded support. Table 3.8 provides the algorithm for this case.

**Table 3.8** Transformed rejection combined with the IoD method

| |
|---|
| 1. Set $i = 1$. Let $N$ be the number of desired samples from $p_o(x)$ |
| 2. Find a monotonic transformation $h(y)$ so that<br>$q(\theta) = p_o^{-1}(h^{-1}(\theta))\left|\dot{h}^{-1}(\theta)\right|$ is bounded with bounded support |
| 3. Draw samples $\theta' \sim \pi(\theta)$ and $u' \sim \mathcal{U}([0,1])$ |
| 4. If $u' \leq \frac{q(\theta')}{L\pi(\theta')}$, then draw $v' \sim \mathcal{U}([0,1])$. Set $x^{(i)} = v' h^{-1}(\theta')$<br>and $i = i + 1$ |
| 5. Otherwise, if $u' > \frac{q(\theta')}{L\pi(\theta')}$, discard $\theta'$ and go back to step 3 |
| 6. If $i > N$, then stop. Else go back to step 3 |

### Bounded Target pdf $p_o(x)$ with Unbounded Support

Let us consider a monotonically decreasing and bounded target pdf, $p(x)$, with unbounded support $\mathbb{R}^+$, such that $p(0) = 1$ and $p(x) \to 0$ when $x \to \infty$. This implies that the inverse target pdf, $p_o^{-1}(y)$, is unbounded but has a bounded support, $\mathcal{D}_Y = (0, 1]$. The density of $\Theta = h(Y)$, with $Y \sim p_o^{-1}(y)$, is

$$q(\theta) = p_o^{-1}\big(h^{-1}(\theta)\big)\big|\dot{h}^{-1}(\theta)\big| \qquad \text{for} \quad \theta \in (0, 1]. \tag{3.43}$$

Now, since $p_o^{-1}(y) \to \infty$ when $y \to y^* = 0$, a necessary condition to obtain a bounded pdf $q(\theta)$ is

$$\lim_{\theta \to h(y^*)} \big|\dot{h}^{-1}(\theta)\big| = \lim_{y \to y^*} \big|\dot{h}(y)\big|^{-1} = 0.$$

Once more, focusing on the limit of interest in this case,

$$\begin{aligned} L_3 &= \lim_{\theta \to h(y^*)} q(\theta) \\ &= \lim_{\theta \to h(y^*)} p_o^{-1}\big(h^{-1}(\theta)\big)\big|\dot{h}^{-1}(\theta)\big| \\ &= \lim_{\theta \to h(y^*)} \frac{p_o^{-1}\big(h^{-1}(\theta)\big)}{\big|\dot{h}(y)\big|_{y=h^{-1}(\theta)}} \\ &= \lim_{y \to y^*} \frac{p_o^{-1}(y)}{\big|\dot{h}(y)\big|}, \end{aligned}$$

we realize that a necessary and sufficient condition to obtain a pdf $q(\theta)$ bounded with a bounded support is that $1/\big|\dot{h}(y)\big|$ is an infinitesimal of equal or higher order than $1/p_o^{-1}(y)$ at $y = y^* = 0$.

### Unbounded Target pdf $p_o(x)$ with Bounded Support

Consider now an unbounded and monotonically decreasing target pdf, $p_o(x)$, with a vertical asymptote at $x = x^* = 0$ (i.e., $\lim_{x \to x^*} p_o(x) = \infty$), but bounded support, $\mathcal{D} = (0, b]$. Hence, the inverse target pdf, $p_o^{-1}(y)$, is monotonically decreasing and bounded ($0 < p_o^{-1}(y) \leq b$), but has an unbounded support $\mathcal{D}_Y = \mathbb{R}^+$. Now, considering a continuous and monotonic transformation, $h : \mathbb{R}^+ \to (0, 1)$, the r.v. $\Theta = h(Y)$ with $Y \sim p_o^{-1}(y)$ has pdf

$$q(\theta) = p_o^{-1}\big(h^{-1}(\theta)\big)\big|\dot{h}^{-1}(\theta)\big| \qquad \text{for} \quad \theta \in (0, 1). \tag{3.44}$$

Again, although the first term, $p_o^{-1}\big(h^{-1}(\theta)\big)$, is bounded, $q(\theta)$ may be unbounded, since the second term is unbounded in general. Indeed, $h(y)$ must reach a horizontal asymptote when $y \to \infty$, so that $h^{-1}(\theta)$ has a vertical asymptote either at $\theta = \theta^* = 1$ (when $h^{-1}(\theta)$ is increasing) or at $\theta = \theta^* = 0$ (when $h^{-1}(\theta)$ is decreasing), implying that

$$\lim_{\theta \to \theta^*} \left|\dot{h}^{-1}(\theta)\right| = \lim_{y \to h^{-1}(\theta^*)} \left|\dot{h}(y)\right|^{-1} = \infty.$$

To obtain a bounded pdf $q(\theta)$, the limit of interest is

$$\begin{aligned} L_4 &= \lim_{\theta \to \theta^*} q(\theta) \\ &= \lim_{\theta \to \theta^*} p_o^{-1}\left(h^{-1}(\theta)\right)\left|\dot{h}^{-1}(\theta)\right| \\ &= \lim_{\theta \to \theta^*} \frac{p_o^{-1}\left(h^{-1}(\theta)\right)}{\left|\dot{h}(y)\right|_{y=h^{-1}(\theta)}} \\ &= \lim_{y \to h^{-1}(\theta^*)} \frac{p_o^{-1}(y)}{\left|\dot{h}(y)\right|}. \end{aligned}$$

Hence, a necessary and sufficient condition for having $L_4 < \infty$ is that $p_o^{-1}(y)$ is an infinitesimal of equal or higher order than $\left|\dot{h}(y)\right|$ at $y = h^{-1}(\theta^*)$.

## 3.8 Examples

In this section, we present some relevant RS schemes that serve both to illustrate the range of application of the RS technique and also to introduce some methodologies that will be used within other sampling algorithms in subsequent chapters. Furthermore, all the examples given in this section are non-intuitive (i.e., non-trivial) applications of the RS principle and provide a solution for relevant sampling problems, such as drawing from a mixture of pdfs with negative coefficients or drawing from a target pdf expressed as a sequence of functions (where the target may also be unknown analytically).

### *3.8.1 RS for Generating Order Statistics*

Assume that we wish to generate $n$ ordered samples $x_1 < x_2 < \ldots < x_n$ from $p_o(x) \propto p(x)$, and consider a proposal $\pi(x)$ and a constant $L$ such that $L\pi(x) \geq p(x)$. The following RS scheme can be used [13, Chap. 5]:

1. Choose a value $m > n$.
2. Generate ordered samples $x_1 < x_2 < \ldots < x_m$ from $\pi(x)$.

3. Draw $m$ i.i.d. uniform r.v.'s, $u_1, \ldots, u_m \sim \mathcal{U}([0,1])$.
4. For $i = 1, \ldots, m$, if $u_i > \frac{p_o(x_i)}{L\pi(x_i)}$, then delete $x_i$. We obtain a subset of $m^*$ ordered samples with $m^* < m$.
5. If $m^* < n$, then repeat from 2.
6. If $m^* \geq n$, then delete randomly chosen $m^* - n$ samples and stop.

### 3.8.2 Mixtures with Negative Coefficients

Consider a target density of the form

$$p_o(x) = \sum_{i=1}^{\infty} \alpha_i h_i(x), \tag{3.45}$$

where the functions $h_i(x)$ are normalized pdfs[4] and the constants $\alpha_i \in \mathbb{R}$ are real numbers such that

$$\sum_{i=1}^{\infty} \alpha_i = 1. \tag{3.46}$$

We remark that the $\alpha_i$, $i = 1, \ldots, +\infty$, can be either positive or negative. Hence, it is not straightforward to draw from a target expressed as in Eq. (3.45) (recall Sect. 2.3.4). However, as long as it is possible to find a decomposition

$$\alpha_i = \alpha_i^+ - \alpha_i^-, \quad \alpha_i^+ \geq 0, \quad \alpha_i^- \leq 0,$$

such that

$$\sum_{i=1}^{+\infty} \alpha_i^+ < +\infty, \tag{3.47}$$

we can design a valid algorithm based on the RS principle [4]. Indeed, setting

$$\pi(x) \propto \sum_{i=1}^{\infty} \alpha_i^+ h_i(x),$$

[4]We assume normalized functions for simplicity of presentation. Unnormalized, integrable functions can be handled as well.

the RS inequality is satisfied, i.e.,

$$p_o(x) \leq L\pi(x) = \sum_{i=1}^{+\infty} \alpha_i^+ h_i(x),$$

where $L = \int_{\mathcal{D}} \pi(x)dx = \sum_{i=1}^{+\infty} \alpha_i^+ < +\infty$, since we have assumed that $\int_{\mathcal{D}} h_i(x)dx = 1$, $\forall i \in \mathbb{N}$. Hence, a simple RS algorithm consists of the following steps:

1. Draw $x' \sim \pi(x) \propto \sum_{i=1}^{\infty} \alpha_i^+ h_i(x)$ using the technique described in Sect. 2.3.4 (e.g., using the inversion method). Note that it is not necessary to compute $L = \sum_{i=1}^{+\infty} \alpha_i^+$ explicitly. However, it is necessary to be able to compute the majorating series $L\pi(x)$ shown above.
2. Generate $u' \sim \mathcal{U}([0, 1])$.
3. If $u' \leq \frac{p_o(x')}{L\pi(x')}$, then accept $x'$. Otherwise, reject $x'$.

Clearly, this method is also valid for a finite mixture. Let us observe that, since $\sum_{i=1}^{\infty} \alpha_i = 1$, then $L = \sum_{i=1}^{+\infty} \alpha_i^+ \geq 1$. The acceptance rate of the algorithm is $\hat{a} = 1/L$.

*Example 3.7* Let us consider

$$p_o(x) = \frac{3}{4}(1 - x^2), \quad -1 \leq x \leq 1,$$

as the target pdf. Note that, it can be rewritten as

$$p_o(x) = \frac{3}{2}\left(\frac{1}{2}\right) - \frac{1}{2}\left(\frac{3}{2}x^2\right), \quad -1 \leq x \leq 1.$$

Then, we can identify $\alpha_1^+ = \frac{3}{2}$, $\alpha_2^+ = 0$, $\alpha_1^- = 0$ and $\alpha_2^- = -\frac{1}{2}$. Consequently, the proposal, $\pi(x) \propto \alpha_1^+ h_1(x) + \alpha_2^+ h_2(x)$, is a uniform pdf in $[-1, 1]$

$$\pi(x) = \frac{1}{2}, \quad -1 \leq x \leq 1,$$

where $h_1(x) = \frac{1}{2}$, $h_2(x) = \frac{3x^2}{2}$, and $L = \alpha_1^+ = \frac{3}{2}$. The RS ratio is then

$$\frac{p_o(x)}{L\pi(x)} = 1 - x^2,$$

and the algorithm turns out to be simply the following:

1. Draw $x' \sim \mathcal{U}([-1, 1])$,
2. Draw $u' \sim \mathcal{U}([0, 1])$,
3. If $u' \leq 1 - (x')^2$, then accept $x'$. Otherwise, reject $x'$.

The acceptance rate is $1/L = 2/3$.

### 3.8.3 Pdfs Expressed as Sequences of Functions

Consider a target pdf $p_o(x) \propto p(x)$ that we are not able to evaluate analytically, but we know two sequences of functions, $\varphi_n(x)$ and $\psi_n(x)$, such that

$$\begin{aligned} &\lim_{n\to+\infty} \varphi_n(x) = p(x), \\ &\lim_{n\to+\infty} \psi_n(x) = p(x), \\ &\varphi_n(x) \leq p(x) \leq \psi_n(x), \qquad \text{for } n = 1, 2, \ldots, \infty. \end{aligned} \tag{3.48}$$

Namely, $p(x)$ can be expressed using two sequences of functions that converge to it from above and from below, respectively. Furthermore, we assume known a proposal pdf $\pi(x)$ and a bound $L$, such that

$$L\pi(x) \geq p(x),$$

or, alternatively,

$$L\pi(x) \geq \psi_n(x) \geq p(x), \quad \forall n \in \mathbb{N}.$$

In both cases the algorithm that we show below is valid. The first case is preferrable in terms of acceptance rate. However, in the second case we could also construct a sequence of upper bounds, $\{L_n\}_{n=0}^{+\infty}$, such that $L_n\pi(x) \geq \psi_n(x)$ to improve the acceptance rate.

An algorithm for sampling from $p_o(x) \propto p(x)$ in this case based on the RS principle is described in Table 3.9 [13, Chap. 5]. Note that $\varphi_n(x)$ is used as a squeeze function. Indeed, the proposed sample $x'$ is accepted immediately if $u'L\pi(x') \leq \varphi_n(x')$, where $u' \sim \mathcal{U}([0, 1])$. Otherwise, if $\varphi_n(x') \leq u'L\pi(x') \leq \psi_n(x')$, a better approximation of the target is needed (i.e., $n$ has to be increased) and the RS tests are repeated. Finally, if $u'L\pi(x') > \psi_n(x')$ the sample $x'$ is definitely rejected.

**Table 3.9** RS for a target pdf expressed as a series

| |
|---|
| 1. Set $i = 1$. Let $N$ be the number of desired samples from $p_o(x)$ |
| 2. Draw samples $x' \sim \pi(x)$ and $u' \sim \mathcal{U}([0, 1])$ |
| 3. Set $n = 1$ |
| 4. If $u' \leq \frac{\varphi_n(x')}{L\pi(x')}$, then set $x^{(i)} = x'$, $i = i + 1$ and go to step 6 |
| 5. If $u' \leq \frac{\psi_n(x')}{L\pi(x')}$, then update $n = n + 1$ and repeat from 4<br>Otherwise, if $u' > \frac{\psi_n(x')}{L\pi(x')}$, reject $x'$ |
| 6. If $i > N$, then stop. Else, go back to step 2 |

**Table 3.10** RS for a target expressed as a series with a known upper bound of the remainder

| |
|---|
| 1. Set $i = 1$. Let $N$ be the number of desired samples from $p_o(x)$ |
| 2. Draw samples $x' \sim \pi(x)$ and $u' \sim \mathcal{U}([0, 1])$ |
| 3. Set $w' = u'L\pi(x')$, $n = 0$, and $s' = 0$ |
| 4. Update $n = n + 1$ and $s' = s' + s_n(x')$ |
| 5. If $\lvert w' - s' \rvert \leq R_{n+1}(x')$, then repeat from step 4 |
| 6. If $w' > s'$, then reject $x'$ and repeat from step 2 |
| 7. If $w' \leq s'$, then set $x^{(i)} = x'$ and $i = i + 1$ |
| 8. If $i > N$, then stop. Else, go back to step 2 |

An interesting variant occurs when the target can be expressed as a series and the remainder can be upper-bounded. This case lends itself naturally to be handled with this scheme. Namely, assume that the target is given by

$$p_o(x) = \sum_{i=1}^{+\infty} s_i(x) \leq L\pi(x), \tag{3.49}$$

and an upper bound, $R_{n+1}(x)$, of the remainder (i.e., the excess error obtained truncating the series at the $n$th term of the sum) is known, i.e.,

$$\left| \sum_{i=n+1}^{+\infty} s_i(x) \right| \leq R_{n+1}(x). \tag{3.50}$$

The resulting algorithm [13, Chap. 5] is given in Table 3.10.

## 3.9 Monte Carlo Estimation via RS

Rejection samplers can be used as random number generators for simulating complex systems or, alternatively, for Monte Carlo integration. Consider, for instance, the approximation of the integral

$$I = \int_{\mathcal{D}} f(x) p_o(x) dx. \tag{3.51}$$

Using RS we can generate a set of samples, $\{x^{(i)}, i = 1, \ldots, N_a\}$, and then compute the approximation

$$\hat{I}_{N_a} = \frac{1}{N_a} \sum_{i=1}^{N_a} f(x^{(i)}) \approx I. \tag{3.52}$$

When $N_a \to \infty$, $\hat{I}_{N_a} \to I$ under mild assumptions [22, 24, 37]. In order to obtain $N_a$ samples via RS we need to generate $N_{tot} \geq N_a$ samples, since $N_r = N_{tot} - N_a$ samples are rejected. However, these rejected samples can be reused to design better estimators, as shown in the next Sect. 3.9.1. Alternatively, RS can be mixed with other Monte Carlo approaches (such as importance sampling and MCMC techniques) as shown in Sect. 3.9.2. Below, we compare the RS and importance sampling approaches considering the same target and proposal functions.

## *Rejection Sampling as a Special Case of Importance Sampling*

Let us consider an extended target density defined as

$$p^*(x, y) \propto \begin{cases} L\pi(x), & \text{if } x \in \mathcal{D}, \text{ and } y \in \left[0, \dfrac{p(x)}{L\pi(x)}\right], \\ 0, & \text{otherwise.} \end{cases} \tag{3.53}$$

Note that, marginalizing w.r.t. the variable $y$, we have

$$p^*(x) = \int_0^{\frac{p(x)}{L\pi(x)}} p^*(x, y)dy \propto p(x),$$

i.e., the marginal pdf is $p^*(x) = p_o(x)$. Let us define the extended proposal pdf

$$\pi^*(x, y) \propto \begin{cases} \pi(x), & \text{if } x \in \mathcal{D}, \text{ and } y \in \left[0, \dfrac{p(x)}{L\pi(x)}\right], \\ 0, & \text{otherwise.} \end{cases} \tag{3.54}$$

Now, drawing $N_{tot}$ pairs $(x^{(i)}, y^{(i)})$, $i = 1, \ldots, N_{tot}$, from the extended proposal $\pi^*(x, y)$, the associated importance weights are

$$w_{RS}^{(i)} = w(x^{(i)}, y^{(i)}) = \frac{p^*(x^{(i)}, y^{(i)})}{\pi^*(x^{(i)}, y^{(i)})} \propto \begin{cases} L, & \text{if } x \in \mathcal{D}, \text{ and } y \in \left[0, \dfrac{p(x)}{L\pi(x)}\right], \\ 0, & \text{otherwise.} \end{cases}$$

and the corresponding importance sampling (IS) estimator is [24, 37]

$$\hat{I}^{RS}_{N_{tot}} = \frac{1}{\sum_{n=1}^{N_{tot}} w^{(n)}_{RS}} \sum_{i=1}^{N_{tot}} w^{(i)}_{RS} f(x^{(i)}), \tag{3.55}$$

$$= \frac{1}{N_a} \sum_{i=1}^{N_a} f(x^{(i)}) = \hat{I}_{N_a}. \tag{3.56}$$

Namely, $\hat{I}^{RS}_{N_{tot}}$ is equivalent to the RS estimator $I_{N_a}$ in Eq. (3.52) [9]. Note that the number $N_a$ of the samples with $w^{(i)}_{RS} = L > 0$ coincides with the number of the auxiliary samples $y^{(i)}$ such that $y^{(i)} \in \left[0, \frac{p(x)}{L\pi(x)}\right]$. The standard IS estimator obtained with the same samples $x^{(i)} \sim \pi(x)$, $i = 1, \ldots, N$, is

$$\hat{I}^{IS}_{N_{tot}} = \frac{1}{\sum_{n=1}^{N_{tot}} w^{(n)}_{IS}} \sum_{i=1}^{N_{tot}} w^{(i)}_{IS} f(x^{(i)}), \quad \text{with } w^{(i)}_{IS} = \frac{p(x^{(i)})}{\pi(x^{(i)})}. \tag{3.57}$$

It is possible to show that the estimator $\hat{I}^{RS}_{N_{tot}}$ is less efficient than the estimator $\hat{I}^{IS}_{N_{tot}}$ (i.e., with less variance) [9]. Therefore, considering the same proposal pdf, the standard IS approach is preferable (although an additional study about the bias is required). However, in the following, we discuss how it is possible to improve the RS estimators by reusing the rejected samples or combining RS with other Monte Carlo schemes.

### 3.9.1 Recycling Rejected Samples

The rejected samples can be "recycled," with a slight computational overhead, to provide a better approximation of the target integral [8]. More specifically, let $p_o(x)$ be the target pdf and let $\{y^{(1)}, \ldots, y^{(N_r)}\}$ be the set of samples discarded in the RS test. As shown in Sect. 3.2.2, the rejected samples are distributed as

$$q_r(y) = \frac{1}{L-1}(L\pi(y) - p_o(y)),$$

where $\pi(x)$ is the proposal pdf used in the RS technique. Resorting to the importance sampling principle we can build an estimator using only the rejected samples

$$\hat{I}_{N_r} = \frac{1}{N_r} \sum_{i=1}^{N_r} \frac{p_o(y^{(i)})}{q_r(y^{(i)})} f(y^{(i)}) = \frac{1}{N_r} \sum_{i=1}^{N_r} \frac{(L-1)p_o(y^{(i)})}{L\pi(y^{(i)}) - p_o(y^{(i)})} f(y^{(i)}). \tag{3.58}$$

Moreover, we can define another estimator as a convex combination of $\widehat{I}_{N_a}$ and $\widehat{I}_{N_r}$, namely

$$\widehat{I}_{N_{tot},1} = \frac{N_a}{N_a + N_r}\widehat{I}_{N_a} + \frac{N_r}{N_a + N_r}\widehat{I}_{N_r}. \tag{3.59}$$

All these $\widehat{I}_{N_a}$, $\widehat{I}_{N_r}$, and $\widehat{I}_{N_{tot},1}$ are unbiased estimators of $I$ [8, 37]. It appears intuitive that $\widehat{I}_{N_r}$ provides the worst performance among them and that $\widehat{I}_{N_{tot},1}$ is the best estimator (in terms of its variance). A theoretical comparison of these three estimators is given in [8]. Another possible (but biased) estimator is

$$\widehat{I}_{N_{tot},2} = \frac{N_a}{N_a + N_r}\widehat{I}_{N_a} + \frac{N_r}{N_a + N_r}\left(\frac{1}{S_{N_r}}\sum_{i=1}^{N_r} w(y^{(i)})f(y^{(i)})\right), \tag{3.60}$$

where

$$w(y^{(i)}) = \frac{(L-1)p_o(y^{(i)})}{L\pi(y^{(i)}) - p_o(y^{(i)})} \quad \text{and} \quad S_{N_r} = \sum_{j=1}^{N_r} w(y^{(j)}).$$

Although $\widehat{I}_{N_{tot},2}$ is biased, it can be shown that $\widehat{I}_{N_{tot},2}$ outperforms $\widehat{I}_{N_{tot},1}$ under certain conditions [8].

### 3.9.2 RS with a Generic Constant $L > 0$

In the previous sections, we have already remarked that finding a suitable constant $L$ such that

$$L\pi(x) \geq p(x), \quad \forall x \in \mathcal{D},$$

can be a non-trivial task. In this section, we assume that we are using a generic positive constant $L$ such that the RS condition $L\pi(x) \geq p(x)$ is not necessarily fulfilled for every $x \in \mathcal{D}$. In this case, a sample $x'$ in the region $\{x \in \mathcal{D} : L\pi(x) < p(x)\}$ will always be accepted by the RS test. Indeed, the accepted samples using the RS technique are distributed as $q(x) \propto \min\{L\pi(x), p(x)\}$ (see Sect. 3.2.3).

Let us consider again the problem of the approximation of the integral

$$I = \int_{\mathcal{D}} f(x)p_o(x)dx \tag{3.61}$$

using samples generated by an RS method with a non-suitable, generic constant $L > 0$. In this case, the accepted samples, $\{x^{(i)}, i = 1, \ldots, N_a\}$, have a pdf $q(x) \propto$

$\min\{L\pi(x), p(x)\}$, and the estimator

$$\hat{I}_{N_a} = \frac{1}{N_a} \sum_{i=1}^{N_a} f(x^{(i)}),$$

is not adequate, i.e., $\hat{I}_{N_a} \nrightarrow I$ as $N_a \to \infty$. In this section, we consider two possible solutions: (a) combining the RS method and importance sampling and (b) combining the RS method with Markov Chain Monte Carlo (MCMC) algorithms. With some modifications, these schemes could also be used jointly with the previous considerations in Sect. 3.9.1 for recycling the rejected samples.

### Rejection Control

The technique described below, called *rejection control* (RC) [24], combines the rejection and the importance sampling techniques. Given a sample $x^{(i)} \sim \pi(x)$, accepted in the RS test with a generic positive constant $L$, the rejection control works as follows:

- If the accepted sample $x^{(i)}$ belongs to the region $\mathcal{A}_{true} = \{x \in \mathcal{D} : L\pi(x) \geq p(x)\}$, it is assigned the weight $w(x^{(i)}) = 1$.
- Otherwise, if the accepted sample $x^{(i)}$ belongs to the region $\mathcal{A}_{false} = \{x \in \mathcal{D} : L\pi(x) < p(x)\}$, then it is assigned the weight $w(x^{(i)}) = \frac{p(x^{(i)})}{L\pi(x^{(i)})}$.

The final estimator, using the accepted samples and the RC weights, is

$$\hat{I}_{N_a,RC} = \frac{1}{\sum_{i=1}^{N_a} w(x^{(i)})} \sum_{i=1}^{N_a} w(x^{(i)}) f(x^{(i)}). \tag{3.62}$$

Note that the RC method is equivalent to using a *modified* proposal density in an importance sampling scheme. Indeed, let us consider the proposal pdf

$$\pi^*(x) \propto \min\{L\pi(x), p(x)\}, \tag{3.63}$$

which is the density of the accepted samples with a generic $L$ (see Sect. 3.2.3). The weights of the RC method become

$$w(x^{(i)}) = \frac{p(x^{(i)})}{\min\{L\pi(x^{(i)}), p(x^{(i)})\}},$$

where $x^{(i)} \sim \pi^*(x)$. The proposal in Eq. (3.63) is clearly closer to the target $p_o(x) \propto p(x)$ than $\pi(x)$, thus providing a better estimation of the integral, as stated in the following theorem.

**Theorem 3.4 ([24])** *The rejection control technique reduces the discrepancy between proposal and target densities w.r.t. a standard importance sampling scheme. As a consequence, it is possible to show that*

$$\text{var}_{\pi^*}\left[\frac{p(x)}{\pi^*(x)}\right] \leq \text{var}_{\pi}\left[\frac{p(x)}{\pi(x)}\right]. \qquad \square$$

## Rejection Sampling Chains

When a suitable constant $L > 0$ for an RS scheme is unknown, an MCMC method can be applied to guarantee that the generated samples are distributed according to the target, $p_o(x) \propto p(x)$ [39, 40] (see Chap. 7). For a generic value of $L$, the density $\pi^*(x)$, given by Eq. (3.63) can be used as the proposal pdf in an independent *Metropolis-Hastings* (MH) algorithm [14, 24, 31, 32, 37] (see Chap. 7). Given a sample $x' \sim \pi^*(x)$ and a previous state $x_{t-1}$ of the Markov chain, the probability of accepting a new state in the MH method is

$$\alpha(x_{t-1}, x') = \min\left[1, \frac{p(x')\pi^*(x_{t-1})}{p(x_{t-1})\pi^*(x')}\right]. \tag{3.64}$$

The convergence of the chain to the target $p_o(x)$ is guaranteed as $t \to +\infty$. Table 3.11 describes the algorithm: an RS scheme is used to drive an independent MH algorithm.

Observe that, in Table 3.11, we have used the definition of $\pi^*(x) \propto \min\{L\pi(x), p(x)\}$ in the expression of $\alpha(x_{t-1}, x')$. Note also that in this case the resulting samples are correlated.

**Table 3.11** Rejection sampling chains

| |
|---|
| 1. Set $t = 1$, choose $L > 0$ and an initial state $x_0$. Let $N$ be the maximum number of iterations |
| 2. Draw samples $x' \sim \pi(x)$ and $u' \sim \mathcal{U}([0, 1])$ |
| 3. If $u' > \frac{p(x')}{L\pi(x')}$ then repeat from step 2 |
| 4. Otherwise, if $u' \leq \frac{p(x')}{L\pi(x')}$, set $x_t = x'$ with probability $\alpha(x_{t-1}, x') = \min\left[1, \frac{p(x')\min\{L\pi(x_{t-1}),p(x_{t-1})\}}{p(x_{t-1})\min\{L\pi(x'),p(x')\}}\right]$, or $x_t = x_{t-1}$ with probability $1 - \alpha(x_{t-1}, x')$ |
| 5. Update $t = t + 1$. If $t > N$, then stop. Else, go back to step 2 |

### Alternative Expression of the Acceptance Probability $\alpha$

In order to compare an RS chain with a standard MH algorithm using an independent proposal, it is interesting to observe that (as noted by L. Tierney in [40]), defining the set

$$\mathcal{C} = \{x \in \mathcal{D} : p(x) \leq L\pi(x)\},$$

the probability $\alpha(x, y)$ of accepting a new state $y$ given a previous state $x$ can be rewritten as

$$\alpha(x, y) = \begin{cases} 1, & \text{for } x \in \mathcal{C}, \\ \frac{L\pi(x)}{p(x)}, & \text{for } x \notin \mathcal{C}, y \in \mathcal{C}, \\ \min\left[1, \frac{p(y)\pi(x)}{p(x)\pi(y)}\right], & \text{for } x \notin \mathcal{C}, y \notin \mathcal{C}. \end{cases} \tag{3.65}$$

Equations (3.65) and (3.64) are completely equivalent. Indeed, if $x \in \mathcal{C}$, the modified proposal in Eq. (3.63) becomes

$$\pi^*(x) \propto \min\{L\pi(x), p(x)\} = p(x), \tag{3.66}$$

by definition of $\mathcal{C}$ and, substituting (3.66) into (3.64) we obtain

$$\begin{aligned} \alpha(x, y) &= \min\left[1, \frac{p(y)\pi^*(x)}{p(x)\pi^*(y)}\right] \\ &= \min\left[1, \frac{p(y)p(x)}{p(x)\min\{L\pi(y), p(y)\}}\right] \\ &= \min\left[1, \frac{p(y)}{\min\{L\pi(y), p(y)\}}\right] \\ &= 1, \end{aligned} \tag{3.67}$$

since the denominator is always smaller than the numerator. The first part of Eq. (3.65) is hence proved.

Moreover, if $x \notin \mathcal{C}$ and $y \in \mathcal{C}$, then we have $\pi^*(x) \propto L\pi(x)$ and $\pi^*(y) \propto p(y)$. Thus, replacing $\pi^*(x)$ and $\pi^*(y)$ in Eq. (3.64), the acceptance probability becomes

$$\begin{aligned} \alpha(x, y) &= \min\left[1, \frac{p(y)\pi^*(x)}{p(x)\pi^*(y)}\right] \\ &= \min\left[1, \frac{p(y)L\pi(x)}{p(x)p(y)}\right] \end{aligned} \tag{3.68}$$

$$= \min\left[1, \frac{L\pi(x)}{p(x)}\right]$$
$$= \frac{L\pi(x)}{p(x)},$$

since $L\pi(x) < p(x)$ when $x \notin \mathcal{C}$. Finally, considering $x \notin \mathcal{C}$ and $y \notin \mathcal{C}$, then $\pi^*(x) \propto L\pi(x)$ and $\pi^*(y) \propto L\pi(y)$, and the $\alpha$ in Eq. (3.64) can be rewritten in this case as

$$\begin{aligned} \alpha(x, y) &= \min\left[1, \frac{p(y)\pi^*(x)}{p(x)\pi^*(y)}\right] \\ &= \min\left[1, \frac{p(y)L\pi(x)}{p(x)L\pi(y)}\right] \\ &= \min\left[1, \frac{p(y)\pi(x)}{p(x)\pi(y)}\right]. \end{aligned} \tag{3.69}$$

Hence, Eqs. (3.65) and (3.64) are equivalent.

## 3.10 Summary

In this chapter, we have introduced the basic theory of the rejection sampling (RS) technique. We have shown that the RS method is an important tool for Monte Carlo algorithms, since it is a universal sampler that can be used to generate random samples from arbitrary target densities. Indeed, its applicability is only limited by the ability of finding a suitable upper bound of the ratio between the target and the proposal pdfs. We have also discussed performance and computational cost. The main figure of merit of an RS scheme is the acceptance rate, which depends strongly on the discrepancy between the shape of the proposal and the target pdfs. In general, the acceptance rate is improved by constructing more elaborate proposal functions but, in turn, such construction is often associated to computationally complex procedures.

In Sect. 3.4, we have introduced a little-known and interesting variant of the standard RS method: the band rejection (BR) approach. BR provides better performance compared to the standard RS scheme, but can only be applied for target pdfs with bounded support. Another interesting variant, described in Sect. 3.5, is the acceptance-complement method, which can be applied when the target can be expressed as a discrete mixture of two pdfs.

Strip methods have been briefly introduced in Sect. 3.6.1. They are RS algorithms in which the proposal pdf is a finite mixture of piece-wise uniform densities with disjoint support. The combination of the RS technique with the inversion method has been presented in Sect. 3.6.2 whereas the joint use of the RS method with a transformation of a random variable has been described in Sect. 3.7.1. We have also shown the power and versatility of RS methodologies with some relevant examples (Sect. 3.8), such as sampling a mixture of pdfs with negative coefficients or drawing from a target that can only be expressed as a series.

Finally, the efficient use of the RS technique for the important problem of numerical integration has been tackled in Sect. 3.9. First, we have described how the rejected samples can be recycled to improve the approximation of the target integral. Then, we have analyzed how proper estimators can be designed when the RS scheme is run using a generic, non-adequate, constant $L > 0$. In this case, we have considered two possibilities: (a) the combination of the RS method with the importance sampling approach and (b) the use of the RS technique to drive an MCMC algorithm.

## References


1. J.H. Ahrens, Sampling from general distributions by suboptimal division of domains. Grazer Math. Berichte **319**, 20 (1993)
2. J.H. Ahrens, A one-table method for sampling from continuous and discrete distributions. Computing **54**(2), 127–146 (1995)
3. N.C. Beaulieu, C. Cheng, Efficient Nakagami-*m* fading channel simulation. IEEE Trans. Veh. Technol. **54**(2), 413–424 (2005)
4. A. Bignami, A. De Matteis, A note on sampling from combinations of distributions. J. Inst. Math. Appl. **8**(1), 80–81 (1971)
5. C. Botts, W. Hörmann, J. Leydold, Transformed density rejection with inflection points. Stat. Comput. **23**, 251–260 (2013)
6. J.C. Butcher, Random sampling from the normal distribution. Comput. J. **3**, 251–253 (1960)
7. B.S. Caffo, J.G. Booth, A.C. Davison, Empirical supremum rejection sampling. Biometrika **89**(4), 745–754 (2002)
8. G. Casella, C.P. Robert, Post-processing accept-reject samples: recycling and rescaling. J. Comput. Graph. Stat. **7**(2), 139–157 (1998)
9. R. Chen, Another look at rejection sampling through importance sampling. Stat. Probab. Lett. **72**, 277–283 (2005)
10. R.C.H. Cheng, The generation of gamma variables with non-integral shape parameter. J. R. Stat. Soc. Ser. C Appl. Stat. **26**, 71–75 (1977)
11. J. Dagpunar, *Principles of Random Variate Generation* (Clarendon Press, Oxford/New York, 1988)
12. L. Devroye, Random variate generation for unimodal and monotone densities. Computing **32**, 43–68 (1984)
13. L. Devroye, *Non-Uniform Random Variate Generation* (Springer, New York, 1986)
14. W.R. Gilks, N.G. Best, K.K.C. Tan, Adaptive rejection Metropolis sampling within Gibbs sampling. Appl. Stat. **44**(4), 455–472 (1995)
15. J.A. Greenwood, Moments of time to generate random variables by rejection. Ann. Inst. Stat. Math. **28**, 399–401 (1976)
16. W. Hörmann, The transformed rejection method for generating Poisson random variables. Insur. Math. Econ. **12**(1), 39–45 (1993)
17. W. Hörmann, A note on the performance of the Ahrens algorithm. Computing **69**, 83–89 (2002)

18. W. Hörmann, G. Derflinger, The transformed rejection method for generating random variables, an alternative to the ratio of uniforms method. Commun. Stat. Simul. Comput. **23**, 847–860 (1994)
19. W. Hörmann, J. Leydold, G. Derflinger, *Automatic Nonuniform Random Variate Generation* (Springer, New York, 2003)
20. R.A. Kronmal, A.V. Peterson, A variant of the acceptance-rejection method for computer generation of random variables. J. Am. Stat. Assoc. **76**(374), 446–451 (1981)
21. R.A. Kronmal, A.V. Peterson, An acceptance-complement analogue of the mixture-plus-acceptance-rejection method for generating random variables. ACM Trans. Math. Softw. **10**(3), 271–281 (1984)
22. P.K. Kythe, M.R. Schaferkotter, *Handbook of Computational Methods for Integration* (Chapman and Hall/CRC, Boca Raton, 2004)
23. J. Leydold, J. Janka, W. Hörmann, Variants of transformed density rejection and correlation induction, in *Monte Carlo and Quasi-Monte Carlo Methods 2000* (Springer, Heidelberg, 2002), pp. 345–356
24. J.S. Liu, *Monte Carlo Strategies in Scientific Computing* (Springer, New York, 2004)
25. J.S. Liu, R. Chen, W.H. Wong, Rejection control and sequential importance sampling. J. Am. Stat. Assoc. **93**(443), 1022–1031 (1998)
26. D. Luengo, L. Martino, Almost rejectionless sampling from Nakagami-$m$ distributions ($m \geq 1$). IET Electron. Lett. **48**(24), 1559–1561 (2012)
27. I. Lux, Another special method to sample some probability density functions. Computing **21**, 359–364 (1979)
28. G. Marsaglia, The exact-approximation method for generating random variables in a computer. Am. Stat. Assoc. **79**(385), 218–221 (1984)
29. G. Marsaglia, W.W. Tsang, The ziggurat method for generating random variables. J. Stat. Softw. **8**(5), 1–7 (2000)
30. L. Martino, J. Míguez, A novel rejection sampling scheme for posterior probability distributions, in *IEEE International Conference on Acoustics, Speech and Signal Processing (ICASSP)* (2009)
31. L. Martino, J. Read, D. Luengo, Independent doubly adaptive rejection metropolis sampling within Gibbs sampling. IEEE Trans. Signal Process. **63**(12), 3123–3138 (2015)
32. L. Martino, H. Yang, D. Luengo, J. Kanniainen, J. Corander, A fast universal self-tuned sampler within Gibbs sampling. Dig. Signal Process. **47**, 68–83 (2015)
33. L. Martino, D. Luengo, Extremely efficient acceptance-rejection method for simulating uncorrelated Nakagami fading channels. Commun. Stat. Simul. Comput. (2018, to appear)
34. W.K. Pang, Z.H. Yang, S.H. Hou, P.K. Leung, Non-uniform random variate generation by the vertical strip method. Eur. J. Oper. Res. **142**, 595–609 (2002)
35. W.H. Payne, Normal random numbers: using machine analysis to choose the best algorithm. ACM Trans. Math. Softw. **4**, 346–358 (1977)
36. J.G. Proakis, *Digital Communications*, 4th edn. (McGraw-Hill, Singapore, 2000)
37. C.P. Robert, G. Casella, *Monte Carlo Statistical Methods* (Springer, New York, 2004)
38. M. Sibuya, Further consideration on normal random variable generator. Ann. Inst. Stat. Math. **14**, 159–165 (1962)
39. L. Tierney, Exploring posterior distributions using Markov Chains, in *Computer Science and Statistics: Proceedings of IEEE 23rd Symposium on the Interface* (1991), pp. 563–570
40. L. Tierney, Markov chains for exploring posterior distributions. Ann. Stat. **22**(4), 1701–1728 (1994)
41. M.D. Troutt, W.K. Pang, S.H. Hou, *Vertical Density Representation and Its Applications* (World Scientific, Singapore, 2004)
42. I. Vaduva, On computer generation of gamma random variables by rejection and composition procedures. Math. Oper. Stat. Ser. Stat. **8**, 545–576 (1977)

43. J. von Neumann, Various techniques in connection with random digits, in *Monte Carlo Methods*, ed. by A.S. Householder, G.E. Forsythe, H.H. Germond. National Bureau of Standards Applied Mathematics Series (U.S. Government Printing Office, Washington, DC, 1951), pp. 36–38
44. C.S. Wallace, Transformed rejection generators for gamma and normal pseudo-random variables. Aust. Comput. J. **8**, 103–105 (1976)
45. Q.M. Zhu, X.Y. Dang, D.Z. Xu, X.M. Chen, Highly efficient rejection method for generating Nakagami-*m* sequences. IET Electron. Lett. **47**(19), 1100–1101 (2011)

# Chapter 4
# Adaptive Rejection Sampling Methods

**Abstract** This chapter is devoted to describing the class of the adaptive rejection sampling (ARS) schemes. These (theoretically) universal methods are very efficient samplers that update the proposal density whenever a generated sample is rejected in the RS test. In this way, they can produce i.i.d. samples from the target with an increasing acceptance rate that can converge to 1. As a by-product, these techniques also generate a sequence of proposal pdfs converging to the true shape of the target density. Another advantage of the ARS samplers is that, when they can be applied, the user only has to select a set of initial conditions. After the initialization, they are completely automatic, self-tuning algorithms (i.e., no parameters need to be adjusted by the user) regardless of the specific target density. However, the need to construct a suitable sequence of proposal densities restricts the practical applicability of this methodology. As a consequence, ARS schemes are often tailored to specific classes of target distributions. Indeed, the construction of the proposal is particularly hard in multidimensional spaces. Hence, ARS algorithms are usually designed only for drawing from univariate densities.

In this chapter we discuss the basic adaptive structure shared by all ARS algorithms. Then we look into the performance of the method, characterized by the acceptance probability (which increases as the proposal is adapted), and describe various extensions of the standard ARS approach which are aimed either at improving the efficiency of the method or at covering a broader class of target pdfs. Finally, we consider a hybrid method that combines the ARS and Metropolis-Hastings schemes.

## 4.1 Introduction

The main limitation of RS methods is the difficulty of finding a proposal function $\pi(x)$ and a bound $L \geq p(x)/\pi(x)$, such that the envelope function, $L\pi(x) \geq p(x)$, is actually "close" enough to the target density, which is required in order to attain good acceptance rates. One way to tackle this difficulty is by constructing the proposal $\pi(x)$ adaptively.



L. Martino et al., *Independent Random Sampling Methods*, Statistics and Computing, https://doi.org/10.1007/978-3-319-72634-2_4

In this chapter, different adaptive accept/reject sampling schemes are described. Whenever they are applicable, adaptive RS techniques should be preferred in practical applications, because the resulting algorithms are *automatic* (i.e., *black-box*) schemes: they are almost completely self-tuning algorithms (no parameters need to be selected, except for a set of initial points) regardless of the target distribution, efficient (after a usually short learning period they produce samples from the target pdf with probability close to 1) and informative (they generate an approximation of the target density).

We divide the presentation of the adaptive rejection sampling (ARS) algorithms into two parts: we describe first the adaptive structure shared by all of the ARS techniques in Sect. 4.2 and then, in Sect. 4.3, we introduce several different specific constructions of the proposal pdf, provided so far in the literature. It is important to remark that the adaptive approach of ARS, described in Sect. 4.2, defines a universal sampling scheme, applicable theoretically to any generic multimodal and multidimensional target pdf. However, important limitations to its applicability are due to the need to design a suitable construction procedure. For this reason, all ARS schemes proposed in the literature are tailored to specific classes of target distributions (see Sect. 4.3).

The performance and computational cost of the standard ARS schemes are discussed in Sect. 4.4. Alternative adaptive approaches are introduced in Sect. 4.5. These methods do not guarantee that the acceptance rate becomes closer and closer to 1, as guaranteed using the standard ARS scheme of Sect. 4.2.2, but allow the user to control the computational cost. For this reason, these alternative schemes may ease the design of ARS methods for drawing from multidimensional target densities.

Finally, Sect. 4.6 describes the adaptive rejection Metropolis sampling (ARMS) method, that combines the ARS and Metropolis-Hastings (MH) algorithms. ARMS is a universal and efficient sampler: it can be applied for drawing virtually from any target pdf, since it does not impose any restriction on the proposal (unlike ARS, which requires a proposal construction mechanism ensuring that $Lp(x) \geq \pi(x)$ for all $x \in \mathcal{D}$), and it also improves the performance of the conventional MH algorithm, since the proposal is adaptively updated. Another advantage w.r.t. a standard MH method is that the ARMS procedure can be used as a black-box algorithm, exactly the same as the ARS scheme. On the other hand, the ARMS method is an MCMC sampler, implying that the generated samples are correlated (unlike ARS schemes, which produce i.i.d. samples).

## 4.2 Generic Structure of an Adaptive Rejection Sampler

In this section, we outline the general structure of an adaptive rejection sampler, following the basic concepts and ideas that were introduced in [5] and [9]. First, in Sect. 4.2.1 we describe the conditions needed to build an adequate proposal pdf that can be used within an ARS scheme. Then, we describe a generic ARS algorithm in Sect. 4.2.2. Note that, in the sequel we write $\pi(x)$ to denote a normalized proposal density, whereas $\bar{\pi}(x)$ is a non-normalized function proportional to $\pi(x)$, i.e., $\pi(x) \propto \bar{\pi}(x)$.

### 4.2.1 *Proposal Densities*

Consider a target pdf $p_o(x) \propto p(x)$ with $x \in \mathcal{D}$. Let us define the set of support points

$$\mathcal{S}_t \triangleq \{s_1, \ldots, s_{m_t}\},$$

where $s_i \in \mathcal{D}$, $i = 1, \ldots, m_t$. The variable $t \in \mathbb{N}$ denotes the iteration index ($t = 0, 1, 2, \ldots$) and the number of points $m_t$ can grow with the iteration index $t$. The sets of support points are the basis for the construction of suitable proposal functions that can be used within an ARS framework. In particular, we aim at obtaining a sequence of non-negative functions, $\{\bar{\pi}_t(x)\}_{t=0}^{+\infty}$, such that:

1. $\bar{\pi}_t(x) \geq p(x), \quad \forall x \in \mathcal{D}, \quad \forall t \in \mathbb{N}$.
2. It is possible to draw samples exactly from $\pi_t(x) \propto \bar{\pi}_t(x)$, $\forall t \in \mathbb{N}$.
3. If $|\mathcal{S}_t| = m_t \to \infty$, then $\bar{\pi}_t(x)$ converges towards $p(x)$, i.e.,

$$\lim_{m_t \to \infty} \bar{\pi}_t(x) = p(x), \tag{4.1}$$

   almost everywhere in $\mathcal{D}$.

Examples of this kind of constructions can be found in Sect. 4.3. Conditions 1 and 2 enable us to apply the RS principle for drawing from $p_o(x) \propto p(x)$ using $\pi_t(x) \propto \bar{\pi}_t(x)$ as a proposal density. Condition 3 is needed in order to ensure the efficiency of the resulting RS scheme and, at the same time, keep the computational cost of the resulting algorithm bounded.

### 4.2.2 *Generic Adaptive Algorithm*

Let us assume that a suitable construction of the sequence $\{\bar{\pi}_t(x)\}_{t=0}^{+\infty}$ is available. Then, the adaptive rejection sampling (ARS) scheme consists of the following steps:

1. Set $t = 0$ and choose an initial set of support points, $\mathcal{S}_0 = \{s_1, \ldots, s_{m_0}\}$ with $s_i \in \mathcal{D}$ and $i = 1, \ldots, m_0$. Let $N$ be the number of desired samples distributed according to $p_o(x)$.

2. Build $\pi_t(x) \propto \bar{\pi}_t(x)$ using a suitable procedure given the set $\mathcal{S}_t$.
3. Draw $x' \sim \pi_t(x)$ and $u' \sim \mathcal{U}([0, 1])$.
4. If $u' \leq p(x')/\bar{\pi}_t(x')$, then accept $x^{(i)} = x'$, set $i = i + 1$, $\mathcal{S}_{t+1} = \mathcal{S}_t$ and $t = t + 1$.
5. If $u' > p(x')/\bar{\pi}_t(x')$, then reject $x'$, set $\mathcal{S}_{t+1} = \mathcal{S}_t \cup \{x'\}$ and $t = t + 1$.
6. If $i > N$ stop. Otherwise, go to step 2.

Note that when a candidate sample is accepted in step 4, the set of support points remains unchanged (i.e., $m_{t+1} = m_t$). On the other hand, when the candidate sample is rejected, in step 5, it is added to the set of support points (i.e., $m_{t+1} = m_t + 1$).

### Essential Features

When a proposed sample is rejected, it is incorporated to the set of support points for the next iterations, $\mathcal{S}_{t+1}$. If the proposal construction approach used is properly chosen, the proposal $\pi_{t+1}(x)$ is then improved w.r.t. $\pi_t(x)$ and becomes closer to the target pdf, $p_o(x)$. As a consequence, the acceptance rate increases quickly, tending to 1 asymptotically. Therefore, at the same time, the probability of rejecting a sample, which is identical to the probability of adding a new point in $\mathcal{S}_{t+1}$, vanishes to zero. This maintains the computational cost of the construction procedure bounded.

The use of information provided by the *rejected* samples to improve the proposal is a key feature: if a sample $x'$ is rejected this means that the discrepancy between the proposal and the target is high at $x'$. Then, it looks appropriate to incorporate this local information to the construction of a better proposal density, which is closer to the target. A more detailed analysis of the performance and computational cost of the algorithm is given in Sect. 4.4.

### Parameters of ARS Algorithms

Given a target pdf $p_o(x) \propto p(x)$, the difference between ARS schemes lies in the procedure followed for the construction of the sequence of proposal pdfs. Once this procedure is selected, the unique parameters of the algorithm are the initial support points, i.e., $\mathcal{S}_0 = \{s_1, \ldots, s_{m_0}\}$, which can be chosen by the user (using some prior information available about the target). Otherwise, the algorithm is completely self-tuning. It is also interesting to note that there is a trade-off between initial performance and computational cost. Indeed, in order to improve the initial performance (i.e., to obtain a higher acceptance rate in the first iterations), a larger number of initial points $m_0$ could be used, but this also implies an increase of the initial computational cost.

### Advantages and Limitations

The ARS technique can be considered a *universal* and *black-box* algorithm. Indeed, assuming an adequate construction of the proposals is feasible, there are no further restrictions imposed on the target density $p_o(x)$. The algorithm only requires being able to evaluate the target $p_o(x) \propto p(x)$, up to a normalizing constant. Moreover, it ensures high acceptance rates after the initial iterations.

The adaptive rejection sampler described previously can also be applied for drawing samples from multidimensional target densities. However, the applicability of the adaptive RS scheme in practice is limited by the ability to find an appropriate procedure to build the sequence of proposals, satisfying the three conditions described in Sect. 4.2.1. Unfortunately, a general procedure is not available, although in the literature it is possible to find constructions valid for several important families of target distributions, as we describe in the sequel.

## 4.3 Constructions of the Proposal Densities

The main problem when using an RS method is to choose, or design, a "good" envelope function $\bar{\pi}(x) \geq p(x)$, where "good" means that $\bar{\pi}(x)$ and $p(x)$ are similar in some quantifiable sense (e.g., we may ask that $\int_{\mathcal{D}} (\bar{\pi}(x) - p(x))\, dx < \epsilon$ for some $\epsilon > 0$).

We can find a specific envelope function for some simple examples or after a deep analytical study of the target distribution. In the previous chapter, we have considered envelope functions of the type $\bar{\pi}(x) = L\pi(x)$, by keeping the proposal density fixed and then finding an upper bound $L \geq \frac{p(x)}{\pi(x)}$. However, in this case even when using the best bound $L \geq \sup_{x\in\mathcal{D}} \frac{p(x)}{\pi(x)}$ the acceptance rate can be very low.

In this section, we consider construction procedures that are completely automatic and provide an acceptance rate that tends to 1 asymptotically. Indeed, the procedures presented in the sequel enable the construction of a sequence of proposal densities, $\{\pi_t(x)\}_{t\in\mathbb{N}}$, $\pi_t(x) \propto \bar{\pi}_t(x)$, tailored to different families of target densities. The most appealing feature of these constructions is that each time we draw a sample from a proposal $\pi_t$ and it is rejected, we can use this sample to build an improved proposal, $\pi_{t+1}$, as required in an adaptive rejection sampler.

### 4.3.1 Standard Adaptive Rejection Sampling

The original adaptive rejection sampling (ARS) scheme was introduced in [9] and, hence, we refer to this method as standard, or conventional, ARS algorithm. Unfortunately, the procedure for the construction of proposal densities in this standard method is only suitable for target pdfs which are log-concave (and, hence, unimodal), which is a stringent limitation for many practical applications.

In order to describe the technique, let us assume that we want to draw from the target pdf $p_o(x) \propto p(x) \geq 0$ with support in $\mathcal{D} \subseteq \mathbb{R}$. The standard ARS procedure can be applied when $\log[p(x)]$ is concave, i.e., when the *potential function*[1]

$$V(x) \triangleq -\log[p(x)], \quad x \in \mathcal{D} \subseteq \mathbb{R}, \tag{4.2}$$

is convex. The basic idea is to partition the domain $\mathcal{D}$ into several intervals and construct an envelope function locally on each of these pieces. Let

$$\mathcal{S}_t = \{s_1, s_2, \ldots, s_{m_t}\} \subset \mathcal{D}$$

be the set of support points available at the $t$th iteration of the algorithm, sorted in ascending order

$$s_1 < \ldots < s_{m_t}.$$

From $\mathcal{S}_t$ we build a piecewise-linear lower hull of $V(x)$, denoted $W_t(x)$, formed by segments of linear functions tangent to $V(x)$ at the support points $s_k \in \mathcal{S}_t$. If we denote as $w_k(x)$ the linear function tangent to $V(x)$ at $s_k$, then we can define

$$W_t(x) \triangleq \max\{w_1(x), \ldots, w_{m_t}(x)\} \leq V(x) \quad \forall x \in \mathcal{D}. \tag{4.3}$$

Figure 4.1 illustrates the construction of $W_t(x)$ with three support points for the convex potential function $V(x) = x^2$. It is apparent that $W_t(x) \leq V(x)$ by construction, therefore $\bar{\pi}_t(x) = \exp(-W_t(x))$ is an envelope function for $p(x)$, i.e.,

$$\bar{\pi}_t(x) = \exp(-W_t(x)) \geq p(x) = \exp(-V(x)). \tag{4.4}$$

Once $W_t(x)$ is built, we can use it to obtain a piecewise-exponential proposal density

$$\pi_t(x) = c_t \exp(-W_t(x)) \propto \bar{\pi}_t(x), \tag{4.5}$$

[1] We assign a name to the function $V(x)$ to ease the treatment, so that we can refer directly to it. The name *potential function* recalls a physical interpretation. In statistical mechanics, for instance, the potential energy function $V$ is central to the evaluation of many thermodynamic properties, where $V$ is used as log-density [31]. To be specific, the estimation of the thermodynamic properties demands the computation of integrals involving the function $\exp(-V)$. This interpretation of the log-pdf as a potential is also evoked explicitly in the Hamiltonian MCMC techniques [32].

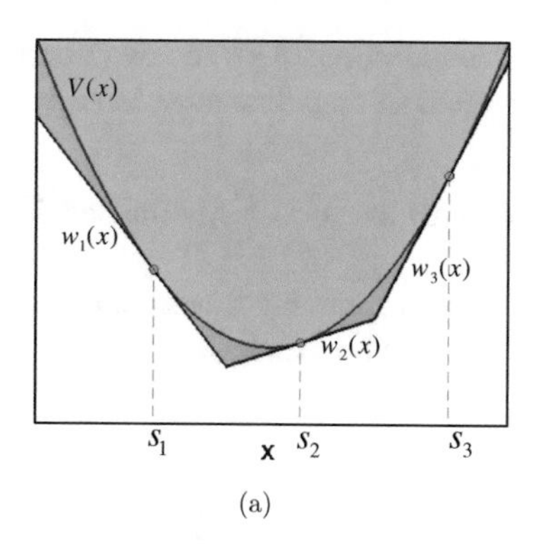


(a)

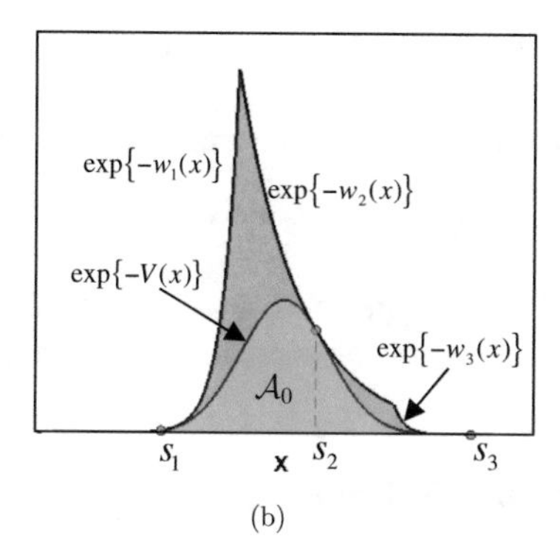


(b)

**Fig. 4.1** **(a)** Example of construction of the piecewise linear function $W_t(x)$ with three support points $\mathcal{S}_t = \{s_1, s_2, s_{m_t=3}\}$, as carried out by the original ARS technique [9]. The function $W_t(x) \triangleq \max[w_1(x), w_2(x), w_3(x)]$ is formed by segments of linear functions tangent to the potential $V(x) = x^2$ at the support points in $\mathcal{S}_t$. **(b)** The corresponding envelope function $\exp(-W_t(x))$ and the target function $p(x) = \exp(-V(x))$

where $c_t = \left(\int \exp(-W_t(x))dx\right)^{-1}$ is the normalizing constant. In Sect. 4.2.1 we stated three conditions that the proposal densities in an ARS scheme should satisfy for the method to be efficient. The first condition was that $\bar{\pi}(x) \geq p(x)$ and this is already fulfilled. In the sequel we show that the standard ARS scheme also satisfies the other two conditions (as long as the target is log-concave), namely (a) it is possible to draw samples from $\pi_t(x)$ and (b) it is possible to improve the proposal pdf along the iterations.

### Drawing from $\pi_t(x)$

We can draw from $\pi_t(x)$ easily. Let $\xi_i$, $i = 1, \ldots, m_t - 1$, denote the intersection points between consecutive tangent lines, i.e., the abscissae such that $w_i(\xi_i) = w_{i+1}(\xi_i)$, $i = 1, \ldots, m_t - 1$. Moreover, let be $\mathcal{I}_0 = (-\infty, \xi_1]$, $\mathcal{I}_i = (\xi_i, \xi_{i+1}]$ for $i = 1, \ldots, m_t - 1$ and $\mathcal{I}_{m_t} = (\xi_{m_t}, +\infty)$. First, we calculate the area

$$\omega_k = \int_{\mathcal{I}_k} \bar{\pi}_t(x)dx, \quad k = 0, \ldots, m_t,$$

below each piece of $\bar{\pi}_t(x) = \exp(-W_t(x))$. Then we obtain the normalized weights

$$\bar{\omega}_k = \frac{\omega_k}{\sum_{k=0}^{m_t} \omega_k} = \frac{\omega_k}{c_t}. \tag{4.6}$$

where $c_t = \sum_{k=0}^{m_t} \omega_k$ is the normalizing constant of $\bar{\pi}_t(x)$, i.e., $\pi_t(x) = c_t \bar{\pi}_t(x)$. Note that the $\omega_k$'s can be calculated exactly, as we only need to integrate functions of the form $\exp\{-\lambda x\}$ (for a constant $\lambda$) in different intervals. It is important to note that these intervals are defined by the intersection points between two contiguous tangent lines $w_i(x)$ and $w_{i+1}(x)$ (then, these intersection points must be calculated). Hence, in order to draw a sample from $\pi_t(x)$:

1. first we randomly choose a piece according to the probability masses $\bar{\omega}_k$, $k = 0, \ldots, m_t$,
2. and then we generate a sample $x'$ from the corresponding truncated exponential pdf using, for instance, the inversion method (see Sect. 2.4.1).

### Improving the Proposal pdf

It is apparent that the previous procedure allows us to build better envelope functions $\bar{\pi}_t(x)$ and, as a consequence, better proposal pdfs $\pi_t(x)$ just using a greater number of support points in the construction.

The adaptive structure in Sect. 4.2.2 suggests that when a sample $x'$ from $\pi_t(x)$ is rejected in an RS scheme we incorporate $x'$ into the set of support points, i.e., $\mathcal{S}_{t+1} = S_t \cup \{x'\}$ and $m_{t+1} = m_t + 1$. Then, we compute a refined lower hull, $W_{t+1}(x)$, and a new proposal density $\pi_{t+1}(x) = c_{t+1} \exp(-W_{t+1}(x))$ that is closer to the target pdf. Figure 4.2 illustrates different constructions of the envelope function $\bar{\pi}_t(x) = \exp(-W_t(x))$ using 2, 3, 4, and 5 support points, respectively.

### Initial Conditions

The minimum number of initial support point is $m_0 = 2$, $\mathcal{S}_0 = \{s_1, s_2\}$ where $s_1 < s_2$. To obtain a proper initial proposal $\pi_0(x)$ the derivatives of $V(x)$ at $s_1, s_2 \in \mathcal{D}$ must have different signs. This also ensures that the mode of $p(x)$ is contained in $[s_1, s_2]$. Clearly more initial support points can be used.

*Example 4.1* Take as target pdf a standard Gaussian density, i.e.,

$$p_o(x) \propto p(x) = \exp\{-x^2/2\}.$$

In this case the potential function is

$$V(x) = \frac{1}{2}x^2,$$

which is convex (i.e., $p(x)$ is log-concave). The first derivative of the potential function is $\frac{dV}{dx} = x$. Hence, we can apply the ARS method using the construction in Eq. (4.3). We draw $T = 500$ i.i.d. samples from $p_o(x)$ via ARS using $\mathcal{S}_0 = \{-1.3, 2\}$ as the initial set of support points. The procedure is repeated for 10,000 different runs.

Figure 4.3a shows the acceptance rate (averaged over the 10,000 runs) for the $i$th sample, with $1 \leq i \leq 500$. We can observe that the acceptance rate converges

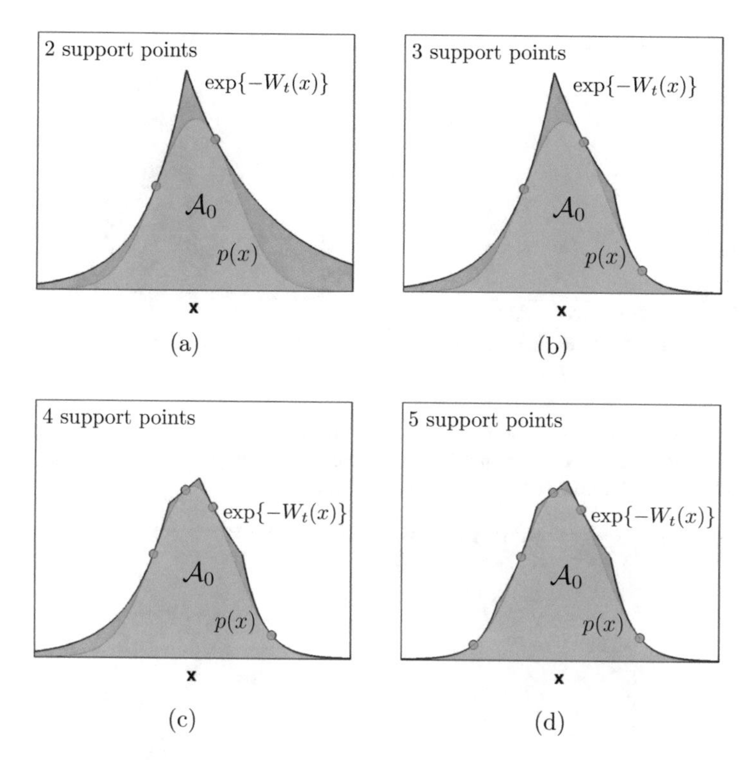


**Fig. 4.2** Examples of construction of the envelope function $\bar{\pi}_t(x) = \exp(-W_t(x))$ with different number of support points. With more support points the envelope function $\bar{\pi}_t(x) = \exp(-W_t(x))$ approaches $p(x)$

quickly to 1, so the ARS method virtually becomes a direct sampler for $p_o(x)$. This is obtained with a bounded computational cost. Indeed, the (empirical) mean final number of support points used in the ARS scheme (i.e., contained in $\mathcal{S}_T$ with $T = 500$) is $m_T = 15.5$. This means that, we discard on average just 13.5 proposed samples for drawing 500 samples and we can obtain an acceptance rate close to 1 just using on average less than 16 support points.

Figure 4.3b shows the histogram of values of $m_T$ for the set of 10,000 simulations. The minimum and maximum value of this empirical distribution is $m_T = 9$ and $m_T = 23$ support points, respectively. Since the acceptance rate after accepting 500 samples is close to 1, the probability of rejecting the following samples will be negligible and, as a consequence, the probability of adding a new support point is almost zero.

As shown in Fig. 4.3a, the acceptance rate is varying with the iterations. However, since for drawing 500 samples we need on average 513.5 iterations, the mean acceptance rate of the overall ARS algorithm is 97.7%.

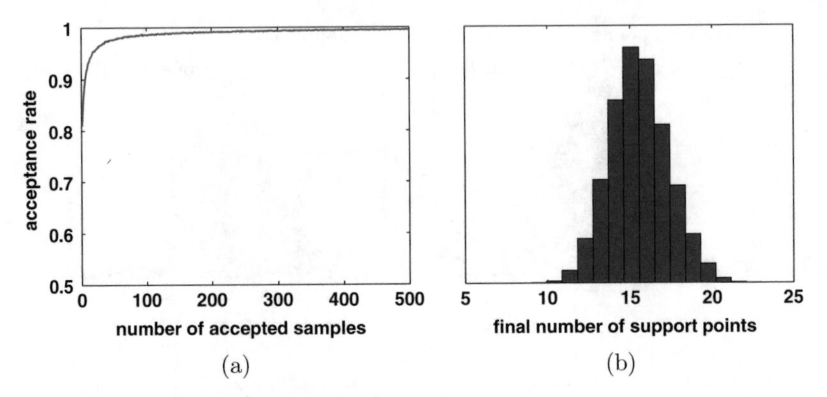


**Fig. 4.3** **(a)** The curve of acceptance rates (averaged over 10,000 runs) as a function of the number of accepted samples. **(b)** Histogram of the final number of support points after accepting 500 samples for 10,000 different runs of the ARS algorithm

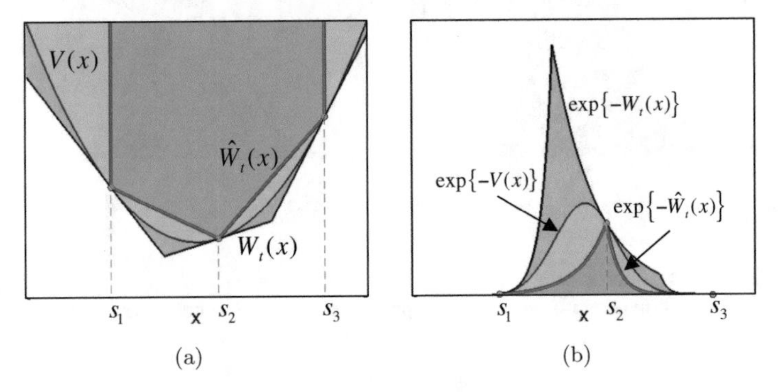


**Fig. 4.4** **(a)** Example of construction of the squeeze function $\exp(-\hat{W}_t(x))$ with three support points $\mathcal{S}_t = \{s_1, s_2, s_{m_t=3}\}$. **(b)** The corresponding hat and squeeze functions $\exp(-W_t(x))$, $\exp(-\hat{W}_t(x))$, together with the target function $p(x) = \exp(-V(x))$

## Applying the Squeeze Principle

If the target density is computationally expensive to evaluate, it is also possible to construct a squeeze function

$$\exp(-\hat{W}_t(x)) \leq p(x) = \exp(-V(x)),$$

for all $x \in \mathcal{D}$ (see Sect. 3.3.2). In order to construct $\hat{W}_t(x)$ in such a way that it is also piecewise linear, we can use the secant lines passing through the points $(s_k, V(s_k))$ and $(s_{k+1}, V(s_{k+1}))$ where $s_k, s_{k+1} \in \mathcal{S}_t$ are support points. Obviously, as illustrated in Fig. 4.4, this construction is possible only in the finite domain $[\min(\mathcal{S}_t), \max(\mathcal{S}_t)]$, hence, we set $\hat{W}_t(x) \to +\infty$ for any $x \notin [\min(\mathcal{S}_t), \max(\mathcal{S}_t)]$. It is straightforward to see that with this construction

$$\exp(-W_t(x)) \geq \exp(-V(x)) \geq \exp(-\hat{W}_t(x)), \tag{4.7}$$

so we can also apply the squeeze technique.

### 4.3.2 *Derivative-Free Constructions for Log-Concave pdfs*

A variation of the standard ARS algorithm that avoids the need to compute derivatives of $V(x)$ and lends itself to a simpler automatic implementation has been proposed in [8].

Given the set of support points $\mathcal{S}_t = \{s_1, \ldots, s_{m_t}\}$, here we denote with $w_k(x)$ the secant line passing through the points $(s_k, V(s_k))$ and $(s_{k+1}, V(s_{k+1}))$, for $k = 1, \ldots, m_t - 1$. Whereas for $k \in \{-1, 0, m_t, m_{t+1}\}$ we set

$$w_k(x) \rightarrow -\infty, \tag{4.8}$$

as infinite constant values. Consider also the definition of the intervals $\mathcal{I}_0 = (-\infty, s_1]$, $\mathcal{I}_j = [s_j, s_{j+1}]$, $j = 1, \ldots, m_t$ and $\mathcal{I}_{m_t+1} = [s_{m_t}, +\infty)$. Then, the piecewise linear function $W_t(x)$ is constructed as

$$W_t(x) \triangleq \max[w_{k-1}(x), w_{k+1}(x)] \text{ for } x \in \mathcal{I}_k,\ k = 0, \ldots, m_t. \tag{4.9}$$

Figure 4.5 illustrates the construction of $W_t(x)$ using the derivative-free ARS algorithm with 4 and 5 support points.

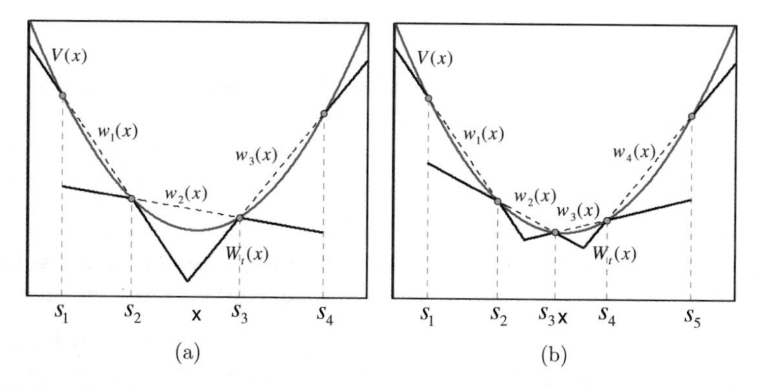


**Fig. 4.5** Example of construction of the piecewise linear function $W_t(x)$, as carried out by the derivative-free ARS technique. The function $W_t(x)$ is composed by pieces of secant lines passing through $(s_k, V(s_k))$ and $(s_{k+1}, V(s_{k+1}))$, $k = 1, \ldots, m_t - 1$, as described in Eqs. (4.8) and (4.9). **(a)** Construction with four support points $\mathcal{S}_t = \{s_1, s_2, s_3, s_{m_t=4}\}$. **(b)** Construction with five support points $\mathcal{S}_t = \{s_1, s_2, s_3, s_4, s_{m_t=5}\}$

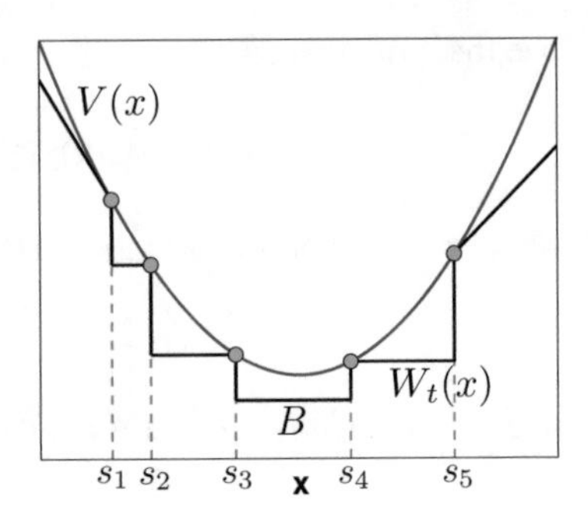


**Fig. 4.6** Example of stepwise construction (with the exception of the tails in $\mathcal{I}_0$ and $\mathcal{I}_{m_t}$) of $W_t(x)$ with five support points

## Stepwise Proposal pdfs

Given a log-concave pdf $p_o(x) \propto p(x)$, assume that an upper bound $B \geq p(x)$ is available. Assume also that the mode $x^*$ of $p(x)$ is contained in the $k$th interval,[2] $\mathcal{I}_k = [s_k, s_{k+1}]$. In this case, a simpler alternative construction is possible. Indeed, we can define

$$W_t(x) = \begin{cases} w_1(x), & x \in \mathcal{I}_0, \\ \min[V(s_j), V(s_{j+1})], & x \in \mathcal{I}_j, \text{ with } j \neq k \\ B, & x \in \mathcal{I}_k, \\ w_{m_t-1}(x), & x \in \mathcal{I}_{m_t}, \end{cases} \tag{4.10}$$

where $j = 1, \ldots, m_t$, $w_1(x)$ is the secant line passing through the points $(s_1, V(s_1))$ and $(s_2, V(s_2))$ and $w_{m_t-1}(x)$ is the secant line passing through the points $(s_{m_t-1}, V(s_{m_t-1}))$ and $(s_{m_t}, V(s_{m_t}))$. Figure 4.6 depicts an example with five support points. The proposal pdf is, again,

$$\pi_t(x) \propto \bar{\pi}_t(x) = \exp(-W_t(x)).$$

Therefore, except for the tails, $\pi_t(x)$ is formed by constant pieces (i.e., a mixture of uniform pdfs). Although the approximation of a curve (the target $p(x)$ in this case) is less accurate using constant functions than using linear functions, this approach has two main advantages:

- it is easier to extend this construction for the multidimensional case,
- and an addition of a new support point just varies the construction in one interval $\mathcal{I}_k$ so that in each step it is not necessary to rebuild the complete proposal.

Observe that this can be considered an adaptive version of the strip methods in Sect. 3.6.1. See Sect. 4.5.2 for an example of application of this construction.

[2]Note that with simple inspections it is always possible to know the interval including the mode (for instance, considering the signs of the slopes of the secant lines passing through the support points).

**Table 4.1** Log-concave densities

| Name | Density $p_o(x) \propto p(x)$ | Parameters |
|---|---|---|
| Gaussian | $p(x) = \exp\left(-\frac{(x-\mu)^2}{2\sigma^2}\right)$ | $\forall \mu, \sigma \in \mathbb{R}$ |
| Exponential power | $p(x) = \exp\left(-\lvert x\rvert^{\alpha}\right)$ | $\alpha \geq 1$ |
| Weibull | $p(x) = x^{a-1}\exp(-x^a)$ | $a \geq 1$ |
| Gamma | $p(x) = x^{a-1}\exp(-x)$ | $a \geq 1$ |
| Beta | $p(x) = x^{a-1}(1-x)^{b-1}$ | $a, b \geq 1$ |
| Planck | $p(x) = x^a/(e^x - 1)$ | $a \geq 1$ |
| Perks | $p(x) = 1/(e^x + e^{-x} + a)$ | $a \geq -2$ |

#### Examples of Log-Concave pdfs

Table 4.1 provides some examples of log-concave pdfs. However, the condition of $\log[p(x)]$ being concave rules out many target pdfs of interest. Indeed, in many practical applications the target is *non*-log-concave or, in general, multimodal and the standard (or derivative-free) ARS techniques cannot be applied. In order to deal with such densities several generalizations of the standard ARS method have been proposed in the literature and we explore some of them in the sequel.

### *4.3.3 Concave-Convex ARS*

Consider a target pdf $p_o(x) \propto p(x) = \exp(-V(x))$ defined in a bounded domain $x \in \mathcal{D}$, where the potential $V(x)$ can be decomposed into a sum of convex, $V_1(x)$, and concave, $V_2(x)$, functions, i.e.,

$$V(x) = V_1(x) + V_2(x). \tag{4.11}$$

In this case, the *concave-convex ARS* (CCARS) method [11, 13] can be applied. The two parts can be analyzed separately in order to obtain two different piecewise linear functions, $W_{t,1}(x)$ and $W_{t,2}(x)$, such that

$$W_{t,i}(x) \leq V_i(x),$$

with $i = 1, 2$. Clearly, the envelope function in this case is

$$\bar{\pi}_t(x) = \exp(-W_{t,1}(x) - W_{t,2}(x)) \geq p(x) = \exp(-V(x)).$$

Figure 4.7 illustrates the procedures to handle the potentials $V_1$ and $V_2$ with different concavity.

Given a set of support points $\mathcal{S}_t = \{s_1, s_2, \ldots, s_{m_t}\} \subset \mathcal{D}$, sorted in ascending order, we already know that we can use pieces of tangent lines at the support points

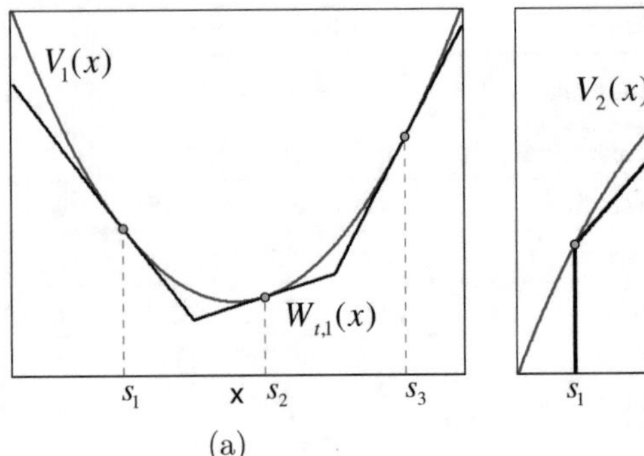


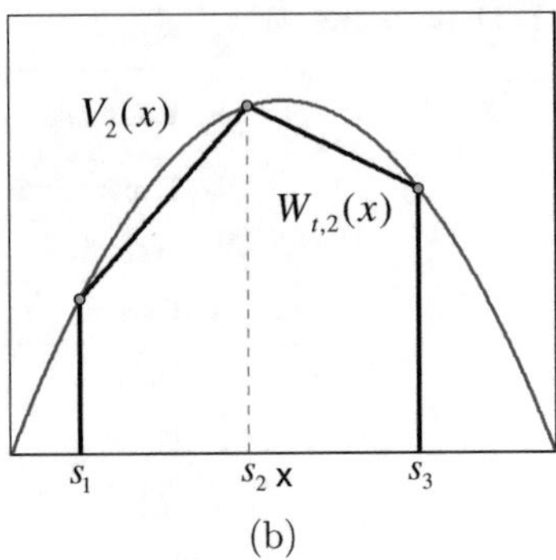


**Fig. 4.7** Application of the concave-convex ARS method. Example of construction with three support points $\mathcal{S}_t = \{s_1, s_2, s_{m_t=3}\}$ of the two piecewise linear functions $W_{t,i}(x)$, $i = 1, 2$, such that $W_{t,i}(x) \leq V_i(x)$. **(a)** Since $V_1(x)$ is convex, we can use pieces of lines tangent to $V_1(x)$ at the support points. **(b)** Since $V_2(x)$ is concave, the function $W_t(x)$ is composed by pieces of secant lines passing through $(s_k, V(s_k))$ and $(s_{k+1}, V(s_{k+1}))$, $k = 1, 2$, to fulfill the inequality $W_{t,2}(x) \leq V_2(x)$

to build a lower hull $W_{t,1}(x)$ for the concave potential $V_1(x)$ as in the standard ARS technique. For the concave potential $V_2(x)$, we can use the secant lines passing through $(s_k, V(s_k))$ and $(s_{k+1}, V(s_{k+1}))$, $k = 1, \ldots, m_t - 1$, to obtain a suitable piecewise linear function $W_{t,2}(x)$ such that $W_{t,2}(x) \leq V_2(x)$. Then the proposal pdf is

$$\pi_t(x) \propto \bar{\pi}_t(x) = \exp(-W_{t,1}(x) - W_{t,2}(x)),$$

that is formed by exponential pieces. Hence, the ARS approach can be applied using $\pi_t(x)$ as proposal and $\bar{\pi}_t(x)$ as envelope function.

*Remark 4.1* It should be noticed that CCARS procedure to build $\pi_t(x)$ is possible only in a finite domain, precisely in the interval $[\min(\mathcal{S}_t), \max(\mathcal{S}_t)]$. However, if the tails of the entire potential $V(x)$ are convex, we can also apply this technique to a target pdf with unbounded domain. Indeed, in this situation we can handle the tails separately using tangent lines in order to build a function $W_t(x) \leq V(x)$ (also in the tails), as in the standard ARS method.

### 4.3.4 *Transformed Density Rejection*

The standard construction given in [9] and described in Sect. 4.3.1 can be applied only when $V(x) = -\log[p(x)]$ is convex. In [6, 16, 18, 21], the authors suggested to replace the $\log(\vartheta)$ function with another monotonically increasing transformation

$$T(\vartheta) : \mathbb{R}^+ \to \mathbb{R}$$

such that $T[p(x)]$ (with $p_o(x) \propto p(x)$) is concave or, equivalently, the corresponding potential function

$$V_T(x) \triangleq -T[p(x)], \tag{4.12}$$

is convex. Equation (4.12) above implies that the target pdf can be expressed as

$$p_o(x) \propto p(x) = T^{-1}[-V_T(x)]. \tag{4.13}$$

Obviously, we go back to the standard ARS construction procedure by choosing $T(\vartheta) = \log(\vartheta)$. This method is known as *transformed density rejection* (TDR) algorithm.

### Algorithm

Let us consider a monotonically increasing transformation $T(\vartheta)$. Given a set of support points $\mathcal{S}_t = \{s_1, s_2, \dots, s_{m_t}\} \subset \mathcal{D}$, the idea is, again, to replace the convex potential $V_T(x)$ with a piecewise-linear function $W_t(x)$, such that $W_t(x) \leq V_T(x)$ and formed by segments of linear functions that, for instance, are tangent to $V_T(x)$ at the points $(s_k, V_T(s_k))$ such that $s_k \in \mathcal{S}_t$. If we let $w_k(x)$ be the linear function tangent to $V_T(x)$ at $s_k$, then the piecewise linear function $W_t(x)$ is defined as $W_t(x) = \max\{w_1(x), \dots, w_{m_t}(x)\}$, exactly as in Sect. 4.3.1. Clearly, the derivative-free procedures in Sect. 4.3.2 can also be applied to $V_T(x)$. Then, the proposal pdf has the form

$$\pi_t(x) \propto \bar{\pi}_t(x) = T^{-1}[-W_t(x)], \tag{4.14}$$

where $\bar{\pi}_t(x) \geq p(x)$ by construction. Moreover, if $T$ is adequately chosen, we can draw easily from $\pi_t(x)$ (as explained in Sect. 4.3.1) and then apply the RS test (using the ratio $p(x)/\bar{\pi}_t(x)$) to generate samples from $p_o(x) \propto p(x)$.

*Remark 4.2* This technique extends the standard ARS algorithm in [9] but it still can be applied only to unimodal target densities, since $T$ is a monotonic (increasing) function.

### Necessary Conditions for $T$

For this procedure, the key is the identification of an adequate transformation $T(\vartheta)$. To be useful, $T : \mathbb{R}^+ \rightarrow \mathbb{R}$ has to satisfy the following conditions:

1. It has to be monotonically increasing.
2. Given the inverse transformation $T^{-1}(z)$, the integral

$$Q_T(z) = \int_{-\infty}^{z} T^{-1}(z')dz',$$

must be bounded for all (finite) possible values of $z < +\infty$ in the domain of $T^{-1}$.

3. It must be possible to compute the integral

$$Q_T(b) - Q_T(a) = \int_a^b T^{-1}(z)dz \tag{4.15}$$

exactly, where $a$ and $b$ can also be non-finite.

4. The composition $(T \circ p)(x) = T[p(x)]$ has to be concave, i.e.,

$$\frac{d^2}{dx^2}T[p(x)] = \left[\frac{d^2T}{d\vartheta^2}\right]_{p(x)} \left(\frac{dp}{dx}\right)^2 + \left[\frac{dT}{d\vartheta}\right]_{p(x)} \frac{d^2p}{dx^2} \leq 0.$$

The satisfaction of the first condition guarantees that the inverse transformation $T^{-1} : \mathbb{R} \to \mathbb{R}^+$ exists and it is monotonically increasing as well. The second condition is required to ensure that the integral of the envelope function $T^{-1}[-W_t(x)]$ in a domain bounded or unbounded in one side, i.e.,

$$\int_a^b T^{-1}[-W_t(x)]dx \leq +\infty,$$

where at least one of values $a, b$ is finite.The third condition assures that we are able to calculate these kind of integral exactly (recall that $W_t(x)$ is a piecewise linear function). This is necessary in order to draw from the resulting piecewise proposal $\pi_t(x)$. The last condition is necessary to allow the construction of $W_t(x)$ using tangent lines such that $W_t(x) \leq V_T(x)$ and, correspondingly, $T^{-1}[-W_t(x)] \geq p(x)$.

### Examples of Suitable $T$-Transformations

In many applications, the most used suitable class of transformations is the family of power functions,

$$T_c(\vartheta) = \text{sign}(c)\vartheta^c \quad \forall \vartheta \in \mathbb{R}^+, \forall c \in \mathbb{R} \setminus \{0, 1\}.$$

Note that $T_c(\vartheta) : \mathbb{R}^+ \to \mathbb{R}^+$ is an increasing function for all values of the parameter $c \in \mathbb{R}$. The authors in [16, 17] complete the family defining

$$T_0(\vartheta) = \log(\vartheta),$$

which is based on the limit [17, Chap. 4]

$$\lim_{\lambda \to 0} \frac{\vartheta^\lambda - 1}{\lambda} = \log(\vartheta).$$

The most commonly used members of this family are, certainly, $T_0(\vartheta) = \log(\vartheta)$ and $T_{-1/2}(\vartheta) = -1/\sqrt{\vartheta}$. Note that the integral of $T^{-1}(z) = z^{1/c}$, for $c > 0$, and $T^{-1}(z) = (-z)^{1/c}$, for $c < 0$ ($c \neq -1$), can be calculated analytically and yield (recalling that $Q_T(z) = \int_{-\infty}^{z} T^{-1}(z')dz'$)

$$
\begin{aligned}
Q_T(z) &= \frac{c}{c+1} z^{\frac{c+1}{c}}, \quad c > 0, \\
Q_T(z) &= -\frac{c}{c+1} (-z)^{\frac{c+1}{c}}, \quad c < 0.
\end{aligned}
\tag{4.16}
$$

Moreover, both functions $Q_T$ can be inverted analytically to obtain $Q_T^{-1}$ and this enables us to use the inversion method to draw from each piece forming the corresponding proposal $\pi_t(x) \propto T_c^{-1}[-W_t(x)]$ (where $W_t(x)$ is a piecewise linear function). The value $c = -1$ is not feasible because $T_{-1}$ does not satisfy the second condition needed for the corresponding proposal $\pi_t(x) \propto T_{-1}^{-1}[-W_t(x)]$ to be a proper pdf.

*Remark 4.3* It is important to be aware that the transformation $T_c(\vartheta)$ is applicable to densities defined in an unbounded domain only for the values $c \in (-1, 0]$ [17, Chap. 4].

*Example 4.2* Consider a Cauchy target pdf, i.e.,

$$
p_o(x) = \frac{1}{\pi(1+x^2)}, \quad x \in \mathbb{R},
$$

and the $T$-transformation

$$
T_{-1/2}(\vartheta) = -\sqrt{\frac{1}{\vartheta}}, \quad \vartheta \in \mathbb{R}^+.
$$

Then, the potential function

$$
V_T(x) = -T \circ p_o(x) = \sqrt{\pi(1+x^2)},
$$

is convex. Using tangent lines we can construct a piecewise linear function $W_t(x) \leq V_T(x)$ using the set of support points $\mathcal{S}_t$. Hence, the proposal $\pi_t(x) \propto T_{-1/2}^{-1}[-W_t(x)]$, where $T_{-1/2}^{-1}(z) = \frac{1}{z^2}$, is formed by pieces of the form

$$
\pi_t(x) \propto \bar{\pi}_t(x) = \frac{1}{(a_i x + b_i)^2}, \quad \mathcal{I}_i = [s_i, s_{i+1}],
$$

where $i = 1, \ldots, |\mathcal{S}_t| = m_t$, and $a_i$, $b_i$ are the coefficients of the $i$th linear piece in $W_t(x)$. We can integrate each piece in order to compute suitable weights, and we can also draw samples from a chosen piece easily using the inversion method.

We remark again that, since $T(\vartheta)$ has to be a monotonically increasing function and $T[p(x)]$ has to be concave, this procedure can be applied only to unimodal target pdfs. Moreover, finding a suitable transformation $T$ is not a always an easy task. For these reasons, different extensions have been investigated in the literature. We summarize and describe them below.

### 4.3.5 *Extensions of TDR*

#### TDR and Transformed Rejection Method

Let us mention that, despite the similarity in the names, the "transformed rejection method" of Sect. 3.7 is rather different from the TDR algorithm. Specifically, in the transformed rejection scheme the transformation is applied to the random variable $X$, with pdf $p_o(x)$, while in TDR the transformation is applied directly to the density $p_o(x)$.

However, both approaches can be used jointly. Indeed, in certain cases, rather than working directly with $p_o(x)$, it is easier to find a suitable transformation $T$ for the transformed density

$$q(y) = p_o(\rho^{-1}(y)) \left| \frac{d\rho^{-1}(y)}{dy} \right|$$

of a r.v. $Y$, obtained as $Y = \rho(X)$ with $X \sim p_o(x)$. Namely, $\rho$ is a transformation applied to the r.v. $X \sim p_o(x)$ whereas $T$ is applied to the pdf $q(y)$. In this case, an ARS scheme can be used to draw a sample $y'$ from $q(y)$ and then we can set $x' = \rho^{-1}(y')$.

A typical example, taken from [17, Chap. 4], is given by the family of pdfs $p_o(x)$ such that the density of $Y = \log X$ (i.e., $\rho(x) = \log(x)$),

$$q(y) = p_o(\exp(y)) \exp(y),$$

is log- concave (i.e., $T(\vartheta) = \log(\vartheta)$). In the literature, pdfs of this class are usually called *log-log-concave* densities [17], but this name does not clarify that the first logarithm function is applied to the r.v. $X \sim p_o(x)$, not directly to the target pdf $p_o(x)$.

*Example 4.3* The Indian buffet process (IBP) is a well-known stochastic process often used in Bayesian nonparametric methodologies [12]. The so-called stick-breaking construction of IBP [34] demands the ability to draw samples from univariate pdfs of the form

$$p_o(x) \propto p(x) = x^{1-\alpha}(1-x)^N \exp\left(\alpha \sum_{i=1}^{N} \frac{1}{i}(1-x)^i\right), \quad x \in [0, b],$$

where $\alpha > 0$, $N \in \mathbb{N}$ and $b > 0$ are all positive constants. If we choose the transformation $Y = \log(X)$, then it is possible to prove that the pdf of $Y$,

$$q(y) \propto \exp((1-\alpha)y)(1-\exp(y))^N \exp\left(\alpha \sum_{i=1}^{N} \frac{1}{i}(1-\exp(y))^i\right),$$

is log-concave and so the ARS algorithm can be applied.

### Extensions for Concave Potentials

The TDR method is not necessarily restricted to the case in which $T$ is monotonically increasing and $V_T(x) = -T[p(x)]$ is convex, although it was originally proposed in this setup [16]. We can extend it in a way similar to the CCARS method of Sect. 4.3.3.

Indeed, we can consider combinations of increasing and decreasing functions $T$ with corresponding concave or convex potentials $V_T$. The procedure to construct the piecewise linear function $W_t(x)$, however, is different depending on the type of $T$ and $V_T$ at hand. This is briefly analyzed in this section.

Let us recall Eq. (4.13), $p_o(x) \propto p(x) = T^{-1}[-V_T(x)]$. So far, we have considered the combination of

(1) a monotonically increasing function $\vartheta = T^{-1}(z)$,
(2) with a concave function $z = -V_T(x)$.

In such case, a piecewise linear function $z = -W_t(x)$ formed by straight lines tangent to $z = -V_T(x)$ can be used to construct an envelope function $T^{-1}[-W_t(x)]$ such that

$$T^{-1}[-W_t(x)] \geq T^{-1}[-V_T(x)].$$

The other cases of interest, depending on to the choice of $T$ and the concavity of $V_T$, are listed below and summarized in Table 4.2.

- If $\vartheta = T^{-1}(z)$ is increasing and $z = -V_T(x) = T[p(x)]$ is convex, a suitable function $z = -W_t(x)$ can be constructed using secant lines. However, the construction is possible only in a bounded domain (see Fig. 4.8a, c).

**Table 4.2** Possible combinations of monotonic transformation $T$ and concavity of $V_T$

| $\vartheta = T^{-1}(z)$ | $z = -V_T(x)$ | $z = -W_t(x)$ | Domain | Figure |
|---|---|---|---|---|
| Increasing | Concave | Tangent lines | Unbounded | 4.8a, b |
| Increasing | Convex | Secant lines | Bounded | 4.8a, c |
| Decreasing | Convex | Tangent lines | Unbounded | 4.9a, c |
| Decreasing | Concave | Secant lines | Bounded | 4.9a, b |

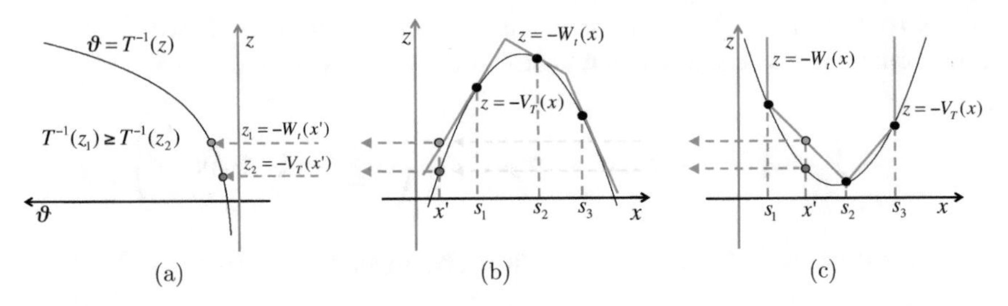


**Fig. 4.8** Example of construction of an adequate piecewise linear function $z = -W_t(x)$, with three support points $\mathcal{S}_t = \{s_1, s_2, s_3\}$, for an increasing function $\vartheta = T^{-1}(z)$. Figures **(a)–(c)** consider the increasing function $\vartheta = T^{-1}(z)$ (w.r.t. the variable $z$), hence, $W_t(x)$ is built to ensure $-W_t(x) \geq -V_T(x)$. In figure **(a)**, the axis associated to the variable $z$ is vertical

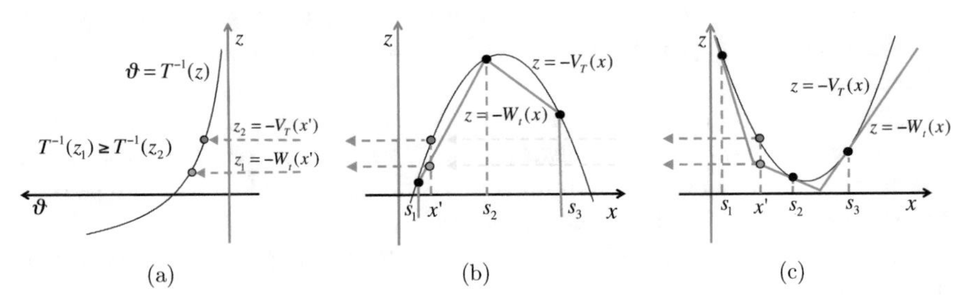


**Fig. 4.9** Example of construction of an adequate piecewise linear function $z = -W_t(x)$, with three support points $\mathcal{S}_t = \{s_1, s_2, s_3\}$, for a decreasing function $\vartheta = T^{-1}(z)$. Figures **(a)–(c)** depict the decreasing function $\vartheta = T^{-1}(z)$, hence we need to build $W_t(x)$ to ensure that $-W_t(x) \leq -V_T(x)$. Note that the axis associated to the independent variable $z$ in figure **(a)** is vertical

- If $\vartheta = T^{-1}(z)$ is decreasing and $z = -V_T(x) = T[p(x)]$ is concave, an adequate function $z = -W_t(x)$ has to be formed by secant lines. Also in this case, the construction is possible only in a bounded domain (see Fig. 4.9a, b).
- If $\vartheta = T^{-1}(z)$ is decreasing and $z = -V_T(x) = T[p(x)]$ is convex, the straight lines tangent to $z = -V_T(x)$ can also be used to build $z = -W_t(x)$. The construction is also possible in an infinite domain (see Fig. 4.9a, c).

Figures 4.8 and 4.9 (jointly with Table 4.2) summarize these different scenarios. Given an arbitrary $x'$, it is possible to see that the value $T^{-1}[-W_t(x')]$ (green point) is always greater than $T^{-1}[-V_T(x')]$ (red point), i.e., $T^{-1}[-W_t(x')] \geq T^{-1}[-V_T(x')]$. Note that, in Fig. 4.8, given a generic $x' \in \mathcal{D}$ and values $z_1 = W_t(x')$ and $z_2 = -V_T(x')$, we have

$$T^{-1}(z_1) \geq T^{-1}(z_2).$$

Since this is true for all $x \in \mathcal{D}$, the inequality $T^{-1}[-W_t(x)] \geq T^{-1}[-V_T(x)]$ is always satisfied. In Fig. 4.8b, c, $W_t(x)$ needs to stay below $V(x)$ whereas, in Fig. 4.9b, c, $W_t(x)$ needs to stay above $V(x)$ for all $x \in \mathcal{D}$.

All the cases above assume a monotonic transformation $T$. A generalization for non-monotonic transformations of the target density is given in Sect. 4.3.6. In the sequel we discuss how it is possible to handle a generic potential function $V_T(x)$ with second derivative with non-constant sign.

#### Potential Functions with Known Inflection Points

Consider a target density $p_o(x)$ that can be written as

$$p_o(x) \propto p(x) = T^{-1}[-V_T(x)],$$

where the potential $V_T(x)$ can possibly be non-convex (it may present several minima) but we assume that the positions of all its inflection points are known. The TDR algorithm can be extended to handle this case.

Indeed, we can find a partition of the support $\mathcal{D} = \cup_{i=1}^{n} \mathcal{D}_i$, $\mathcal{D}_i \cap \mathcal{D}_j = \emptyset$, $i \neq j$, where within each $\mathcal{D}_i$ the function $V_T(x)$ has a second derivative with constant sign. Therefore, in each interval $\mathcal{D}_i$ where $V_T(x)$ is convex, we use tangent lines to build $W_t(x)$. Alternatively, if $V_T(x)$ is concave in $\mathcal{D}_i$, the function $W_t(x)$ is composed by secant lines.

Clearly, since this procedure can be applied to non-convex potentials $V_T(x)$, it could be applied to tackle multimodal target pdfs. In general, however, for complicated target densities it is not straightforward to study analytically the second derivative of the potential $V_T(x)$. Furthermore, even if the inflection points are known, we need that the tails of the potential be convex, as in Sect. 4.3.3, in order to build a proper proposal.

Recently, another approach has been studied [3] that requires only knowledge of an interval where an inflection point is located, but not exactly the position of the inflection point. To apply this method the potential has to be three-times differentiable, though, and it can be used only with target pdfs with bounded domain.

### 4.3.6 *Generalized Adaptive Rejection Sampling*

In this section, we describe another generalization of the adaptive rejection sampling algorithm proposed in [26, 27]. Let us consider a target pdf $p_o(x)$, $x \in \mathcal{D} \subseteq \mathbb{R}$, that can be written as

$$p_o(x) \propto p(x) = \exp\left(-V(x; \mathbf{g})\right) = \exp\left(-\sum_{i=1}^{n} \bar{V}_i(g_i(x))\right), \tag{4.17}$$

where the potential function has the form

$$V(x; \mathbf{g}) \triangleq \sum_{i=1}^{n} \bar{V}_i(g_i(x)), \tag{4.18}$$

and $\mathbf{g} = [g_1, \ldots, g_n]^\top$ is a vector of real functions. We assume that

1. the functions $\bar{V}_i(\vartheta_i)$, for $i = 1, \ldots, n$ (hereafter called *marginal potentials*), are convex with a minimum at $\mu_i$ and
2. the nonlinearities $g_i(x)$, $i = 1, \ldots, n$, are either convex or concave (i.e., they have a second derivative with constant sign).

The potential $V(x; \mathbf{g})$ in Eq. (4.18) is, in general, a non-convex function. Moreover, in general it is impossible to study analytically the first and second derivatives of the potential $V(x; \mathbf{g})$ of Eq. (4.18) in order to calculate the stationary or inflection points.

In the simplest case when $n = 1$ the pdf in Eq. (4.17) is reduced to the form

$$p_o(x) \propto \exp(-\bar{V}(g(x))), \tag{4.19}$$

where $\bar{V}(\vartheta)$ is a convex function while $g(x)$ can be either a convex or a concave function. It is apparent that if $g(x)$ is a linear function then we go back to the standard ARS framework described in Sect. 4.3.1. More generally, given

$$p_o(x) \propto \exp\left(-\sum_{i=1}^{n} \bar{V}_i(g_i(x))\right),$$

if all $g_i(x)$ are linear functions, the potential $V(x; \mathbf{g}) \triangleq \sum_{i=1}^{n} \bar{V}_i(g_i(x))$ is convex, so that we go back again the standard ARS case. Indeed, in general, replacing each nonlinearity $g_i(x)$ with a linear function $r_i(x)$, $i = 1, \ldots, n$, i.e., we have $\mathbf{r} = [r_1, \ldots, r_n]^\top$, it is straightforward to prove that the *modified potential*

$$V(x; \mathbf{r}) = \sum_{i=1}^{n} \bar{V}_i(r_i(x)) \tag{4.20}$$

is convex. This result follows from each term $\bar{V}_i(r_i(x))$ being convex, namely

$$\begin{aligned} \frac{d^2 \bar{V}_i(r_i(x))}{dx^2} &= \frac{d\bar{V}_i}{d\vartheta} \frac{d^2 r_i}{dx^2} + \left(\frac{dr_i}{dx}\right)^2 \frac{d^2 \bar{V}_i}{d\vartheta^2} \\ &= 0 + \left(\frac{dr_i}{dx}\right)^2 \frac{d^2 \bar{V}_i}{d\vartheta^2} \geq 0 \end{aligned} \tag{4.21}$$

where we have used that

$$\frac{d^2 r_i}{dx^2} = 0,$$

because $r_i$ is linear, and the convexity of the marginal potentials $\bar{V}_i(\vartheta)$, $i = 1, \ldots, n$. The argument above suggests that a generalized ARS method can be designed based on constructing suitable piecewise-linear approximations of the nonlinearities $g_i$.

### The GARS Algorithm

Let us recall the set of support points at the $t$th iteration,

$$\mathcal{S}_t \triangleq \{s_1, s_2, \ldots, s_{m_t}\} \subset \mathcal{D},$$

and sort them in ascending order, $s_1 < \ldots < s_{m_t}$. From the points in $\mathcal{S}_t$ we construct the closed intervals $\mathcal{I}_k = [s_k, s_{k+1}]$ for $k = 1, \ldots, m_t - 1$, together with two semi-open intervals $\mathcal{I}_0 = (-\infty, s_1]$ and $\mathcal{I}_{m_t} = [s_{m_t}, +\infty)$.

For each interval $\mathcal{I}_k$, $k = 0, \ldots, m_t$, the GARS method proceeds in two steps. Consider the interval $\mathcal{I}_k \subset \mathcal{D}$. First, every nonlinearity $g_i(x)$ is replaced by a suitable linear function $r_{i,k}(x)$. In this way we generate a modified potential $V(x, \mathbf{r}_k)$ in $\mathcal{I}_k$, with

$$\mathbf{r}_k(x) = [r_{1,k}(x), \ldots, r_{n,k}(x)]^\top,$$

that lies below the original one, i.e.,

$$V(x, \mathbf{r}_k) \leq V(x, \mathbf{g}), \quad \forall x \in \mathcal{I}_k.$$

Second, we construct a linear function $W_t(x)$ that is tangent at an (arbitrary) point $x_k^* \in \mathcal{I}_k$ to the modified potential $V(x, \mathbf{r}_k)$. The two steps are described in detail below.

1. GARS builds suitable linear functions $r_{i,k}(x)$ such that

$$\bar{V}_i(r_{i,k}(x)) \leq \bar{V}_i(g_i(x)), \quad \forall x \in \mathcal{I}_k, \tag{4.22}$$

for every $i = 1, \ldots, n$ and $k = 0, \ldots, m_t$. As a consequence, substituting $\mathbf{g}$ by $\mathbf{r}_k$ into the functional $V(x; \cdot)$, we obtain the inequality

$$V(x; \mathbf{r}_k) = \sum_{i=1}^{n} \bar{V}_i(r_{i,k}(x)) \leq V(x; \mathbf{g}) = \sum_{i=1}^{n} \bar{V}_i(g_i(x)),$$

$\forall x \in \mathcal{I}_k$. Note that, if all the $r_{i,k}(x)$ are adequately built, $\exp\{-V(x; \mathbf{r}_k)\}$ is already an envelope function for $p(x)$, i.e.,

$$\exp\{-V(x; \mathbf{r}_k)\} \geq \exp\{-V(x; \mathbf{g})\} = p(x), \quad \forall x \in \mathcal{I}_k.$$

However, in general it is not possible to draw from $\pi^*(x) \propto \exp\{-V(x; \mathbf{r})\}$ and we need to seek further simplifications. Full details on the construction of $\mathbf{r}_k$ are provided in the next subsection.

2. The modified potential $V(x; \mathbf{r})$ is convex in $\mathcal{I}_k$ as we have shown previously [see Eq. (4.21)]. Therefore, we can choose a straight line tangent to $V(x; \mathbf{r}_k)$ at an arbitrary point $x_k^* \in \mathcal{I}_k$ to build a linear function

$$W_t(x) = w_k(x), \quad x \in \mathcal{I}_k,$$

such that $W_t(x) \leq V(x; \mathbf{r})$ for all $x \in \mathcal{I}_k$, exactly as in a standard ARS method. Thus,

$$\begin{aligned} \bar{\pi}_t(x) = \exp(-W_t(x)) &\geq \exp(-V(x; \mathbf{r}_k)) \\ &\geq \exp(-V(x; \mathbf{g})) = p(x) \quad \forall x \in \mathcal{I}_k, \end{aligned} \tag{4.23}$$

is an envelope function for $p(x) \propto p_o(x)$. The built proposal pdf,

$$\pi_t(x) \propto \bar{\pi}_t(x) = \exp\{-W_t(x)\}, \quad \forall x \in \mathcal{I}_k,$$

is an exponential pdf within $\mathcal{I}_k$, since $W_t(x) = w_k(x)$ is linear in this interval, exactly as in the standard ARS technique in Sect. 4.3.1.

Figure 4.10 illustrates the construction of the piecewise linear function $W_t(x)$ using the proposed technique for the non-convex potential $V(x; \mathbf{g}) = 16 - 8x^2 + x^4$ with three support points, $\mathcal{S}_t = \{s_1, s_2, s_{m_t=3}\}$. Indeed, this potential can be rewritten as

$$V(x; \mathbf{g}) = 16 - 8x^2 + x^4 = (4 - x^2)^2.$$

so that we can interpret it as a composition of functions $(\bar{V}_1 \circ g_1)(x)$, where $\bar{V}_1(\vartheta) = \vartheta^2$ and $g_1(x) = 4 - x^2$ (i.e., $n = 1$ and the vector of nonlinearities $\mathbf{g} = g_1$ is scalar). The dashed line shows the modified potentials $V(x; \mathbf{r}_k)$, $k = 0, \ldots, m_t = 3$. The function $W_t(x)$ consists of segments of linear functions $w_k(x)$ tangent to the modified potentials $V(x; \mathbf{r}_k)$ at arbitrary points $x_k^* \in \mathcal{I}_k$, with $k = 0, \ldots, m_t = 3$.

Using the GARS construction for an ARS scheme, when a sample $x'$ drawn from $\pi_t(x) \propto \exp(-W_t(x))$ is rejected, $x'$ is incorporated as a support point in the new set $\mathcal{S}_{t+1} \triangleq \mathcal{S}_t \cup \{x'\}$ and, as a consequence, a refined lower hull $W_{t+1}(x)$ is constructed yielding a better approximation of the potential function $V(x; \mathbf{g})$.

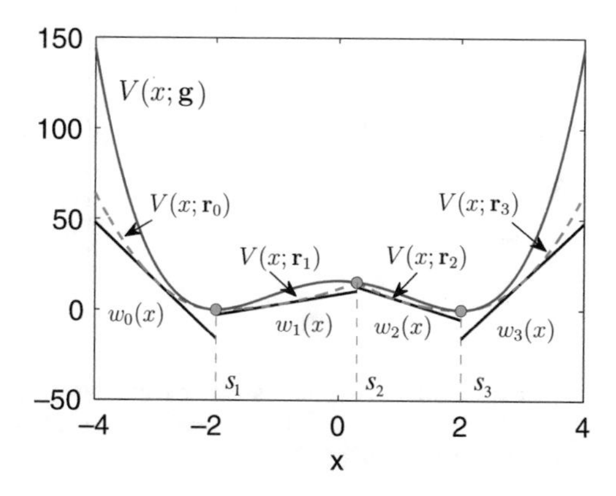


**Fig. 4.10** Example of construction of the piecewise linear function $W_t(x)$ with three support points $\mathcal{S}_t = \{s_1, s_2, s_{m_t=3}\}$, as carried out by the GARS technique. The potential $V(x; \mathbf{g})$ is shown with solid line. The modified potential $V(x; \mathbf{r}_k)$, for $x \in \mathcal{I}_k$, is depicted with dashed lines. The piecewise linear function $W_t(x)$ (depicted with solid lines) consists of segments of linear functions $w_k(x)$ tangent to the modified potential $V(x; \mathbf{r}_k)$ at arbitrary points $x_k^* \in \mathcal{I}_k$, with $k = 0, \ldots, 3$

### Construction of the Linear Functions $r_{i,k}$

The GARS algorithm relies on the ability to obtain linear functions $r_{i,k}(x)$ such that $\bar{V}_i(r_{i,k}(x)) \leq \bar{V}_i(g_i(x))$ for $i = 1, \ldots, n$ and $k = 0, \ldots, m_t$. In order to build suitable linear functions we need to introduce the set of *simple estimates* corresponding to the nonlinearity $g_i(x)$ as

$$\mathcal{X}_i \triangleq \{x_i \in \mathbb{R} : \; g_i(x_i) = \mu_i\}, \tag{4.24}$$

where $\mu_i$ is the position of the minimum of the marginal potential $\bar{V}_i$, i.e.,

$$\mu_i = \min_{\vartheta} \bar{V}_i(\vartheta).$$

The background of the name "simple estimates" is clarified in a later section, when discussing the applicability of GARS.

We recall that each function $g_i(x)$ is assumed to have a second derivative with constant sign, hence the equation $\mu_i = g_i(x_i)$ can yield zero ($|\mathcal{X}_i| = 0$, i.e., $\mathcal{X}_i$ is empty), one ($|\mathcal{X}_i| = 1$), or two ($|\mathcal{X}_i| = 2$) simple estimates. Clearly, if $g_i(x)$ is a monotonic function then $|\mathcal{X}_i| \leq 1$. Figure 4.11 displays the three possible cases for a generic concave $g_i(x)$.

We assume that all the simple estimates in $\mathcal{X}_i$, $i = 1, \ldots, n$, are included in the initial set of support points $\mathcal{S}_0$ ($t = 0$), i.e.,

$$\mathcal{X}_i \subset \mathcal{S}_0, \quad \text{for } i = 1, \ldots, n.$$

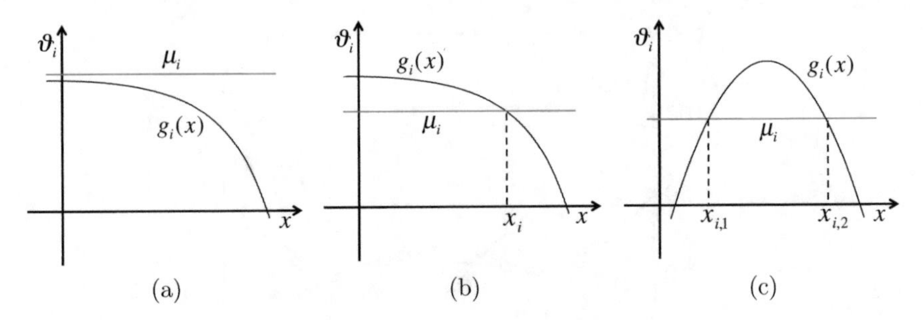


**Fig. 4.11** Example of the three possible cases for a concave nonlinearity $g_i(x)$. **(a)** The set $\mathcal{X}_i$ is empty, $|\mathcal{X}_i| = 0$. **(b)** There exists one simple estimate $x_i$, i.e., $|\mathcal{X}_i| = 1$. **(c)** The nonlinearity $g_i(x)$ is a non-monotonic function and $|\mathcal{X}_i| = 2$ ($\mathcal{X}_i = \{x_{i,1}, x_{i,2}\}$)

We remark that this condition is needed for the construction of suitable linear functions $r_{i,k}(x)$, $i = 1, \ldots, n$ and $k = 0, \ldots, m_t$.

*Remark 4.4* It is possible to build the functions $r_{i,k}(x)$ adequately without knowing analytically the simple estimates and the positions of the minima $\mu_i$ as well [23, Chap. 5]. However, for the sake of simplicity, here we assume that we are able to compute the simple estimates.

*Example 4.4* Consider the bimodal target pdf

$$p_o(x) \propto p(x) = \exp\left(-(4 - x^2)^2\right), \quad x \in \mathbb{R},$$

where the potential is $V(x; \mathbf{g}) = (4 - x^2)^2 = \bar{V}_1 \circ g_1(x)$ with $\bar{V}_1(\vartheta) = \vartheta^2$ and $g_1(x) = 4 - x^2$ ($n = 1$). In this case, we have $\mu_1 = 0$ and the set of simple estimates is obtained by solving the equation $4 - x^2 = 0$, i.e.,

$$\mathcal{X}_1 = \{x_{i,1} = -2, x_{i,2} = 2\}.$$

It is easy to see that the inequality (4.22) is satisfied for the class of marginal potential functions $\bar{V}_i$ (convex with a minimum at $\mu_i$) if

$$|\mu_i - r_{i,k}(x)| \leq |\mu_i - g_i(x)| \text{ and} \tag{4.25}$$

$$(\mu_i - r_{i,k}(x))(\mu_i - g_i(x)) \geq 0 \tag{4.26}$$

jointly, $\forall x \in \mathcal{I}_k$, where $\mu_i = \arg\min_{\vartheta} \bar{V}_i(\vartheta)$. Indeed, if $\mu_i \leq a \leq b$ then $\bar{V}_i(a) \leq \bar{V}_i(b)$ because $\bar{V}_i$ is increasing in $(\mu_i, +\infty)$ whereas for $b \leq a \leq \mu_i$ we have also $\bar{V}_i(a) \leq \bar{V}_i(b)$ because $\bar{V}_i$ is decreasing in $(-\infty, \mu_i)$. Figure 4.12 illustrates the latter inequalities. We can see that the green points are always closer to the minimum $\mu_i$ than the red points, and so they always have a smaller potential value.

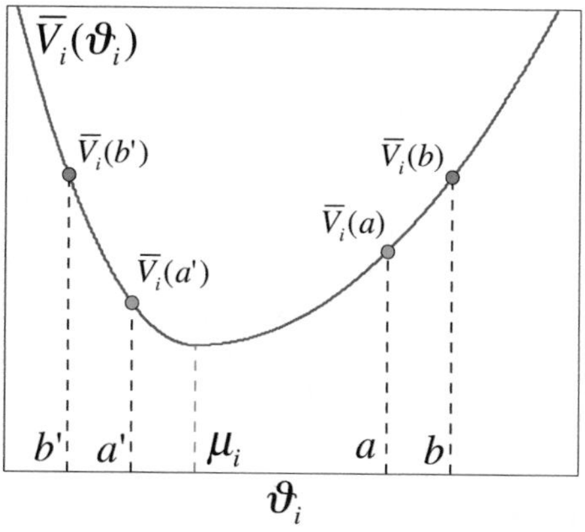

**Fig. 4.12** An example of marginal potential $\bar{V}_i(\vartheta_i)$. Since we assume that $\bar{V}_i$ is convex with a minimum at $\mu_i$, we have always $\bar{V}_i(a) \leq \bar{V}_i(b)$ if $b \geq a \geq \mu_i$ or $\bar{V}_i(a') \leq \bar{V}_i(b')$ if $b' \leq a' \leq \mu_i$

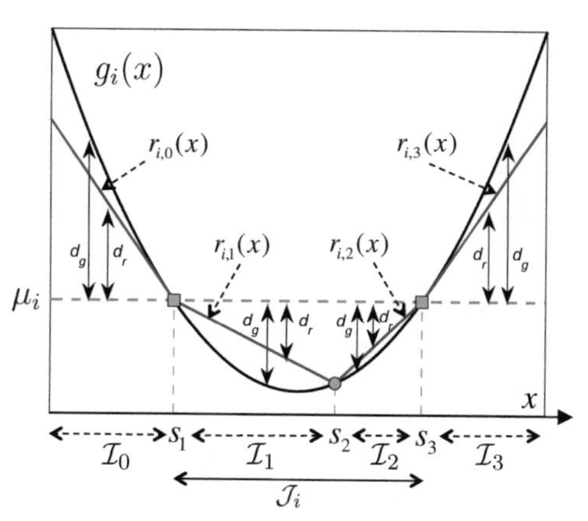


**Fig. 4.13** Example of construction of the linear function $r_{i,k}(x)$ in order to replace a convex nonlinearity $g_i(x)$ in different intervals $\mathcal{I}_k$, using $m_t = 3$ support points, $\mathcal{S}_t = \{s_1 = x_{i,1}, s_2, s_3 = x_{i,2}\}$

Figure 4.13 depicts the basic idea of how to construct the linear functions $r_{i,k}(x)$, $k = 0, \ldots, 3$, for three support points $\mathcal{S}_t = \{s_1 = x_{i,1}, s_2, s_3 = x_{i,2}\}$ ($s_1$ and $s_3$ coincide with the two simple estimates). We seek a linear function $r_{i,k}(x)$ such that the absolute difference $d_r = |\mu_i - r_{i,k}(x)|$ is always less than the distance $d_g = |\mu_i - g_i(x)|$, i.e., $d_r \leq d_g$ in an interval $\mathcal{I}_k$. Therefore, in the intervals $\mathcal{I}_0 = [-\infty, s_1]$ and $\mathcal{I}_3 = [s_3, +\infty]$ we should use tangent straight lines while in $\mathcal{I}_1 = [s_1, s_2]$ and $\mathcal{I}_2 = [s_2, s_3]$ we should use the linear functions passing through the two support points.

In general, take a non-monotonic function $g_i(x)$ and assume that the set of simple estimates $\mathcal{X}_i$ is not empty. In such case, we can define the interval

$$\mathcal{J}_i = [x_{i,1}, x_{i,2}],$$

limited by the simple estimates associated to the function $g_i(x)$. Clearly, $\mathcal{J}_i$ can be empty or a semi-open interval when $g_i(x)$ is a monotonic function [23, Chap. 5].

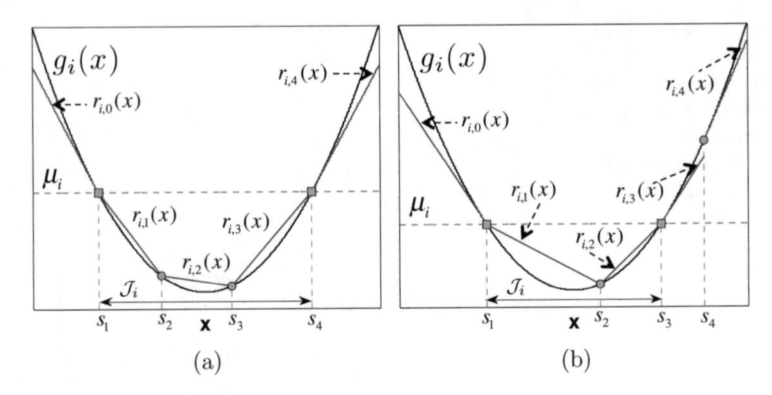


**Fig. 4.14** Examples of construction of the appropriate linear functions $r_{i,k}(x)$, with $k = 0, \ldots, m_t = 4$ for a non-monotonic convex nonlinearity $g_i(x)$. The interval defined by the simple estimates $\mathcal{J}_i = [x_{i,1}, x_{i,2}]$ is indicated by solid double arrows

Then, the procedure in Fig. 4.13 can be summarized as

1. if $\mathcal{I}_k \subset \mathcal{J}_i$ (i.e., $\mathcal{I}_k \cap \mathcal{J}_i = \mathcal{I}_k$), use secant lines,
2. otherwise, if $|\mathcal{I}_k \cap \mathcal{J}_i| = 0$, use tangent lines.

Recall that we are assuming that all the simple estimates are contained in the initial set $\mathcal{S}_0$ so that we have only two possibilities: either the $\mathcal{I}_k$ is completely contained in $\mathcal{J}_i$ or the intersection between $\mathcal{I}_k$ and $\mathcal{J}_i$ is just one support point.

Figure 4.14 displays two examples of construction with four support points. Specifically, it shows the construction of the linear functions $r_{i,k}(x)$ when $g_i(x)$ is non-monotonic and convex with $m_t = 4$ support points. In Fig. 4.14a the intervals $\mathcal{I}_0$ and $\mathcal{I}_4$ are not contained in $\mathcal{J}_i = [x_{i,1}, x_{i,2}]$, hence we use two tangent lines to build $r_{i,0}(x)$ and $r_{i,4}(x)$. Since $\mathcal{I}_1, \mathcal{I}_2, \mathcal{I}_3 \subseteq \mathcal{J}_i$, we use secant lines for $r_{i,1}(x)$, $r_{i,2}(x)$ and $r_{i,3}(x)$. In Fig. 4.14b the intervals $\mathcal{I}_0$, $\mathcal{I}_3$, and $\mathcal{I}_4$ are not contained in $\mathcal{J}_i$, hence we use tangent lines for $r_{i,0}(x)$, $r_{i,3}(x)$ and $r_{i,4}(x)$. Since $\mathcal{I}_1, \mathcal{I}_2 \subseteq \mathcal{J}_i$, we use secant lines for $r_{i,1}(x)$ and $r_{i,2}(x)$.

### Applicability

Densities of the form of Eq. (4.17) appear naturally in statistical inference problems [1, 4, 7, 19, 33] where it is desired to draw from the posterior pdf $p(x|\mathbf{y})$ with $\mathbf{y} = [y_1, y_2, \ldots, y_n] \in \mathbb{R}^n$, of a random variable $X$ given a collection of observations

$$\begin{cases} Y_1 = \bar{g}_1(X) + \Theta_1, \\ \vdots \\ Y_n = \bar{g}_n(X) + \Theta_n, \end{cases} \tag{4.27}$$

where $\Theta_1, \ldots, \Theta_n$ are independent "noise" variables. In fact, writing the noise pdfs as $p(\vartheta_i) \propto \exp\{-\bar{V}_i(\vartheta_i)\}$ (with a mode at $\vartheta_i^* = \mu_i$), $i = 1, \ldots, n$, the likelihood function can be expressed as

$$p(\mathbf{y}|x) \propto \exp\left\{ -\sum_{i=1}^{n} \bar{V}_i(y_i - \bar{g}_i(x)) \right\}. \tag{4.28}$$

Therefore, denoting $g_i(x) = y_i - \bar{g}_i(x)$ and writing the prior pdf as $p(x) \propto \exp\{-\bar{V}_{n+1}(g_{n+1}(x))\}$, the potential function is

$$\begin{aligned} V(x; \mathbf{g}) &= -\log[p(x|\mathbf{y})] \\ &= -\log[p(\mathbf{y}|x)p(x)] = \sum_{i=1}^{n+1} \bar{V}_i(g_i(x)). \end{aligned} \tag{4.29}$$

Since we are assuming that each $p(\vartheta_i) \propto \exp\{-\bar{V}_i(\vartheta_i)\}$ has only a mode at $\vartheta_i^* = \mu_i$, if we have only one observation ($n = 1$, hence we have only one equation, for instance, $Y_1 = \bar{g}_1(X) + \Theta_1$) the set of maximum likelihood estimators $\hat{\mathcal{X}}$ of the variable of interest $x$ is

$$\hat{\mathcal{X}} = \{x \in \mathcal{D} : g_1(x) = \mu_1\}, \tag{4.30}$$

where $g_1(x) = y_1 - \bar{g}_1(x)$. Note that Eq. (4.30) is exactly the definition of the "simple estimates" [see Eq. (4.24)] for the first nonlinearity $g_1(x)$, hence the choice of the name.

*Example 4.5* The standard ARS algorithm can be interpreted as a method for sampling from pdfs of the form $p_o(x) \propto \exp\{-h(x)\}$, where $h(x)$ is a convex function. From a similar perspective, the proposed GARS algorithm can handle target pdfs of the form

$$\begin{aligned} p_o(x) &\propto h(x) \exp\left(-h(x)\right) \\ &= \exp\left(-h(x) + \log[h(x)]\right), \end{aligned} \tag{4.31}$$

where the function $h(x)$ can be either convex or concave. In fact, in this case we can write $-\log[p_o(x)]$ as a composition of two functions, $\bar{V}_1 \circ g_1$, where $\bar{V}_1(\theta_1) \triangleq \vartheta_1 - \log[\vartheta_1]$ (which is convex with a minimum at $\mu_1 = 1$) and $g_1(x) \triangleq h(x)$.

*Example 4.6* Consider a generic polynomial potential of fourth order,

$$V(x; \mathbf{g}) = a_0 + a_1 x + a_2 x^2 + a_3 x^3 + a_4 x^4, \tag{4.32}$$

with $a_4 > 0$. This can always be written as

$$\begin{aligned} V(x;\mathbf{g}) &= \kappa + (\alpha + \beta x + \gamma x^2)^2 + (\delta + \eta x)^2 \\ &= \kappa + \bar{V}_1(g_1(x)) + \bar{V}_2(g_2(x)), \end{aligned} \tag{4.33}$$

where $\kappa, \alpha, \beta, \gamma, \eta, \delta$ are real constants, $\bar{V}_i(\vartheta_i) = \vartheta_i^2$, $i = 1, 2$, $g_1(x) \triangleq \alpha+\beta x+\gamma x^2$ is a second-order polynomial and $g_2(x) \triangleq \delta + \eta x$ is linear. Since $\bar{V}_1(\vartheta) = \bar{V}_2(\vartheta)$ are convex, $\frac{d^2 g_1}{dx^2} = \gamma$ is constant and $g_2(x)$ is linear, it is straightforward to apply the GARS procedure. The constants $\kappa$, $\alpha$, $\beta$, $\gamma$, $\eta$, and $\delta$ have to satisfy the following equalities

$$\begin{cases} \gamma^2 = a_4, \\ 2\beta\gamma = a_3, \\ \beta^2 + 2\alpha\gamma + \eta^2 = a_2, \\ 2\alpha\beta + 2\delta\eta = a_1, \\ \alpha^2 + \delta^2 + \kappa = a_0. \end{cases} \tag{4.34}$$

This is a nonlinear system of five equations and six unknowns that can be always solved if we assume $a_4 > 0$. As a consequence, the composition of the potential is always possible.

*Example 4.7* Consider now a polynomial potential of eighth order,

$$V(x;\mathbf{g}) = a_0 + a_1 x + a_2 x^2 + a_3 x^3 + a_4 x^4 + a_6 x^6 + a_8 x^8, \tag{4.35}$$

where the coefficients corresponding to the powers 5, 7 are null, i.e., $a_5 = a_7 = 0$. Moreover, if $a_2, a_6, a_8 > 0$, we can rewrite the polynomial in Eq. (4.35) as

$$\begin{aligned} V(x;\mathbf{g}) &= \kappa + \left(\frac{a_1}{2\sqrt{a_2}} + \sqrt{a_2}x\right)^2 + \left(\frac{a_3}{2\sqrt{a_6}} + \sqrt{a_6}x^3\right)^2 + \left(\frac{a_4}{2\sqrt{a_8}} + \sqrt{a_8}x^4\right)^2 \\ &= \kappa + \bar{V}_1(g_1(x)) + \bar{V}_2(g_2(x)) + \bar{V}_3(g_3(x)), \end{aligned}$$

where

$$\kappa = a_0 - \frac{a_1^2}{2a_2} - \frac{a_3^2}{2a_6} - \frac{a_4^2}{2a_8},$$

and $\bar{V}_i(\vartheta_i) = \vartheta_i^2$, $i = 1, 2, 3$. We observe that $g_1(x) = \frac{a_1}{2\sqrt{a_2}} + \sqrt{a_2}x$ is already linear, $g_2(x) = \frac{a_3}{2\sqrt{a_6}} + \sqrt{a_6}x^3$ is concave when $x < 0$ and convex when $x > 0$ and $g_3(x) = \frac{a_4}{2\sqrt{a_8}} + \sqrt{a_8}x^4$ is always convex. It is important to remark that, although the second derivative of $g_2(x)$ does not have a constant sign, it is possible to apply the GARS procedure because we know the inflection points ($x = 0$ in this case).

#### Relationship Between GARS and TDR

The GARS method can be extended and related with the $T$-transformation technique described in Sect. 4.3.4 (see also [23, Chap. 5]). The TDR technique enables us to draw from target pdfs of the type

$$p_o(x) \propto T^{-1}[g(x)] = (T^{-1} \circ g)(x), \quad x \in \mathcal{D} \subseteq \mathbb{R},$$

where $T^{-1}(z)$ is monotonically increasing and $g(x)$ is concave.[3] The GARS technique can be expressed in a similar form, that extends the applicability of the TDR method to non-monotonic transformations. Indeed, let us assume that we are able to draw from the pdf

$$f(z) \propto H(z), \tag{4.36}$$

$\forall z \in \mathbb{R}$, with a single mode at $\mu$.

*Remark 4.5* The argument in this subsection can also be extended to cases where $f(z)$, and $H(z)$, have several modes. Moreover, $f(z)$ can be an arbitrary function, not a pdf, if the inverse function $f^{-1}$ of $f(z)$ satisfies the conditions discussed in Sect. 4.3.5 for the transformation $T$. However, here we consider $f(z)$ to be a pdf to simplify the treatment.

To take advantage of the TDR technique, consider a target density of the form

$$p_o(x) = f(g(x)) \propto H(g(x)), \tag{4.37}$$

where $g(x)$ is either a concave or a convex function (or a general function with known inflection points, see [23, Chap. 5]). Hence, the target pdf is a distribution generated by a transformation of scale $g(x)$ of the pdf $f(z)$ [20].

*Remark 4.6* The function $H(z)$ is non-monotonic though, differently from the transformation $T$ of Sects. 4.3.4 and 4.3.5.

The first step of the standard GARS procedure can be used to replace the nonlinearity $g(x)$ with suitable linear functions $r_k(x) = a_k x + b_k$, for all $x \in \mathcal{I}_k$ and $k = 0, \ldots, m_t$, such that

$$H(r_k(x)) = H(a_k x + b_k) \geq H(g(x)), \tag{4.38}$$

[3]Table 4.2 in Sect. 4.3.5 summarizes the other three possible cases where the technique is applicable: $T^{-1}(z)$ increasing and $g(x)$ convex, $T^{-1}(z)$ decreasing and $g(x)$ concave and finally $T^{-1}(z)$ decreasing and $g(x)$ convex. Note that in all the four cases $T^{-1}(z)$ is a monotonic function.

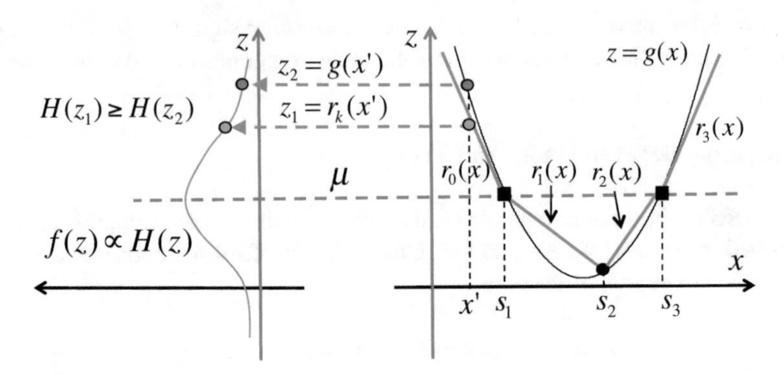


**Fig. 4.15** Example of construction of the linear functions $r_k(x)$, $k = 0, \ldots, m_t = 3$ for a convex nonlinearity $g(x)$ and a function $H(z)$ with a single mode at $\mu$. The simple estimates $s_1$ and $s_3$ are solutions of the equation $\mu = g(x)$

in order to use

$$\pi_t(x) \propto H(r_k(x))$$

as the proposal pdf in an RS scheme. Since $f(z) \propto H(z)$, it is important to remark that $\pi_t(x) \propto f(r_k(x)) \propto H(r_k(x))$ is a linearly-scaled version of the pdf $f(z)$, hence if we can draw from $f$ then we can generate samples also from $\pi_t(x)$. We recall that if the function $H$ is monotonic, then we go back to the scenario in Sect. 4.3.4, where $H(z) = T^{-1}(z)$ (with either $\mu \to +\infty$ if $H$ is increasing, or $\mu \to -\infty$ if $H$ is decreasing).

The construction of the necessary linear functions is very similar to the standard algorithm. Given a set of support points

$$\mathcal{S}_t = \{s_1, \ldots, s_{m_t}\},$$

we aim at finding a linear function $r_k(x)$ in the interval $\mathcal{I}_k = [s_k, s_{k+1}]$ such that $H(r_k(x)) \geq H(g(x))$. An example with $m_t = 3$ support points and a convex nonlinearity $g(x)$ is shown in Fig. 4.15. The simple estimates (shown with squares) are calculated as solutions of the equation $\mu = g(x)$. The straight lines $r_0(x)$ and $r_3(x)$ are tangent to $g(x)$ at $s_1$ and $s_3$, respectively, while $r_1(x)$ and $r_2(x)$ are secant lines. If $\mu \to \pm\infty$, Fig. 4.15 would show the cases also depicted in Figs. 4.8 and 4.9, with $H(z) = T^{-1}(z)$.[4]

[4]The case when the set of simple estimates is empty is similar to the case $\mu \to \pm\infty$, hence the construction of the linear functions is also similar, but it needs a special care. For further details, see [23].

The construction is valid because the inequality in Eq. (4.38) is satisfied. Indeed, arbitrarily choosing $x'$ and taking the values $z_2 = g(x')$ and $z_1 = r_k(x')$ we always have that $H(z_1) \geq H(z_2)$.

Therefore, the GARS method can be interpreted as a technique to use rejection sampling for target densities generated by a transformation of scale [20]. Here, we have considered target pdfs which are scaled versions of a pdf $f(z)$ with a single mode $\mu$, but the GARS technique can also be applied when $f(z)$ has several modes.

## 4.4 Performance and Computational Cost of the ARS Schemes

In Sect. 3.3, we have seen that the computational cost of an RS scheme is determined mainly by the acceptance rate and by the time necessary to draw from the proposal pdf. Here, we study these two aspects in the ARS methodology.

### 4.4.1 Acceptance Rate

In an RS scheme a proposed sample $x'$ is rejected if the discrepancy between the envelope function $\bar{\pi}_t(x)$ and $p(x)$ is high (the probability $\frac{p(x)}{\bar{\pi}_t(x)}$ of accepting $x'$ is small). An ARS scheme uses the rejected $x'$ to improve the construction of the proposal around $x'$. In this way, $\pi_{t+1}(x) \propto \exp(-W_{t+1}(x))$ becomes "closer" to $p(x) \propto p_o(x)$ and it can be expected that the mean acceptance rate becomes higher. This is illustrated, numerically, by Example 4.1.

To be precise, the probability of accepting a sample $x \in \mathcal{D}$ drawn from $\pi_t(x) \propto \bar{\pi}_t(x)$ is

$$a_t(x) = \frac{p(x)}{\bar{\pi}_t(x)},$$

and the acceptance rate at the $t$th iteration, denoted as $\hat{a}_t$, is the expected value of $a_t(x)$ with respect to the proposal density, i.e.,

$$\hat{a}_t = E[a_t(x)] = \int_{\mathcal{D}} a_t(x)\pi_t(x)dx = c_t \int_{\mathcal{D}} \frac{p(x)}{\bar{\pi}_t(x)}\pi_t(x)dx = \frac{c_v}{c_t} \leq 1, \tag{4.39}$$

where $1/c_t$ and $1/c_v$ are the normalizing constants of $\pi_t(x)$ and $p(x)$, namely

$$c_t = \int_{\mathcal{D}} \bar{\pi}_t(x)dx$$

and

$$c_v = \int_{\mathcal{D}} p(x)dx,$$

respectively. Note that $\frac{c_v}{c_t} \leq 1$ because $\bar{\pi}_t(x) \geq p(x)$, $\forall x \in \mathcal{D}$ and $\forall t \in \mathbb{N}$.

From Eq. (4.39), we obtain that $\hat{a}_t = 1$ if, and only if, $c_t = c_v$ or, equivalently, $\hat{a}_t = 1$ if and only if the integral

$$e(t) \triangleq \int_{\mathcal{D}} [\bar{\pi}_t(x) - p(x)] \, dx \tag{4.40}$$

vanishes, i.e., $e(t) = 0$.

The error signal $e(t)$ can be interpreted as a divergence[5] between $\pi_t(x)$ and $p(x)$. In particular if $e(t)$ decreases, the acceptance rate $\hat{a}_t = \frac{c_v}{c_t}$ increases and, since $\bar{\pi}_t(x) \geq p(x)$, $\forall x \in \mathcal{D}$, $e(t) = 0$ if, and only if, $\bar{\pi}_t(x) = p(x)$ almost everywhere. Equivalently, $\hat{a}_t = 1$ if, and only if, $\pi_t(x) = p_o(x)$ almost everywhere.

### 4.4.2 *Probability of Adding a New Support Point*

So far we have seen that the use of more support points improves the proposal pdf $\pi_t(x)$ and, as a consequence, provides better acceptance rates. However, the cost of drawing samples from $\pi_t(x)$ strictly depends on the number of pieces that form the proposal, which, in turn, are determined by the total number of support points (see Sect. 4.3). Then, an increase in the number of points yields higher acceptance rates but also a higher cost in terms of the time needed to draw from $\pi_t(x)$.

Fortunately, the probability of incorporating a new support point in $\mathcal{S}_t$ vanishes quickly as the acceptance rate approaches 1, since in the ARS scheme of Sect. 4.2 the probability of adding a new point coincides exactly with the probability of discarding a sample, that is exactly $1 - a_t(x)$. Since the expected value of $a_t(x)$ converges to 1 then the probability of adding a new support point approaches zero as $t \to +\infty$.

## 4.5 Variants of the Adaptive Structure in the ARS Scheme

The adaptive scheme given in Sect. 4.2.2 presents several advantages but it is certainly not unique. Different possibilities can be considered.

[5] Note that $[\bar{\pi}_t(x) - p(x)] \geq 0$ for all $x \in \mathcal{D}$.

A broad discussion about the optimal distribution of the support points is given in [17, Chap. 4] (see also Sect. 4.5.2 below). Furthermore, the authors in [17, Chap. 4] introduce a *derandomized* version of the ARS scheme, in which support points are chosen according to a deterministic rule: a new point selected deterministically is added where the discrepancy between the envelope function and the target exceeds a threshold (previously chosen by the user).

In the sequel, we describe two alternative structures that maintain a stochastic component in choosing novel support points, i.e., in the construction of the proposal but enable us to control the computational cost. These variants are especially useful to design ARS algorithms for drawing samples from multivariate target pdfs (as shown in Chap. 6).

### 4.5.1 Deterministic Test for Adding New Support Points

In this subsection, we describe a deterministic test for deciding whether to incorporate, or not, a new support point to the proposal. We build on ideas introduced in [29] for similar techniques. The main difference with respect to the derandomized ARS version in [17, Chap. 4] is that the new points are still adaptively (and randomly) chosen as in the standard ARS method.

The first important observation is that in the standard ARS structure of Sect. 4.2.2, the RS test is used to decide, jointly, (a) whether to accept or not the sample $x'$ (with probability $\frac{p(x')}{\bar{\pi}(x')}$) and (b) whether to incorporate or not $x'$ into $\mathcal{S}_t$ (with probability $1-\frac{p(x')}{\bar{\pi}(x')}$). However, we could split this test into two parts: first decide whether to accept the new sample or not and, afterwards, decide whether the sample should be incorporated as a support point or not. This last test can be chosen arbitrarily, designing different algorithms, since it affects only the construction of the proposal but does not interfere with the sampling mechanism, i.e., we can change this test as we wish and the sampling technique still remains valid. For instance, a possible algorithm is outlined below.

1. Choose a value $\epsilon > 0$, an initial set $\mathcal{S}_0 = \{s_1, \ldots, s_{m_0}\}$, $s_i \in \mathcal{D}$, $i = 1, \ldots, m_t$, and set $t = 0$. Let $N$ be the number of desired samples distributed according $p_o(x)$.
2. Build $\pi_t(x) \propto \bar{\pi}_t(x)$ using a suitable procedure (see Sect. 4.3) given the set $\mathcal{S}_t$.
3. Draw $x' \sim \pi_t(x)$ and $u' \sim \mathcal{U}([0,1])$.
4. If $u' \leq p(x')/\bar{\pi}_t(x')$, then accept $x^{(i)} = x'$, set $i = i + 1$. Otherwise, if $u' > p(x')/\bar{\pi}_t(x')$, reject $x'$.
5. If $|\bar{\pi}_t(x')-p(x')| > \epsilon$, then set $\mathcal{S}_{t+1} = \mathcal{S}_t \cup x'$. Otherwise, i.e., $|\bar{\pi}_t(x')-p(x')| \leq \epsilon$, set $\mathcal{S}_{t+1} = \mathcal{S}_t$.
6. If $i < N$ stop. Otherwise, update $t = t + 1$ and go to step 2.

The RS test at step 4 ensures that the accepted samples are distributed exactly according to the target pdf $p_o(x) \propto p(x)$. The second test, at step 5, adds $x'$ to the set of support points if the discrepancy (measured as a the absolute value of the difference) between the envelope function $\bar{\pi}_t(x')$ and the target $p(x)$ exceeds a threshold $\epsilon > 0$.

*Remark 4.7* Since $\epsilon$ is fixed and positive, the main difference with the ARS structure of Sect. 4.2.2 is that here the adaptation could be stopped at some iteration $t^*$, i.e.,

$$\pi_{t^*}(x) = \pi_{t^*+1}(x) = \ldots = \pi_{t^*+\tau}(x) \ldots.$$

if $|\bar{\pi}_{t^*}(x) - p(x)| < \epsilon$ for all $x \in \mathcal{D}$. Namely, we have

$$\mathcal{S}_{t^*} = \mathcal{S}_{t^*+1} = \ldots \mathcal{S}_{t^*+\tau} = \ldots$$

Therefore, we can choose the threshold $\epsilon$ to establish a trade-off between the acceptance rate that we attempt to obtain and the number of support points. With bigger $\epsilon$, on average we add a smaller number of support points but obtain also a smaller acceptance rate. On the contrary, with smaller $\epsilon$ we add on average a greater number of support points but obtain an acceptance rate closer to 1.

The value $\epsilon = 0$ is not admitted, since in this case all proposed samples drawn from $\pi_t(x)$ are included in $\mathcal{S}_t$ and, as a consequence, the computational cost grows without bound. With $\epsilon \to +\infty$ we never update the proposal so that this ARS scheme becomes a simple ("static") RS algorithm.

## Parsimonious Adaptive Rejection Sampling (PARS)

In some cases, it is difficult to choose a proper value of the parameter $\epsilon$ when no a-priori information about the range of values of the function $d(x) = |\bar{\pi}_t(x) - p(x)|$ is available. A variant of the previous technique can be applied which considers a threshold value $\delta \in [0, 1]$. This alternative scheme, called Parsimonious ARS (PARS) [24], is outlined below:

1. Set $t = 0$ and choose an initial set of support points, $\mathcal{S}_0 = \{s_1, \ldots, s_{m_0}\}$ and a threshold value $\delta \in [0, 1]$. Let $N$ be the number of desired samples distributed according to $p_o(x)$.
2. Build $\pi_t(x) \propto \bar{\pi}_t(x)$ using a suitable procedure given the set $\mathcal{S}_t$.
3. Draw $x' \sim \pi_t(x)$ and $u' \sim \mathcal{U}([0, 1])$.
4. If $u' \leq p(x')/\bar{\pi}_t(x')$, then accept $x^{(i)} = x'$, and set $i = i + 1$, $\mathcal{S}_{t+1} = \mathcal{S}_t$.
5. If $\frac{p(x')}{\bar{\pi}_t(x')} \leq \delta$, update $\mathcal{S}_{t+1} = \mathcal{S}_t \cup \{x'\}$. Otherwise, if $\frac{p(x')}{\bar{\pi}_t(x')} > \delta$, set $\mathcal{S}_{t+1} = \mathcal{S}_t$.
6. If $i < N$ stop. Otherwise, update $t = t + 1$ and go to step 2.

Note that if $\delta = 0$, no support points are added so that the method is not adaptive in this case. If $\delta = 1$, all the generated samples from the proposal pdf $\pi(x)$ are incorporated as support points, so that the method becomes computationally demanding.

### 4.5.2 An Adaptive Rejection Sampler with Fixed Number of Support Points

Except for the initial set, the locations of the rest of support points are chosen randomly and adaptively in the ARS algorithm introduced in Sect. 4.2. Clearly, if we fix a number $m$ of points and a given construction procedure of the proposal, the optimal distribution of support points $\hat{\mathcal{S}}^{(m)} = \{\hat{s}_1, \ldots, \hat{s}_m\}$ is the one which minimizes the discrepancy between $\bar{\pi}_t(x)$ and $p(x)$. Namely, using the more explicit notation $\bar{\pi}_t(x|\mathcal{S}^{(m)})$ and denoting the $L_1$ distance between $\bar{\pi}_t(x|\mathcal{S}^{(m)})$ and $p(x)$ as

$$D(\mathcal{S}^{(m)}) = \int_{\mathcal{D}} \left|\bar{\pi}_t(x|\mathcal{S}^{(m)}) - p(x)\right| dx = \int_{\mathcal{D}} (\bar{\pi}_t(x|\mathcal{S}^{(m)}) - p(x))dx,$$

then the optimal set of $m$ support points is such that

$$D(\hat{\mathcal{S}}^{(m)}) = \min_{\mathcal{S}^{(m)}} D(\mathcal{S}^{(m)}).$$

By an argument similar to the discussion in Sect. 4.4.1, it is apparent that the (optimal) set $\hat{\mathcal{S}}^{(m)}$ maximizes the achievable acceptance rate subject to having at most $m$ points and using a prescribed (suitable) construction procedure.

The rest of this subsection is devoted to the description of an alternative adaptive structure that uses a fixed total number $m$ of support points, whose positions are still selected randomly. The parameter $m$ can be selected by the user in order to control the computational cost of the resulting adaptive rejection sampler (note that $m$ can also be seen as the maximum allowed number of support points).

The underlying idea relies on the following observation. Equation (4.39) states that the acceptance rate in an RS scheme is

$$\hat{a}_t = \frac{c_v}{c_t},$$

where $c_t = \int_{\mathcal{D}} \bar{\pi}_t(x)dx$ represents the area below the envelope function $\bar{\pi}_t(x)$ whereas $c_v = \int_{\mathcal{D}} p(x)dx$ is the area below $p(x)$. Clearly, in our problem $c_v$ is fixed. As a consequence, to increase the acceptance rate we have to improve the construction of the envelope function in order to diminish $c_t$ (while, obviously, maintaining $\bar{\pi}_t(x) \geq p(x)$). The following algorithm changes the position of the support points in order to decrease $c_t$.

#### Cheap Adaptive Rejection Sampling (CARS) Algorithm

Consider the intervals $\mathcal{I}_0 = (-\infty, s_1]$, $\mathcal{I}_j = [s_j, s_{j+1}]$, $j = 1, \ldots, m$ and $\mathcal{I}_{m+1} = [s_m, +\infty)$. A possible algorithm with a fixed number of support points, called Cheap

Adaptive Rejection Sampling (CARS) [25], is given below:

1. Set $i = 0$, $t = 0$. Choose $m$ points $\mathcal{S}_0^{(m)} = \{s_1, \ldots, s_m\}$. Let $N$ be the number of desired samples distributed according to $p_o(x) \propto p(x)$.
2. Build $\pi_t(x) \propto \bar{\pi}_t(x|\mathcal{S}_t^{(m)})$ using a suitable procedure given the set $\mathcal{S}_t^{(m)}$.
3. Draw $x' \sim \pi_t(x)$ and $u' \sim \mathcal{U}([0, 1])$.
4. If $u' \leq p(x')/\bar{\pi}_t(x')$, then accept $x^{(i)} = x'$, set $i = i + 1$, $\mathcal{S}_{t+1}^{(m)} = \mathcal{S}_t^{(m)}$, $t = t + 1$ and jump to step 6.
5. If $u' > p(x')/\bar{\pi}_t(x')$, then reject $x'$ and

   (a) Find the interval $\mathcal{I}_j = [s_j, s_{j+1}]$ such that $x' \in \mathcal{I}_j$.
   (b) Using the alternative sets

$$\mathcal{A}_1 = \{s_1, \ldots, s_{j-1}, x', s_{j+1}, \ldots, s_m\},$$

$$\mathcal{A}_2 = \{s_1, \ldots, s_j, x', s_{j+2}, \ldots, s_m\},$$

   compute $c_t^{(1)} = \int_{\mathcal{D}} \bar{\pi}_t(x|\mathcal{A}_1)dx$ and $c_t^{(2)} = \int_{\mathcal{D}} \bar{\pi}_t(x|\mathcal{A}_2)dx$.
   (c) If $c_t^{(1)} \leq c_t^{(2)}$, then set $\mathcal{S}_{t+1}^{(m)} = \mathcal{A}_1$, otherwise $\mathcal{S}_{t+1}^{(m)} = \mathcal{A}_2$.

6. If $i > N$ stop, otherwise go to step 2.

Note that at each step the set $\mathcal{S}_t^{(m)}$ either remains the same or just one support point is substituted. It is apparent that the algorithm produces a sequence of envelope functions with non-increasing areas, i.e.,

$$c_0 \geq c_1 \geq \ldots c_t \geq c_{t+1} \ldots . \geq c_{t+\tau} \geq \ldots . \geq \lim_{t\to\infty} c_t = c_\infty.$$

Therefore, the acceptance rate can be expected to grow as the procedure is iterated but, since the number of points is finite, it will not converge to 1, in general, even if $t \to \infty$. As a matter of fact, there is also no guarantee that the algorithm convergences to the optimal distribution of the support points, i.e., $\mathcal{S}_\infty^{(m)} \neq \hat{\mathcal{S}}^{(m)}$ (the final value $c_\infty$ may not be the minimum possible area). The final stationary locations of the support points, $\mathcal{S}_\infty^{(m)}$, attained by the algorithm depend in general on the initial set $\mathcal{S}_0^{(m)}$. The latter, in turn, can be chosen using the standard ARS approach in Sect. 4.2. In this sense, the value $m$ can be seen as a maximum total number of support points.

### Applicability

The applicability of this technique depends strictly on the effort required to apply step 5 of the algorithm, compared to the gain in terms of computational cost using only $m$ points. A good construction procedure for this algorithm is described in Sect. 4.3.2, where the proposal pdf is a stepwise function (with the exception of the tails). Observe that, with this construction, a change of one support point only varies two pieces of the built proposal (corresponding to the intervals $\mathcal{I}_{j-1}$ and $\mathcal{I}_j$) whereas the rest remains invariant. Therefore, to compute the values $c_t^{(1)}$ and $c_t^{(2)}$ we need to rebuild the proposal and recalculate the areas just in these two pieces.

## 4.6 Combining ARS and MCMC

ARS samplers are very appealing tools for computational inference because of their efficiency. However, the range of target probability distributions to which they can be applied is still limited and, hence, there have been attempts to extend the methodology in the direction of obtaining a fully universal adaptive sampling scheme that can be used whenever the target pdf can be evaluated point-wise.

An obvious path to pursue the latter goal is to investigate the combination of the ARS and MCMC methodologies and, in particular, the authors in [10] have addressed the design of adaptive samplers with intertwined Metropolis-Hastings acceptance tests. The resulting algorithm, called *adaptive rejection Metropolis sampling* (ARMS), broadens the range of applicability of the ARS method (it can be always applied as long as the target can be evaluated) and improves the performance of a standard MH technique. Its basic drawback is that the generated samples form a Markov chain and, therefore, they are no longer independent.

### 4.6.1 Adaptive Rejection Metropolis Sampling

This method, introduced in [10], is a generalization of the standard ARS algorithm that includes a Metropolis-Hastings step and can be applied to any target pdf. Unfortunately, because of the incorporation of MCMC steps, the produced samples are correlated, i.e., they are not statistically independent. This technique is a *rejection sampling chain* [35], as described in Sect. 3.9.2, using an adaptive proposal pdf.

The main idea is relatively simple. Consider a proposal $\pi_t(x) \propto \bar{\pi}_t(x)$ built with some prescribed procedure. In this case, we do not need $\bar{\pi}_t(x)$ to be an envelope function, i.e., we can possibly have $\bar{\pi}_t(x) < p(x)$. Therefore, if we apply an RS scheme using $\pi_t(x) \propto \bar{\pi}_t(x)$, we have seen in Sect. 3.2.3 that the accepted samples are distributed as

$$q(x) \propto \min(p(x), \bar{\pi}_t(x)).$$

**Table 4.3** Adaptive rejection metropolis sampling algorithm (ARMS)

| |
|---|
| 1. Start with $t = 0$, $k = 0$, $m_0 = 2$, $\mathcal{S}_0 = \{s_1, s_2\}$ where $s_1 < s_2$, and<br>Let $N$ be the number of desired iterations of the chain.<br>Choose an arbitrary initial state $x_0$ |
| 2. Build the proposal pdf $\pi_t(x) \propto \bar{\pi}_t(x)$ using $\mathcal{S}_t$ |
| 3. Draw $x'$ from $\pi_t(x)$, and $u'$ from $\mathcal{U}([0, 1])$ |
| 4. If $u' > \frac{p(x')}{\bar{\pi}_t(x)}$, then reject $x'$, set $\mathcal{S}_{t+1} = \mathcal{S}_t \cup \{x'\}$ and update<br>$m_{t+1} = m_t + 1$. Jump to step 8 |
| 5. Otherwise, if $u' \leq \frac{p(x')}{\bar{\pi}_t(x)}$, draw $v'$ from $\mathcal{U}([0, 1])$ |
| 6. If $v' \leq \min\left[1, \frac{p(x')\min(p(x_k),\bar{\pi}_t(x_k))}{p(x_k)\min(p(x'),\bar{\pi}_t(x'))}\right]$ then accept $x^{(i)} = x'$, set $x_c = x'$<br>and $\mathcal{S}_{t+1} = \mathcal{S}_t$, $m_{t+1} = m_t$ and $i = i + 1$ |
| 7. If $v' > \min\left[1, \frac{p(x')\min(p(x_k),\bar{\pi}_t(x_k))}{p(x_k)\min(p(x'),\bar{\pi}_t(x'))}\right]$ then reject $x'$, set $x^{(i)} = x_c$, and<br>$\mathcal{S}_{t+1} = \mathcal{S}_t$, $m_{t+1} = m_t$, $i = i + 1$ |
| 8. If $k > N$ then stop, else increment $t = t + 1$ and go back to step 2 |

The generated samples are correlated

To compensate for this distortion, that appears because $\bar{\pi}_t(x)$ is not an envelope function, a Metropolis-Hastings control test is added to ensure sampling from the target $p_o(x) \propto p(x)$. Moreover, exactly as in other ARS schemes, the samples rejected in the RS test are used to improve the construction of the proposal pdf. Table 4.3 summarizes the algorithm.

The validity of the method should be proved using the theory of MCMC methods with an adaptive proposal pdf [22, Chap. 8]. However, since the current state of the chain is never used for updating the proposal the proof is quite easy [10] (see also an important and related result in [14, 15]). The ARMS method can be considered as an MH algorithm with a proposal independent of the current state and the points in $\mathcal{S}_t$ play the role of auxiliary variables [2, 10].

### 4.6.2 A Procedure to Build Proposal pdfs for the ARMS Algorithm

The authors of [10] suggest a specific procedure to build the proposal that we describe in the sequel. Given a set of support points

$$\mathcal{S}_t = \{s_1, \ldots, s_{m_t}\},$$

with $s_1 < \ldots < s_{m_t}$, consider the intervals $\mathcal{I}_0 = (-\infty, s_1]$, $\mathcal{I}_j = [s_j, s_{j+1}]$, $j = 1, \ldots, m_t$ and $\mathcal{I}_{m_t+1} = [s_{m_t}, +\infty)$ and a potential $V(x) = -\log[p(x)]$. Moreover, let us denote as $L_{j,j+1}(x)$ the straight line passing through the points $(s_j, V(s_j))$ and $(s_{j+1}, V(s_{j+1}))$ for $j = 1, \ldots, m_t - 1$, and also set

$$L_{-1,0}(x) = L_{0,1}(x) = L_{1,2}(x),$$

$$L_{m_t,m_t+1}(x) = L_{m_t+1,m_t+2}(x) = L_{m_t-1,m_t}(x).$$

In [10], a piecewise-linear function $W_t(x)$ is constructed as

$$W_t(x) = \min\left[L_{j,j+1}(x), \max\left[L_{j-1,j}(x), L_{j+1,j+2}(x)\right]\right], \tag{4.41}$$

with $x \in \mathcal{I}_j = (s_j, s_{j+1}]$ and $j = 0, \ldots, m_t$. Hence, the proposal pdf, defined as $\pi_t(x) \propto \bar{\pi}_t(x) = \exp(-W_t(x))$, is formed by exponential pieces (that are easy to draw from).

The advantage of using $W_t(x)$ of the form in Eq. (4.41) is that if $V(x)$ is convex (i.e., $p(x)$ is log-concave) then we have

$$W_t(x) = \max\left[L_{j-1,j}(x), L_{j+1,j+2}(x)\right], \quad x \in \mathcal{I}_j = (s_j, s_{j+1}],$$

that is exactly the construction given in Sect. 4.3.2 [see Eq. (4.9)] and a brief examination shows also that in this case

$$\bar{\pi}_t(x) = \exp(-W_t(x)) \geq p(x),$$

so that the ARMS is reduced to the ARS (generating i.i.d. samples) if the target pdf is log-concave. In [30] it is suggested to use functions $W_t(x)$ formed by polynomials of degree 2 (parabolic pieces), instead of linear functions (polynomials of degree 1) in order to produce a better approximation of the real potential $V(x)$ and improve the performance of the algorithm. More recently, other simpler construction procedures have been proposed and analyzed in [28]. See Chap. 7, for further study of the performance and drawbacks of ARMS-type methods.

## 4.7 Summary

The main drawback of the rejection sampling (RS) method is the difficulty to find an envelope function, $L\pi(x) \geq p(x)$, "similar enough" to the target density in order to attain high acceptance rates. In this chapter, we have described a class of adaptive rejection sampling (ARS) algorithms that adaptively build a sequence of proposal functions that converge, under suitable conditions, toward the target density. These methods are very efficient samplers that update the proposal whenever a generated sample is rejected in the RS test, and produce i.i.d. samples from the target with acceptance rate close to 1. One further advantage of ARS schemes is that, after the selection of initial support points, they are completely automatic, self-tuning algorithms regardless of the specific target density.

The main limitation for the practical use of this family of techniques is the ability to build a suitable sequence of proposal densities. While relatively simple proposal construction procedures are easily available for the univariate case, they are particularly hard to design in multidimensional spaces. In Sect. 4.2, we have outlined the general structure of an ARS algorithm and listed the necessary general conditions that a proposal construction procedure should satisfy for the resulting functions to be used within an ARS scheme. In Sect. 4.3, we have reviewed different procedures, tailored to specific classes of target distributions.

The computational cost of an ARS is essentially a result of its acceptance rate and the effort required to generate a sample from the proposals, as discussed in Sect. 4.4. In order to control the computational cost, we have also looked into two variants of the standard adaptive structure. While related to existing "derandomized" techniques [17, Chap. 4], both alternatives, studied in Sect. 4.5, maintain a stochastic component in the construction of the proposal densities. These approaches to proposal construction prove themselves useful when addressing the problem of drawing samples from multivariate distributions, as shown in Chap. 6.

Finally, in Sect. 4.6, we have described the adaptive rejection Metropolis sampling (ARMS) algorithm. It combines the ARS and Metropolis-Hastings (MH) methodologies, in order to broaden the range of applicability of the standard ARS technique and improve the performance of a standard MH algorithm. It has the disadvantage, though, that the generated samples are correlated, unlike in the ARS method.

## References

1. M.S. Arulumpalam, S. Maskell, N. Gordon, T. Klapp, A tutorial on particle filters for online nonlinear/non-Gaussian Bayesian tracking. IEEE Trans. Signal Process. **50**(2), 174–188 (2002)
2. J. Besag, P.J. Green, Spatial statistics and Bayesian computation. J. R. Stat. Soc. Ser. B **55**(1), 25–37 (1993)
3. C. Botts, A modified adaptive accept-reject algorithm for univariate densities with bounded support. Technical Report, http://williams.edu/Mathematics/cbotts/Research/paper3.pdf (2010)
4. G.E.P. Box, G.C. Tiao, *Bayesian Inference in Statistical Analysis* (Wiley, New York, 1973)
5. L. Devroye, *Non-uniform Random Variate Generation* (Springer, New York, 1986)
6. M. Evans, T. Swartz, Random variate generation using concavity properties of transformed densities. J. Comput. Graph. Stat. **7**(4), 514–528 (1998)
7. J. Geweke, Bayesian inference in econometric models using Monte Carlo integration. Econometrica **24**, 1317–1399 (1989)
8. W.R. Gilks, Derivative-free adaptive rejection sampling for Gibbs sampling. Bayesian Stat. **4**, 641–649 (1992)
9. W.R. Gilks, P. Wild, Adaptive rejection sampling for Gibbs sampling. Appl. Stat. **41**(2), 337–348 (1992)
10. W.R. Gilks, N.G. Best, K.K.C. Tan, Adaptive rejection Metropolis sampling within Gibbs sampling. Appl. Stat. **44**(4), 455–472 (1995)

11. D. Gorur, Y.W. Teh, Concave convex adaptive rejection sampling. J. Comput. Graph. Stat. **20**(3), 670–691 (2011)
12. T.L. Griffiths, Z. Ghahramani, The indian buffet process: an introduction and review. J. Mach. Learn. Res. **12**, 1185–1224 (2011)
13. H. Hirose, A. Todoroki, Random number generation for the generalized normal distribution using the modified adaptive rejection method. Int. Inf. Inst. **8**(6), 829–836 (2005)
14. L. Holden, Adaptive chains. Technical Report Norwegian Computing Center (1998)
15. L. Holden, R. Hauge, M. Holden, Adaptive independent Metropolis-Hastings. Ann. Appl. Probab. **19**(1), 395–413 (2009)
16. W. Hörmann, A rejection technique for sampling from T-concave distributions. ACM Trans. Math. Softw. **21**(2), 182–193 (1995)
17. W. Hörmann, J. Leydold, G. Derflinger, *Automatic Nonuniform Random Variate Generation* (Springer, New York, 2003)
18. W. Hörmann, J. Leydold, G. Derflinger, Inverse transformed density rejection for unbounded monotone densities. Research Report Series, Department of Statistics and Mathematics (Economy and Business), Vienna University (2007)
19. Y. Huang, J. Zhang, P.M. Djurić, Bayesian detection for BLAST. IEEE Trans. Signal Process. **53**(3), 1086–1096 (2005)
20. M.C. Jones, Distributions generated by transformation of scale using an extended cauchy-schlömilch transformation. Indian J. Stat. **72-A**(2), 359–375 (2010)
21. J. Leydold, J. Janka, W. Hörmann, Variants of transformed density rejection and correlation induction, in *Monte Carlo and Quasi-Monte Carlo Methods 2000* (Springer, Heidelberg, 2002), pp. 345–356
22. F. Liang, C. Liu, R. Caroll, *Advanced Markov Chain Monte Carlo Methods: Learning from Past Samples*. Wiley Series in Computational Statistics (Wiley, Chichester, 2010)
23. L. Martino, *Novel Schemes for Adaptive Rejection Sampling* (Universidad Carlos III, Madrid, 2011)
24. L. Martino, Parsimonious adaptive rejection sampling. *IET Electron. Lett.* **53**(16), 1115–1117 (2017)
25. L. Martino, F. Louzada, Adaptive rejection sampling with fixed number of nodes. Commun. Stat. Simul. Comput., 1–11 (2017, to appear)
26. L. Martino, J. Míguez, A generalization of the adaptive rejection sampling algorithm. Stat. Comput. **21**, 633–647 (2010). doi:https://doi.org/10.1007/s11222-010-9197-9
27. L. Martino, J. Míguez, Generalized rejection sampling schemes and applications in signal processing. Signal Process. **90**(11), 2981–2995 (2010)
28. L. Martino, J. Read, D. Luengo, Independent doubly adaptive rejection Metropolis sampling within Gibbs sampling, *IEEE Trans. Signal Process.* **63**(12), 3123–3138 (2015)
29. L. Martino, R. Casarin, F. Leisen, D. Luengo, Adaptive independent sticky MCMC algorithms. EURASIP J. Adv. Signal Process. (2018, to appear)
30. R. Meyer, B. Cai, F. Perron, Adaptive rejection Metropolis sampling using Lagrange interpolation polynomials of degree 2. Comput. Stat. Data Anal. **52**(7), 3408–3423 (2008)
31. J. Michel, The use of free energy simulations as scoring functions. PhD Thesis, University of Southampton (2006)
32. R. Neal, MCMC using Hamiltonian dynamics, Chap. 5, in *Handbook of Markov Chain Monte Carlo* ed. by S. Brooks, A. Gelman, G. Jones, X.-L. Meng (Chapman and Hall/CRC Press, Boca Raton, 2011)
33. J.G. Proakis, *Digital Communications*, 4th edn. (McGraw-Hill, Singapore, 2000)
34. Y.W. Teh, D. Görür, Z. Ghahramani, Stick-breaking construction for the indian buffet process, in *Proceedings of the International Conference on Artificial Intelligence and Statistics* (2007)
35. L. Tierney, Exploring posterior distributions using Markov Chains, in *Computer Science and Statistics: Proceedings of IEEE 23rd Symposium on the Interface* (1991), pp. 563–570

# Chapter 5
# Ratio of Uniforms

**Abstract** This chapter provides a detailed description of the so-called ratio-of-uniforms (RoU) methods. The RoU and the generalized RoU (GRoU) techniques were introduced in Kinderman and Monahan (ACM Trans Math Softw 3(3):257–260, 1977); Wakefield et al. (Stat Comput 1(2):129–133, 1991) as bivariate transformations of the bidimensional region $\mathcal{A}_0$ below the target pdf $p_o(x) \propto p(x)$. To be specific, the RoU techniques can be seen as a transformation of a bidimensional *uniform random variable*, defined over $\mathcal{A}_0$, into another two-dimensional random variable defined over an alternative set $\mathcal{A}$. RoU schemes also convert samples uniformly distributed on $\mathcal{A}$ into samples with density $p_o(x) \propto p(x)$ (which is equivalent to draw uniformly from $\mathcal{A}_0$). Therefore, RoU methods are useful when drawing uniformly from the region $\mathcal{A}$ is comparatively simpler than drawing from $p_o(x)$ itself (i.e., simpler than drawing uniformly from $\mathcal{A}_0$). In general, RoU algorithms are applied in combination with the rejection sampling principle and they turn out especially advantageous when $\mathcal{A}$ is bounded. In this chapter, we present first the basic theory underlying RoU methods, and then study in depth the connections with other sampling techniques. Several extensions, as well as different variants and point of views, are discussed.

## 5.1 Introduction

In the previous chapters, we have seen that the best scenario for a rejection sampling (RS) scheme involves a bounded target pdf taking values on a bounded domain. For this reason, for example, we have studied the transformed rejection method (TRM), which, before applying the RS principle, transforms the r.v. $X$ with $p_o(x) \propto p(x)$ into another r.v. $Y$ with bounded density and bounded domain.

The *ratio of uniforms* (RoU) method is a sampling algorithm that identifies a region $\mathcal{A}$ (bounded, in the case of interest) connected to the target pdf $p_o(x)$ in the sense that if we are able to draw uniformly from $\mathcal{A}$, then we can also draw samples from $p_o(x)$. The most appealing feature of RoU schemes is that we do not need to know the boundary of $\mathcal{A}$ explicitly, because we are always able to check whether a point belongs to $\mathcal{A}$ or not (using certain simple inequalities). The RoU method is



L. Martino et al., *Independent Random Sampling Methods*, Statistics and Computing, https://doi.org/10.1007/978-3-319-72634-2_5

closely related to the TRM as we show later in this chapter. We have divided this material into three main blocks:

1. *Basic concepts:* Sections 5.2 and 5.4 provide the basic theory of RoU and the generalized RoU (GRoU) method, respectively. In Sect. 5.3, we also describe an adaptive RS scheme for the RoU technique. Some relevant properties of GRoU samplers are discussed in detail in Sect. 5.5.
2. *Relationships with other techniques:* In Sects. 5.6 and 5.7 we explain the connections with other sampling techniques, clarifying that the GRoU method is equivalent to a transformation of a r.v. $Y$ distributed according to the inverse density $p^{-1}(y)$ associated to $p_o(x) \propto p(x)$. To show this, we need to introduce an extended version of the standard inverse-of-density (IoD) method (Sect. 5.6.1). This approach is also useful, for instance, to relax certain conditions and extend the range of applications of GRoU.

   Given the previous analysis, in Sect. 5.8 we are able to provide the transformation that used within the GRoU method, achieves a rectangular region $\mathcal{A}$ (clearly, an easy one in order to generate uniform random variates). With the same procedure, other different shapes of $\mathcal{A}$ could be studied. In Sect. 5.8.1, we analyze the relationship between the GRoU scheme and the two standard versions of the IoD method described in Chap. 2.
3. *Extensions:* Section 5.9 is devoted to discuss how we can relax the different assumptions used in the GRoU theorem. In Sect. 5.10, we describe the generalization proposed in [18]. This generalization includes, as particular cases, other extensions independently introduced in the literature and enables us to see the GRoU technique from another point of view, as explained in Sect. 5.10.

Some final considerations are made in Sect. 5.11. Before starting with the description of the standard RoU methodology, we clarify an important point about the notation and the definition of inverse densities.

### 5.1.1 A Remark on Inverse Densities

As in the previous chapters, all pdfs of interest are assumed to be proper, yet most often only known up to a proportionality constant. For instance, we usually write the normalized target pdf as $p_o(x) = \frac{1}{c_v}p(x)$, with $\frac{1}{c_v} > 0$ denoting the normalization constant, and all the subsequent methods are formulated in terms of $p(x)$, which is the unnormalized target function such that

$$\int_{\mathcal{D}} p(x)dx = c_v, \tag{5.1}$$

with $c_v > 0$, but $c_v \neq 1$ in general. The set $\mathcal{D}$ denotes the domain of the r.v. $X$.

In Chap. 2, we have considered sampling methods using the inverse function $p^{-1}(y)$ of $p(x)$. Assume that $p(x)$ is monotonic. Since $p(x)$ is unnormalized, $p^{-1}(y)$

is also unnormalized with the same normalizing constant $\int_{\mathcal{D}_Y} p^{-1}(y)dy = c_v$, where $\mathcal{D}_Y$ is the domain of the variable $y$. Note that the domain $\mathcal{D}_Y$ changes depending on the value of $c_v$. Indeed, we have

$$p_o^{-1}(y) = p^{-1}\left(c_v y\right),$$

where $p_o^{-1}$ is normalized and is the inverse function of the normalized target $p_o(x)$. Although $\frac{1}{c_v}p^{-1}(y)$ and $p_o^{-1}(y)$ are both normalized, they are different pdfs, i.e.,

$$\frac{1}{c_v}p^{-1}(y) \neq p_o^{-1}(y) = p^{-1}\left(c_v y\right).$$

However, the techniques to be discussed in the rest of this chapter do not require the knowledge of the normalizing constant. Sometimes, for the sake of simplicity, we refer to $p(x)$ and $p^{-1}(y)$ as "densities" although they are unnormalized.

## 5.2 Standard Ratio of Uniforms Method

The basic formulation of the RoU method has been provided in [19, 25]. It is a sampling technique that relies on the following result.

**Theorem 5.1** *Let $p(x) \geq 0$ be a pdf known only up to a proportionality constant ($p(x) \propto p_o(x)$). If $(v, u)$ is a sample drawn from the uniform distribution on the set*

$$\mathcal{A} = \left\{(v, u) : 0 \leq u \leq \sqrt{2p(v/u)}\right\}, \tag{5.2}$$

*then $x = \frac{v}{u}$ is a sample from $p_o(x)$.*

*Proof* Given the transformation $(v, u) \to (x, y)$

$$\begin{cases} x = \dfrac{v}{u} \\ y = u \end{cases}, \tag{5.3}$$

and a pair of r.v.'s $(V, U)$ uniformly distributed on $\mathcal{A}$, we can write the joint pdf $q(x, y)$ of the transformed r.v.'s $(X, Y)$ as

$$q(x, y) = \frac{1}{|\mathcal{A}|}|\mathbf{J}^{-1}| \quad \text{for all} \quad 0 \leq y \leq \sqrt{2p(x)}, \tag{5.4}$$

where $|\mathcal{A}|$ is the area of $\mathcal{A}$, and $\mathbf{J}^{-1}$ is the Jacobian of the inverse transformation, i.e.,

$$\mathbf{J}^{-1} = \det\begin{bmatrix} y & x \\ 0 & 1 \end{bmatrix} = y. \tag{5.5}$$

Substituting (5.5) into (5.4) yields

$$q(x, y) = \begin{cases} \frac{1}{|\mathcal{A}|} y, & 0 \leq y \leq \sqrt{2p(x)}, \\ 0, & \text{otherwise.} \end{cases} \quad (5.6)$$

Then, the marginal density of the r.v. $X$ obtained by integrating the pdf $q(x, y)$ coincides with $p_o(x)$. Indeed,

$$\begin{aligned} \int_{-\infty}^{+\infty} q(x, y) dy &= \int_0^{\sqrt{2p(x)}} \frac{y}{|\mathcal{A}|} dy = \\ &= \frac{1}{|\mathcal{A}|} \left[ \frac{y^2}{2} \right]_0^{\sqrt{2p(x)}} = \frac{1}{|\mathcal{A}|} p(x), \end{aligned} \quad (5.7)$$

where the first equality follows from Eq (5.6) and the rest of the calculations are straightforward. Since $p_o(x) \propto p(x)$ and $\int_{-\infty}^{+\infty} q(x, y) dy$ is a proper pdf then, necessarily, $p_o(x) = \frac{1}{|\mathcal{A}|} p(x)$ (and $c_v = |\mathcal{A}|$). □

This theorem provides a way to draw from the target $p_o(x)$. Indeed, if we are able to draw a point $(v', u')$ uniformly on $\mathcal{A}$, then the sample $x' = v'/u'$ is distributed according to $p_o(x)$. Therefore, the efficiency of the RoU method depends on the ease with which we can generate variates uniformly within the region $\mathcal{A}$.

## *Measure of the Region $\mathcal{A}$*

The measure of the region $\mathcal{A}$ is clearly related with the area below $p(x)$. In fact, from Eq. (5.7) we can see that $\frac{1}{|\mathcal{A}|}$ is the normalizing constant of $p(x)$, i.e.,

$$|\mathcal{A}| = c_v = \int_{\mathcal{D}} p(x) dx. \quad (5.8)$$

In the case that $p(x) = p_o(x)$, then $c_v = \int_{\mathcal{D}} p(x) dx = 1$ so that $|\mathcal{A}| = 1$.

### Scaled Versions of the Region $\mathcal{A}$

A simple look to the proof above shows that the region $\mathcal{A}$ could be equivalently redefined as

$$\mathcal{A} = \left\{ (v, u) : 0 \leq u \leq \sqrt{2c_{\mathcal{A}} p(v/u)} \right\},$$

where $c_A$ is a positive constant (in Eq. (5.2), we have $c_A = 1$). In this case, the measure of $\mathcal{A}$ changes, $|\mathcal{A}| = c_A c_v$, but the method is still valid. In the literature, several authors define $\mathcal{A}$ using $c_A = \frac{1}{2}$, i.e.,

$$\mathcal{A} = \left\{(v, u) : 0 \leq u \leq \sqrt{p(v/u)}\right\}. \tag{5.9}$$

For simplicity, in the rest of this section, we assume $\mathcal{A}$ given by Eq. (5.9).

### 5.2.1 Some Basic Considerations

The cases of practical interest are those in which the region $\mathcal{A}$ is bounded. To illustrate how the method works, Fig. 5.1a depicts a (rather arbitrary) two-dimensional bounded set $\mathcal{A}$. Note that, for every angle $\alpha \in (-\pi/2, +\pi/2)$ rad, we can draw a straight line that passes through the origin $(0, 0)$ and contains points $(v_i, u_i) \in \mathcal{A}$ such that $x = \frac{v_i}{u_i} = \tan(\alpha)$, i.e., every point $(v_i, u_i)$ in the straight line with angle $\alpha$ yields the same value of $x$. From the definition of $\mathcal{A}$ and in Eq. (5.3), it follows that $u_i \leq p(x)$ and $v_i = xu_i \leq x\sqrt{p(x)}$. Hence, the boundary of $\mathcal{A}$ is defined parametrically by the system of equations

$$\begin{cases} u_b = \sqrt{p(x)}, \\ v_b = x\sqrt{p(x)}, \end{cases} \tag{5.10}$$

namely, the points $(v_b, u_b)$ that satisfy (5.10) lie on the boundary of $\mathcal{A}$. Thus, the set $\mathcal{A}$ is bounded if, and only if, both functions $\sqrt{p(x)}$ and $x\sqrt{p(x)}$ are bounded. It is easy to show that the function $\sqrt{p(x)}$ is bounded if, and only if, the target density

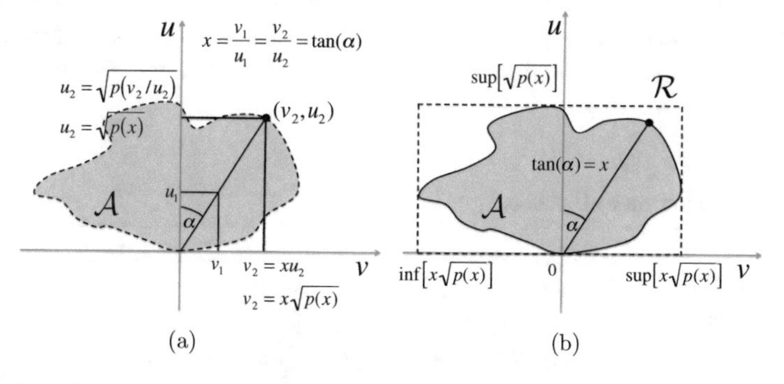


**Fig. 5.1** **(a)** A bounded region $\mathcal{A}$ and the straight line $v = xu$ corresponding to the sample $x = \tan(\alpha)$. Every point in the intersection of the line $v = xu$ and the set $\mathcal{A}$ yields the same sample $x$. The point on the boundary, $(v_2, u_2)$, has coordinates $v_2 = x\sqrt{p(x)}$ and $u_2 = \sqrt{p(x)}$. **(b)** If the two functions $\sqrt{p(x)}$ and $x\sqrt{p(x)}$ are bounded, the set $\mathcal{A}$ is bounded and embedded in the rectangle $\mathcal{R}$

$p_o(x) \propto p(x)$ is bounded, while the function $x\sqrt{p(x)}$ is bounded if, and only if, the tails of $p(x)$ decay as $1/x^2$ or faster. For further details see Sect. 5.5.1.

#### Simplest RoU Scheme

Owing to the definition of the boundary of $\mathcal{A}$, if the supremum of $\sqrt{p(x)}$ and $x\sqrt{p(x)}$, as well as the infimum of $x\sqrt{p(x)}$, can be found, then we can embed the set $\mathcal{A}$ in the rectangular region

$$\mathcal{R} = \left\{(v, u) : 0 \leq u \leq \sup_x \sqrt{p(x)},\ \inf_x x\sqrt{p(x)} \leq v \leq \sup_x x\sqrt{p(x)}\right\}, \tag{5.11}$$

as depicted in Fig. 5.1b. Once the rectangle $\mathcal{R}$ is constructed, it is straightforward to draw uniformly from $\mathcal{A}$ by rejection sampling: simply draw uniformly from $\mathcal{R}$ and then check whether the candidate point belongs to $\mathcal{A}$.

*Remark 5.1* Note that in this rejection procedure we do not need to know the analytical expression of the boundary of the region $\mathcal{A}$. Indeed, Eq. (5.9) provides a way to check whether a point $(v, u)$ falls inside $\mathcal{A}$ or not.

Table 5.1 summarizes this simple accept/reject scheme.

### 5.2.2 *Examples*

Figure 5.2b, d provides two examples in which the region $\mathcal{A}$ corresponds to standard Gaussian and Cauchy densities (shown in Fig. 5.2a, c, respectively). In the case of the standard Cauchy pdf, the region $\mathcal{A}$ is a semi-circle with radius 1 and center in $(0, 0)$. In order to show how RoU works, the pictures also display lines corresponding to $x$ and $y$ constant (dotted line) and the corresponding straight lines in the transformed domain $v - u$. Moreover, Fig. 5.2a depicts the set of points (solid line) that corresponds in the transformed domain $v - u$, to a line having $v$ constant (solid line).

**Table 5.1** Rejection via RoU method

| |
|---|
| 1. Start with $j = 1$ |
| 2. Construct the rectangle $\mathcal{R} \supseteq \mathcal{A}$ |
| 3. Draw a point $(v', u')$ uniformly from the rectangular region $\mathcal{R}$ |
| 4. If $u' \leq \sqrt{p(v'/u')}$, then accept the sample $x^{(j)} = x' = \frac{v'}{u'}$ and set $j = j + 1$ |
| 5. Otherwise, if $u' > \sqrt{p(v'/u')}$, then reject the sample $x' = \frac{v'}{u'}$ |
| 6. If $j > N$ then stop, else go back to step 2 |

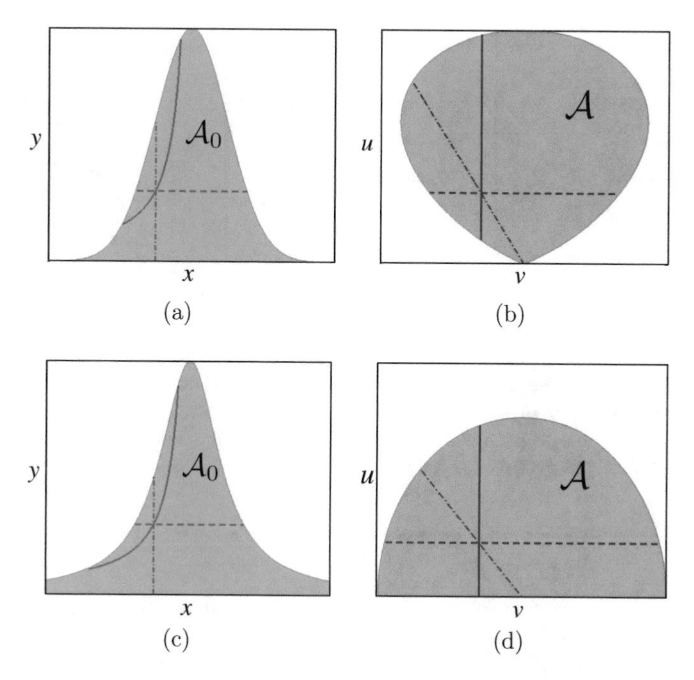


**Fig. 5.2** Examples of the regions $\mathcal{A}$. Each figure shows lines corresponding to $x$ constant (dotted line), $u$ constant (dashed line), and $v$ constant (solid line). **(a)**, **(b)** The standard Gaussian density $p_o(x) \propto \exp\{-x^2/2\}$ and the corresponding region $\mathcal{A}$. **(c)**, **(d)** The standard Cauchy density $p_o(x) \propto 1/(1+x^2)$ and the corresponding region $\mathcal{A}$

In some cases the equation $u = \sqrt{p(v/u)}$ can be solved analytically and the boundary $\mathcal{A}$ can be found explicitly. If we assume a monotonic function $p(x)$, and indicate with $p^{-1}$ its inverse, the boundary can be expressed with the equation

$$v = up^{-1}(u^2). \tag{5.12}$$

In particular, when

$$p_o(x) \propto \frac{\lambda^2}{(\delta x + \beta)^2} \tag{5.13}$$

with $\lambda$, $\delta$, $\beta$ constant values and a compact support, $x \in [a, b]$, the region $\mathcal{A}$ is a triangle, as depicted in Fig. 5.3a, with one vertex at the origin, $\mathbf{v}_1 = (0, 0)$, and the opposite side, $\mathbf{v}_2 - \mathbf{v}_3$, given by the equation $\delta v + \beta u = \lambda$. Figure 5.3b illustrates the particular case with $\delta = 0$, when $p_o(x)$ becomes a uniform distribution and we obtain a triangular region with the side $\mathbf{v}_2 - \mathbf{v}_3$ parallel to the axis $v$. Moreover, if $\beta = 0$ the pdf $p_o(x) \propto \frac{1}{x^2}$, $x \in [a, b]$, is a special case of *Pareto* pdf, also called *reciprocal uniform* density (since we can obtain it by taking the reciprocal of a

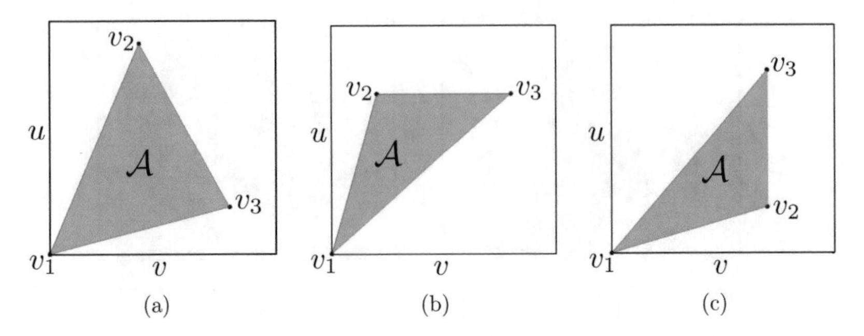


**Fig. 5.3** **(a)** A triangular region $\mathcal{A}$ with a vertex at the origin $\mathbf{v}_1 = (0, 0)$ and where the side $\mathbf{v}_2 - \mathbf{v}_3$ has a generic slope. It corresponds to a density of the form $p_o(x) \propto 1/(\delta x + \beta)^2$ transformed via the RoU method. **(b)** A triangular region $\mathcal{A}$ obtained by transforming a uniform pdf by the RoU method. The side $\mathbf{v}_2 - \mathbf{v}_3$ is parallel to the axis $v$. **(c)** A triangular region $\mathcal{A}$ obtained transforming a reciprocal uniform pdf $p_o(x) \propto 1/x^2, x \in [a, b]$, by the RoU method. The side $\mathbf{v}_2 - \mathbf{v}_3$ is parallel to the axis $u$

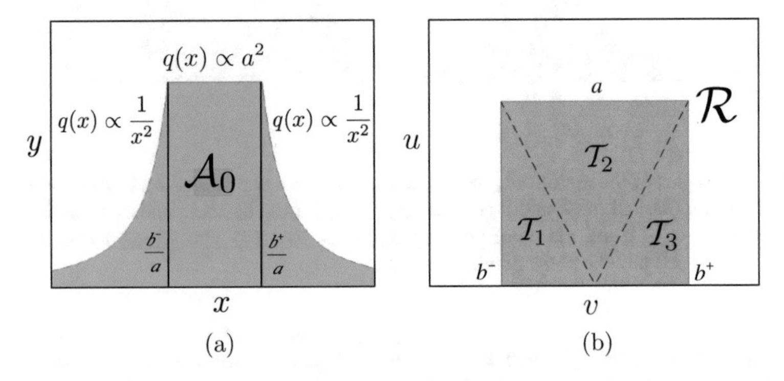


**Fig. 5.4** **(a)** The shape of a table mountain density $q(x)$ defined in Eq. (5.14). **(b)** The region $\mathcal{A}$ obtained with the RoU transformation of the table mountain density is a rectangle, i.e., $\mathcal{A} = \mathcal{R}$. The rectangular region $\mathcal{R}$ can be divided into three non-overlapping triangular parts $\mathcal{R} = \mathcal{T}_1 \cup \mathcal{T}_2 \cup \mathcal{T}_3$

uniform random variable $U$, i.e., $1/U$) and the corresponding region $\mathcal{A}$ is triangular with $\mathbf{v}_1 = (0, 0)$ and the side $\mathbf{v}_2 - \mathbf{v}_3$ parallel to the axis $u$, as shown in Fig. 5.3c.

Another example in which $\mathcal{A}$ has a closed form occurs for the so-called *table mountain density* [13, 16]. In particular, if $\mathcal{A} = [b^-, b^+] \times [0, a]$ is a rectangular region in the $v - u$ domain, then the associated pdf is

$$q(x) \propto \begin{cases} (b^-)^2/x^2 & \text{for} \quad x \in (-\infty, b^-/a] \\ a^2 & \text{for} \quad x \in [b^-/a, b^+/a], \\ (b^+)^2/x^2 & \text{for} \quad x \in [b^+/a, +\infty) \end{cases} \tag{5.14}$$

plotted in Fig. 5.4a (up to a proportionally constant). If we divide the rectangular region $\mathcal{R}$ into three non-overlapping triangular parts, $\mathcal{R} = \mathcal{T}_1 \cup \mathcal{T}_2 \cup \mathcal{T}_3$ as illustrated in Fig. 5.4b, then we can see that each part of $q(x)$ is related to each triangular part $\mathcal{T}_i$, $i = 1, 2, 3$, by comparing Fig. 5.4a, b.

## 5.3 Envelope Polygons and Adaptive RoU

Previously, we have seen that, in the cases of interest, the region $\mathcal{A}$ can be contained within a rectangle $\mathcal{R}$ as illustrated in Fig. 5.1b. This allows the design of a simple RS scheme: draw uniformly a point in $\mathcal{R}$, and accept it if belongs to $\mathcal{A}$ (reject it otherwise). Clearly, there are other possibilities to design an RS technique jointly with the RoU. In the sequel, we provide some examples.

The adaptive rejection sampling idea has been implemented jointly with the RoU method in [20, 21]. Indeed, if the region $\mathcal{A}$ is convex it is possible to construct adaptively a bounding region $\mathcal{P}_t$, such that $\mathcal{A} \subseteq \mathcal{P}_t$, with a polygonal boundary. The underlying idea is that drawing from the polygon $\mathcal{P}_t$ is easier than drawing from $\mathcal{A}$. Indeed, the ability to draw from $\mathcal{P}_t$ readily enables an accept/reject procedure to draw uniformly from $\mathcal{A}$. To be specific consider a set of support points

$$\mathcal{S}_t = \{\mathbf{s}_1, \mathbf{s}_2, \ldots, \mathbf{s}_{m_t}\}$$

where $\mathbf{s}_i = [v_i, u_i]$, $i = 1, \ldots, m_t$, are points on the boundary of $\mathcal{A}$ in the $v - u$ space. The envelope region $\mathcal{P}_t$ can be built using the straight lines tangent at $\mathbf{s}_i$ to the boundary of the convex region $\mathcal{A}$. Figure 5.5 shows an example of bounding set $\mathcal{P}_t$ with polygonal boundary built using $m_t = 5$ support points.

As the next step, note that it is always possible to calculate the first derivative of the boundary of $\mathcal{A}$ if the function $p(x)$ is differentiable, without knowing the explicit equation of the contour. Indeed, the boundary of $\mathcal{A}$ can be described parametrically as

$$\begin{cases} u = u(x) = \sqrt{p(x)} \\ v = v(x) = x\sqrt{p(x)} \end{cases}, \tag{5.15}$$

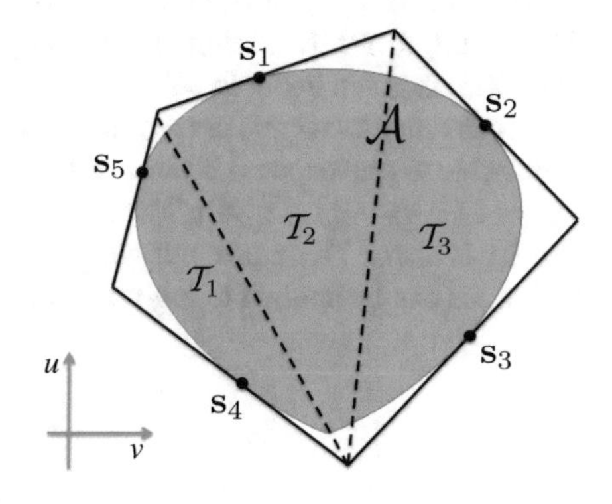


**Fig. 5.5** Example of construction of a bounding polygon $\mathcal{P}_t$ using the tangent lines at the support points $\mathbf{s}_i$, $i = 1, \ldots, m_t = 5$, to the boundary of the convex region $\mathcal{A}$. The polygon can be divided into $m_t - 2 = 3$ non-overlapping triangles, i.e., $\mathcal{P}_t = \cup_{k=1}^{3} \mathcal{T}_k$

in the case of the standard RoU method. Hence, we can use the chain rule for computing the derivative and write

$$\frac{dv}{du} = \frac{dv}{dx}\frac{dx}{du} = \left(\sqrt{p(x)} + \frac{x}{2\sqrt{p(x)}}\frac{dp}{dx}\right)(du/dx)^{-1}$$

$$= \left(\sqrt{p(x)} + \frac{x}{2\sqrt{p(x)}}\frac{dp}{dx}\right)\left(\frac{1}{\frac{1}{2\sqrt{p(x)}}\frac{dp}{dx}}\right)$$

$$= 2\frac{p(x)}{\dot{p}(x)} + x,$$

where $x = \frac{v}{u}$ and $\dot{p} = \frac{dp}{dx}$.

Furthermore, it is straightforward to draw samples uniformly from the polygon $\mathcal{P}_t$ by dividing it into $m_t - 2$ non-overlapping triangular areas $\mathcal{T}_k$, i.e., $\mathcal{P}_t = \cup_{k=1}^{m_t-2}\mathcal{T}_k$ where $\mathcal{T}_i \cap \mathcal{T}_j = \emptyset$ whenever $i \neq j$. Note that it is straightforward to sample uniformly from a triangle $\mathcal{T}_k$ using only two uniform random variables, as will be shown in Sect. 6.6.1.

Therefore, to generate samples uniformly from $\mathcal{P}_t$, we first have to randomly select a triangle with probabilities proportional to the areas $|\mathcal{T}_k|$, $k = 0, \ldots, m_t - 2$, and then draw from the selected triangular subset. For the first step, we define the normalized weights

$$w_k \triangleq \frac{|\mathcal{T}_k|}{\sum_{i=0}^{m_t-2}|\mathcal{T}_i|}, \tag{5.16}$$

and then we choose a triangular piece by drawing an index $k' \in \{0, \ldots, m_t - 2\}$ from the probability distribution $P(k) = w_k$. For the second step, we easily generate a point $(v', u')$ uniformly in the selected triangular region $\mathcal{T}_{k'}$ using the procedure in Sect. 6.6.1 (see also [11]).

If the point $(v', u')$, generated using this two-step procedure, belongs to $\mathcal{A}$, we accept the sample $x' = v'/u'$ and set $m_{t+1} = m_t$, $\mathcal{S}_{t+1} = \mathcal{S}_t$ and $\mathcal{P}_{t+1} = \mathcal{P}_t$. Otherwise, we discard the sample $x' = v'/u'$ and incorporate it into the set of support points, $\mathcal{S}_{t+1} = \mathcal{S}_t \cup \{\mathbf{s}' = (v', u')\}$, so that $m_{t+1} = m_t + 1$ and the region $\mathcal{P}_{t+1}$ is improved by adding another tangent line. An outline of the adaptive RoU algorithm is given in Table 5.2.

Let us remark that this procedure is applicable if the region $\mathcal{A}$ is convex. In turn, it is possible to prove that $\mathcal{A}$ is convex if, and only if, the target pdf $p_o(x)$ is $T$-concave, where $T(x) = -1/\sqrt{x}$ [20]. Moreover, in [15] it is proved that every log-concave density is also a $T$-concave pdf with $T(x) = -1/\sqrt{x}$. Therefore, this adaptive RoU technique can be applied to log-concave target pdfs as well.

**Table 5.2** Adaptive RoU scheme

| |
|---|
| 1. Start with $t = 0, j = 1$ and initialize the set of support points $\mathcal{S}_0 = \{\mathbf{s}_1, \ldots, \mathbf{s}_{m_0}\}$ |
| For every $t \geq 0$: |
| 2. Construct the enveloping polygon $\mathcal{P}_t$ using the lines tangent to the boundary of the convex region $\mathcal{A}$ at $\mathbf{s}_i$, $i = 1, \ldots, m_t$ |
| 3. Construct the triangular regions $\mathcal{T}_k$, $k = 0, \ldots, m_t - 2$, as described in Fig. 5.5 |
| 4. Calculate the area $\|\mathcal{T}_k\|$ of every triangle, and compute the normalizedweights $w_k \triangleq \frac{\|\mathcal{T}_k\|}{\sum_{i=0}^{m_t-2} \|\mathcal{T}_i\|}$, with $k = 0, \ldots, m_t - 2$ |
| 5. Draw an index $k' \in \{0, \ldots, m_t - 2\}$ from the probability distribution $P(k) = w_k$ |
| 6. Generate a point $(v', u')$ uniformly from the region $\mathcal{T}_{k'}$ as explained in Sect. 6.6.1 |
| 7. If $u' \leq \sqrt{p(v'/u')}$, then accept the sample $x^{(j)} = x' = \frac{v'}{u'}$, set $j = j + 1$, $\mathcal{S}_{t+1} = \mathcal{S}_t$ and $m_{t+1} = m_t$ |
| 8. Otherwise, if $u' > \sqrt{p(v'/u')}$, then reject the sample $x' = \frac{v'}{u'}$, set $\mathcal{S}_{t+1} = \mathcal{S}_t \cup \{\mathbf{s}' = (v', u')\}$, and sort $\mathcal{S}_{t+1}$ in ascending order. Finally, update $m_{t+1} = m_t + 1$ |
| 9. If $j > N$ then stop, else go back to step 2 |

## 5.4 Generalized Ratio of Uniforms Method

A more general version of the standard RoU method proposed in [19] can be established using the following theorem from [32].

**Theorem 5.2** *Let $g(u) : \mathbb{R}^+ \to \mathbb{R}^+$ be a strictly increasing differentiable function such that $g(0) = 0$ and let $p(x) \geq 0$ be a function proportional to a target pdf $p_o(x)$. Assume that $(v, u) \in \mathbb{R}^2$ is a sample drawn from the uniform distribution on the set*

$$\mathcal{A}_g = \left\{ (v, u) \in \mathbb{R}^2 : 0 \leq u \leq g^{-1}\left[ c_A\, p\left( \frac{v}{\dot{g}(u)} \right) \right] \right\}, \tag{5.17}$$

*where $c_A > 0$ is a positive constant and $\dot{g} = \frac{dg}{du}$. Then $x = \frac{v}{\dot{g}(u)}$ is a sample from $p_o(x)$.*

*Proof* Given the transformation $(v, u) \in \mathbb{R}^2 \to (x, z)$

$$\begin{cases} x = \dfrac{v}{\dot{g}(u)}, \\ z = u \end{cases} \tag{5.18}$$

and a pair of rv's $(V, U)$ uniformly distributed on $\mathcal{A}_g$, we can write the joint pdf $q(x, y)$ of the transformed rv's $(X, Z)$ as

$$q(x, z) = \frac{1}{|\mathcal{A}_g|} |\mathbf{J}^{-1}| \quad \text{for all} \quad 0 \leq z \leq g^{-1}[c_A p(x)], \tag{5.19}$$

where $|\mathcal{A}_g|$ denotes the area of $\mathcal{A}_g$, and $\mathbf{J}^{-1}$ is the Jacobian of the inverse transformation, namely,

$$\mathbf{J}^{-1} = \det \begin{bmatrix} \dot{g}(z) & x\ddot{g}(z) \\ 0 & 1 \end{bmatrix} = \dot{g}(z). \tag{5.20}$$

Since we assume $\dot{g}(z) \geq 0$ (i.e., $g$ is increasing), then $|\mathbf{J}^{-1}| = |\dot{g}(z)| = \dot{g}(z)$ and substituting (5.20) into (5.19) yields

$$q(x, z) = \begin{cases} \dfrac{1}{|\mathcal{A}_g|}\dot{g}(z) & \text{for} \quad 0 \leq z \leq g^{-1}[c_{\mathcal{A}} p(x)], \\ 0, & \text{otherwise.} \end{cases} \tag{5.21}$$

Hence, integrating $q(x, z)$ w.r.t. $z$ yields the marginal pdf of the rv $X$,

$$\begin{aligned} q(x) &= \int_{-\infty}^{+\infty} q(x, z) dz \\ &= \int_0^{g^{-1}[c_{\mathcal{A}} p(x)]} \frac{1}{|\mathcal{A}_g|}\dot{g}(z) dz \\ &= \frac{1}{|\mathcal{A}_g|}\Big[g(z)\Big]_0^{g^{-1}[c_{\mathcal{A}} p(x)]} \\ &= \frac{c_{\mathcal{A}}}{|\mathcal{A}_g|} p(x) - \frac{1}{|\mathcal{A}_g|} g(0), \end{aligned} \tag{5.22}$$

where the first equality follows from Eq. (5.21) and the remaining calculations are straightforward. Since we have also assumed $g(0) = 0$, it turns out that

$$q(x) = \frac{c_{\mathcal{A}}}{|\mathcal{A}_g|} p(x) = p_o(x). \qquad \square$$

Since $p_o(x)$ is normalized, the measure of $\mathcal{A}_g$ is

$$|\mathcal{A}_g| = c_{\mathcal{A}} c_v = c_{\mathcal{A}} \int_{\mathcal{D}} p(x) dx. \tag{5.23}$$

Moreover, choosing $g(u) = \frac{1}{2}u^2$, we come back to the standard RoU method. Again, the theorem above provides a way to generate samples from $p_o(x)$. Indeed, if we are able to draw uniformly a point $(v', u')$ from $\mathcal{A}_g$, then the sample $x' = v'/\dot{g}(u')$ is distributed according to $p_o(x) \propto p(x)$. Also in this case, the efficiency of the method depends on the ease with which we can generate points uniformly within the region $\mathcal{A}_g$. For this reason, the cases of practical interest are those in which the region $\mathcal{A}_g$ is bounded. Moreover, observe that if $g(u) = u$ and $c_{\mathcal{A}} = 1$ we come back to *the fundamental theorem of simulation* described in Sect. 2.4.3, since $\mathcal{A}_g$ becomes exactly the region $\mathcal{A}_0$.

Other interesting generalizations of the RoU method can be found in [18] and we discuss them in Sect. 5.10. Related work and further developments involving ratios of r.v.'s can be found in [1–3, 7, 12, 24, 27, 30, 31]. The possible combination with MCMC algorithms is discussed in [14].

## 5.5 Properties of Generalized RoU Samplers

### 5.5.1 Boundary of $\mathcal{A}_g$

In the boundary of the region $\mathcal{A}_g$ the coordinates of the point $(v, u)$ satisfy $u = g^{-1}[c_{\mathcal{A}} p(x)]$ and, since $v = x\dot{g}(u)$ for the transformation (5.18), $v = x\dot{g}[g^{-1}(c_{\mathcal{A}} p(x))]$. Therefore, the contour of $\mathcal{A}_g$ can be described parametrically by the pair of equations

$$\begin{cases} u = g^{-1}[c_{\mathcal{A}} p(x)], \\ v = x\dot{g}[g^{-1}(c_{\mathcal{A}} p(x))], \end{cases} \tag{5.24}$$

where $x$ plays the role of a parameter. Hence, if the two functions $g^{-1}[c_{\mathcal{A}} p(x)]$ and $x\dot{g}[g^{-1}(c_{\mathcal{A}} p(x))]$ are bounded, the region $\mathcal{A}_g$ is embedded in the rectangular region

$$\mathcal{R}_g = \Big\{(v, u) \in \mathbb{R}^2 : \ 0 \le u \le \sup_x g^{-1}[c_{\mathcal{A}} p(x)], \\ \inf_x x\dot{g}[g^{-1}(c_{\mathcal{A}} p(x))] \le v \le \sup_x x\dot{g}[g^{-1}(c_{\mathcal{A}} p(x))]\Big\}, \tag{5.25}$$

and it is straightforward to design a rejection sampler in the same vain as for the standard RoU scheme.

### 5.5.2 How to Guarantee that $\mathcal{A}_g$ is Bounded

In this section, we discuss in detail the conditions to be imposed on $g(u)$ and $p(x)$ to ensure that the region $\mathcal{A}_g$ is bounded. In order to handle the general case, we assume that $p(x)$ is defined in an unbounded domain $\mathcal{D} \equiv \mathbb{R}$. As shown above in Eqs. (5.24) and (5.25), the region $\mathcal{A}_g$ is bounded (rather trivially) if the two functions $g^{-1}[c_{\mathcal{A}} p(x)]$ and $x\dot{g}[g^{-1}(c_{\mathcal{A}} p(x))]$ are bounded.

**First Function: $u(x) = g^{-1}[c_A p(x)]$**

Since $g$ is assumed to be a strictly increasing ($\dot{g} > 0$) and continuous function, then $g^{-1}$ is also increasing. As a consequence, the function $u = g^{-1}[c_A p(x)]$ is bounded if, and only if, $p(x)$ is bounded, for all $x \in \mathcal{D}$. Recall that $p_o(x) \propto p(x)$ so that clearly $p(x) \geq 0$ and, denoting $M = \max_{x \in \mathbb{R}} p(x)$, we have

$$0 \leq u \leq g^{-1}(c_A M).$$

**Second Function: $v(x) = x\dot{g}[g^{-1}(c_A p(x))]$**

First, observe that we can write

$$v = x\dot{g}(u),$$

where $u(x) = g^{-1}[c_A p(x)]$ is the first function above. Since $\dot{g} > 0$, for the second factor, $\dot{g}(u)$, to be bounded, $u$ needs to be bounded. We have seen above that $u = g^{-1}[c_A p(x)]$ is bounded when $p(x)$ is bounded. Furthermore, to ensure that $v(x)$ is bounded, we also need that the limits

$$\lim_{x \to \pm\infty} x\dot{g}(u) = \lim_{x \to \pm\infty} x\dot{g}[g^{-1}(c_A p(x))] \tag{5.26}$$

be finite, which occurs whenever

$$\lim_{x \to \pm\infty} \dot{g}[g^{-1}(c_A p(x))] = 0, \tag{5.27}$$

$$\dot{g}\left(g^{-1}(c_A p(x))\right) \leq \frac{\xi}{x}, \tag{5.28}$$

for some constant $\xi$ and sufficiently large $x$. We can also reformulate these conditions in other forms, which will turn out more useful later in this chapter. First, we rewrite them in terms of the variable $u = g^{-1}[c_A p(x)]$ and then in terms of the variable $y = p(x)$.

- Let us write $u(x) = g^{-1}[c_A p(x)]$ and recall that $\lim_{x \to \pm\infty} p(x) = 0$, $g^{-1}(0) = 0$ (since we have assumed $g(0) = 0$). Then we can readily obtain

  $$\lim_{x \to \pm\infty} u(x) = \lim_{x \to \pm\infty} g^{-1}[c_A p(x)] = 0,$$

  and by a simple change of variable,

  $$\lim_{x \to \pm\infty} \dot{g}[g^{-1}(c_A p(x))] = \lim_{u \to 0} \dot{g}(u) = 0. \tag{5.29}$$

  We also require that $\dot{g}(u(x)) \leq \frac{\xi}{x}$ for some constant $\xi$ and sufficiently large $x$.

- For the sake of simplicity, let assume $c_A = 1$. If we let $y = p(x)$ then $u(x) = g^{-1}(p(x))$ can be rewritten as $u(y) = g^{-1}(y)$ and, hence, $g(u(y)) = y$. Then, we can rewrite $\dot{g}[g^{-1}(p(x))]$ as

$$\dot{g}[g^{-1}(p(x))] = \left.\frac{dg}{du}\right|_{u=g^{-1}(y)} = \frac{1}{\left.\frac{dg^{-1}}{dy}\right|_{y}}, \tag{5.30}$$

where we have used the derivative of the inverse function. Hence, since $y = p(x) \to 0$ for $x \to \pm\infty$, and $g^{-1}(0) = 0$ by assumption, we have

$$\begin{aligned}\lim_{x\to\pm\infty} \dot{g}[g^{-1}(p(x))] &= \lim_{u\to 0} \dot{g}(u) \\ &= \lim_{y\to 0} \frac{1}{\left.\frac{dg^{-1}}{dy}\right|_{y}} \\ &= \lim_{y\to 0} \frac{1}{\frac{dg^{-1}}{dy}} = \lim_{y\to 0} \frac{1}{\dot{g}^{-1}(y)} = 0,\end{aligned} \tag{5.31}$$

where we have used the limit of the composition of functions [23]. Moreover, we require $\frac{1}{\dot{g}^{-1}(y(x))} \leq \frac{\xi}{x}$ for some constant $\xi > 0$ and sufficiently large $x$. If $y = p(x)$ is invertible, this condition becomes

$$\frac{1}{\dot{g}^{-1}(y)} \leq \frac{\xi}{p^{-1}(y)},$$

for some constant $\xi > 0$ and sufficiently large $p^{-1}(y)$. Finally, note that we can rewrite the limit in Eq. (5.31) as

$$\lim_{y\to 0} \frac{dg^{-1}}{dy} = \infty. \tag{5.32}$$

## Summary

The region $\mathcal{A}_g$ generated by GRoU is bounded if the following assumptions are jointly satisfied:

(1) The target $p(x)$ is bounded (i.e., if $p(x)$ is monotonic, $x = p^{-1}(y)$ has finite support).
(2) The equality $\lim_{u\to 0} \frac{dg}{du} = 0$, or equivalently $\lim_{y\to 0} \frac{dg^{-1}}{dy} = \infty$, holds.
(3) It is also necessary that $\dot{g}(u(x)) \leq \frac{\xi}{x}$ for some constant $\xi > 0$ and sufficiently large $x$ or, equivalently, $\frac{1}{\dot{g}^{-1}(y)} \leq \frac{\xi}{p^{-1}(y)}$, for some constant $\xi > 0$ and sufficiently large $p^{-1}(y)$.

Moreover, we recall that the Theorem 5.2 makes additional assumptions on the function $g(u)$:

(4) $g(u)$ must be increasing,
(5) $g(u): \mathbb{R}^+ \rightarrow \mathbb{R}^+$,
(6) $g(0) = 0$.

In Sect. 5.9, we show how some conditions can be relaxed. Below we provide a family of suitable functions $g(u)$.

### 5.5.3 Power Functions

A family of transformations $g(u)$ that turns out suitable for its use within the GRoU framework is the set of power functions with the following form

$$g(u) = \frac{u^{r+1}}{(r+1)}, \quad u \geq 0, \tag{5.33}$$

with $r \geq 0$ and $c_A = \frac{1}{r+1}$ [6]. Note that the first derivative $\dot{g}(u) = u^r$ is strictly increasing for $u \geq 0$. The region $\mathcal{A}_g$ defined in Eq. (5.17) becomes

$$\mathcal{A}_g = \mathcal{A}_r = \left\{ (v, u) : 0 \leq u \leq \left[ p\left(\frac{v}{u^r}\right) \right]^{\frac{1}{r+1}} \right\},$$

that we denote $\mathcal{A}_r$, since with $r = 1$ we obtain the same set as with the standard RoU method in Eq. (5.2) (with $c_A = \frac{1}{r+1}$) and the region $\mathcal{A}_0$ (delimited by the pdf $p_o(x)$, see Fig. 2.2) defined in Eq. (2.39) with $r = 0$. In other words, the region $\mathcal{A}_r$ is bounded if the functions $[p(x)]^{1/(r+1)}$ and $x[p(x)]^{r/(r+1)}$ are both bounded. This occurs, in turn, when $p(x)$ is bounded and its tails decay as $1/x^{(r+1)/r}$ or faster. Hence, by choosing $r > 1$ we can handle pdfs with heavier tails than with the standard RoU method.

It is interesting to analyze the probability of acceptance, $p_A(r)$, for a point drawn uniformly from the rectangle $\mathcal{R}_r$, defined in Eq. (5.25) using $g(u) = \frac{u^{r+1}}{(r+1)}$ and $c_A = \frac{1}{r+1}$. This probability is given by the ratio between the two areas, i.e.,

$$p_A(r) = \frac{|\mathcal{A}_r|}{|\mathcal{R}_r|}, \tag{5.34}$$

and

$$|\mathcal{A}_r| = c_A c_v = \frac{1}{r+1} \int_{\mathcal{D}} p(x)dx,$$

as shown in Eq. (5.23). Moreover, if we define

$$a(r) = \sup_x [p(x)]^{\frac{1}{r+1}},$$

$$b^-(r) = \inf_x x[p(x)]^{\frac{r}{r+1}},$$

$$b^+(r) = \sup_x x[p(x)]^{\frac{r}{r+1}},$$

the area of the bounding rectangle can be rewritten as $|\mathcal{R}_r| = a(r)[b^+(r) - b^-(r)]$. Substituting this expression into Eq. (5.34), we obtain

$$p_A(r) = \frac{\int_{\mathcal{D}} p(x)dx}{(r+1)a(r)[b^+(r) - b^-(r)]}. \tag{5.35}$$

In some cases, it is possible to analytically obtain the optimal value of $r$ in order to maximize the acceptance probability $p_A(r)$ in Eq. (5.35). We now show an example involving a standard Gaussian pdf.

*Example 5.1* Consider a standard Gaussian density, i.e., $p_o(x) \propto p(x) = \exp\{-x^2/2\}$ with $x \in \mathbb{R}$. In this case, we know that

$$\int_{\mathbb{R}} p(x)dx = (2\pi)^{1/2}$$

and

$$a(r) = \sup[p(x)]^{1/(r+1)} = 1. \tag{5.36}$$

Moreover, we can find the first derivative of the function $\phi(x) \triangleq x[p(x)]^{r/(r+1)} = x \exp\left\{-\frac{r}{2(r+1)}x^2\right\}$ w.r.t. $x$ and then write

$$\frac{d\phi}{dx} = \left(1 - \frac{r}{r+1}x^2\right) \exp\left\{-\frac{r}{2(r+1)}x^2\right\}. \tag{5.37}$$

The solutions of $\frac{d\phi}{dx} = 0$ are $x_{1,2} = \pm\sqrt{\frac{r+1}{r}}$, where $\phi(x_1) = b^-(r)$ and $\phi(x_2) = b^+(r)$ in Eq. (5.35). Namely, we obtain

$$b^-(r) = \inf_x x[p(x)]^{r/(r+1)} = -\left(\sqrt{\frac{r+1}{r}}\right) \exp\{-1/2\}, \tag{5.38}$$

and

$$b^+(r) = \sup_x x[p(x)]^{r/(r+1)} = \left(\sqrt{\frac{r+1}{r}}\right)\exp\{-1/2\}. \tag{5.39}$$

Substituting (5.36), (5.38) and (5.39) into (5.35) yields

$$p_A(r) = \frac{(2\pi)^{1/2}}{(r+1)\left[(\frac{r+1}{r})^{1/2}\exp\{-1/2\} + (\frac{r+1}{r})^{1/2}\exp\{-1/2\}\right]},$$

which reduces to

$$p_A(r) = \frac{(2\pi re)^{1/2}}{2(r+1)^{3/2}}, \tag{5.40}$$

after some straightforward calculations. The maximization of $p_A(r)$ in (5.40) w.r.t. $r$ yields $\min_r p_A(r) = 0.795$, which is attained for $r^* = \frac{1}{2}$. Note that, for the standard RoU method ($r = 1$), we have $p_A(1) = 0.731$.

## 5.6 Connections Between GRoU and Other Classical Techniques

Consider a monotonic density $p_o(x) \propto p(x)$. For the sake of simplicity, sometimes we refer to $p(x)$ and $p^{-1}(y)$ as densities although they are unnormalized. In this section, we aim at proving the following result.

**Proposition 5.1** *The GRoU method can be interpreted as a combination of*

- *the transformed rejection method applied to a random variable $Y$ distributed according to the inverse density $p^{-1}(y)$,*
- *with the extended inverse-of-density method, that we will introduce in Sect. 5.6.1.*

We first introduce an extension of the inverse-of-density method of Sect. 2.4.4, and then investigate the connection between GRoU and transformed rejection sampling.

### 5.6.1 *Extended Inverse-of-Density Method*

Consider, for simplicity, a monotonic function $p(x)$. The standard inverse-of-density (IoD) method of Sect. 2.4.4 provides the relationship between a r.v. $Y$ distributed as a pdf proportional to $p^{-1}(y)$ and the r.v. $X$ with a pdf proportional to $p(x)$. In the sequel we refer to $p^{-1}(y)$ and $p(x)$ as densities although they are not normalized, in general.

In this section, we study the connection between a transformed random variable $U = h(Y)$, (where $h$ is a monotonic function and $Y$ is distributed according to $p^{-1}(y)$), and the random variable $X$ with density $p(x)$. In this case, we know that the density of $U$ is

$$q(u) = p^{-1}(h^{-1}(u))\left|\frac{dh^{-1}}{du}\right|. \tag{5.41}$$

Denoting as $\mathcal{A}_h$ the area below $q(u)$, our goal is now to find the relationship between the pair $(U, V) \sim \mathcal{U}(\mathcal{A}_h)$, i.e., uniformly distributed on $\mathcal{A}_h$, and the r.v. $X$ with density $p(x)$. Figure 5.6b illustrates an example of a possible pdf $q(u)$, the region $\mathcal{A}_h$ and a random point $(u', v')$ drawn uniformly from $\mathcal{A}_h$. Firstly, let us observe that:

1. The r.v. $U$ in the random vector $(U, V) \sim \mathcal{U}(\mathcal{A}_h)$ has pdf $q(u)$, because of the fundamental theorem of simulation (see Sect. 2.4.3).
2. Because of the relationship $U = h(Y)$, if we are able to draw a sample $u'$ from $q(u)$, then we can easily generate a sample $y'$ from $p^{-1}(y)$ by simply taking

$$y' = h^{-1}(u'). \tag{5.42}$$

3. By the inverse-of-density method (Sect. 2.4.4), we also know that

$$x' = z'p^{-1}(y'), \tag{5.43}$$

where $z' \sim \mathcal{U}([0, 1])$ and $y' \sim p^{-1}(y)$, is distributed according to the target density $p_o(x) \propto p(x)$. This relationship is also illustrated in Fig. 5.6a.

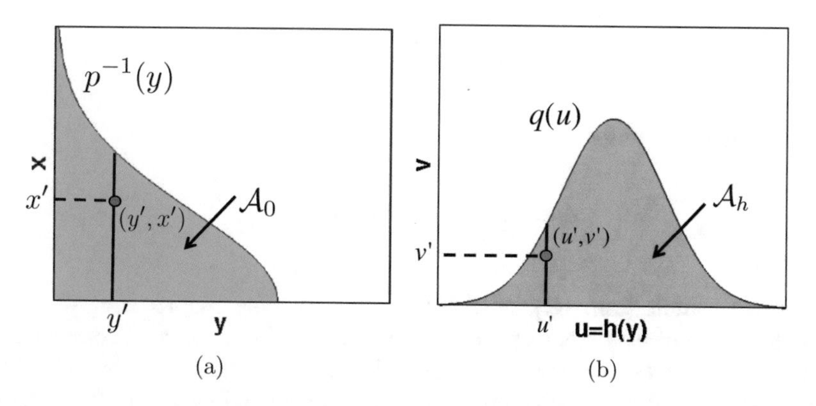


**Fig. 5.6** **(a)** Given a point $(x', y')$ uniformly distributed on $\mathcal{A}_0$, $y'$ has pdf $p^{-1}(y)$ while $x'$ is distributed as $p(x)$, as stated by the fundamental theorem of simulation and the inverse-of-density method. **(b)** Given a transformation $U = h(Y)$ with pdf $q(u)$, and a point $(u', v')$ uniformly distributed on the region $\mathcal{A}_h$ below $q(u)$, then the sample $x' = v'\dot{h}(h^{-1}(u'))$ has density $p(x)$

Then, replacing Eq. (5.42) into Eq. (5.43), we obtain the expression

$$x' = z'p^{-1}(y') = z'p^{-1}(h^{-1}(u')), \tag{5.44}$$

where $z' \sim \mathcal{U}([0,1])$, $u' \sim q(u)$ and $y' = h^{-1}(u') \sim p^{-1}(y)$. Equation (5.44) above connects samples related to the r.v.'s $U$ and $X$. However, we are looking for a relationship involving also the random variable $V$.

To draw a point $(u', v')$ uniformly on $\mathcal{A}_h$, we have seen in Sect. 2.4.4 that we can first draw a sample $u'$ from $q(u)$ and then $v'$ uniformly from the interval $[0, q(u')]$, i.e., $v' \sim \mathcal{U}([0, q(u')])$. Therefore, the sample $v'$ can also be expressed as

$$v' = z'q(u'), \tag{5.45}$$

where $z' \sim \mathcal{U}([0,1])$. Substituting the expression of $q(u)$ in Eq. (5.41) into Eq. (5.45), we obtain

$$v' = z'p^{-1}(h^{-1}(u'))\left|\frac{dh^{-1}}{du}\right|_{u=u'}. \tag{5.46}$$

Furthermore, recalling the previous expression of $x' = z'p^{-1}(h^{-1}(u'))$ in Eq. (5.44), we can easily recognize this term in Eq. (5.46),

$$v' = \underbrace{z'p^{-1}(h^{-1}(u'))}_{x'}\left|\frac{dh^{-1}}{du}\right|_{u=u'}, \tag{5.47}$$

hence

$$v' = x'\left|\frac{dh^{-1}}{du}\right|_{u=u'}. \tag{5.48}$$

Thus, we can also write

$$x' = \frac{v'}{\left|\frac{dh^{-1}}{du}\right|_{u=u'}} = v'|\dot{h}(h^{-1}(u'))|, \tag{5.49}$$

which is a sample from $p(x)$. In (5.49) we have used the notation $\dot{h} = \frac{dh}{dx}$ for the first derivative of $h(x)$. Let us recall that the sample $v'$ is the second component of a random vector $(u', v') \sim \mathcal{U}(\mathcal{A}_h)$: since $\mathcal{A}_h$ represents the area below $q(u)$, $v'$ is distributed according to the inverse pdf $q^{-1}(v)$ corresponding to $q(u)$ (Sect. 2.4.4), where

$$q^{-1}(v) = \{v \in \mathbb{R} : \quad v = q(u), \quad u \in \mathcal{Q} \subseteq \mathbb{R}\},$$

denotes the generalized inverse function of $q(u)$ (and $\mathcal{Q}$ is the support of $q(u)$).

Equation (5.49) links a uniform random point $(U, V) \in \mathcal{A}_h$, as illustrated in Fig. 5.6b, with the r.v. $X$. This relationship is the basis of the extended IoD (E-IoD) method. If we are able to draw points $(u', v')$ uniformly from $\mathcal{A}_h$ we can generate a sample $x'$ from the density $p(x)$ using Eq. (5.49), as formalized by the following proposition.

**Proposition 5.2** *Let $Y$ be a r.v. with a monotonic pdf $p^{-1}(y)$, and let $U = h(Y)$ be another (transformed) r.v., where $h(y)$ is a monotonic transformation. Let us denote with $q(u)$ the density of $U$ and let $\mathcal{A}_h$ be the area below $q(u)$. If we are able to draw a point $(u', v')$ uniformly from the region $\mathcal{A}_h$, then*

$$x' = \frac{v'}{\left|\frac{dh^{-1}}{du}\right|_{u=u'}} = \frac{v'}{\left|\dot{h}^{-1}(u')\right|}, \tag{5.50}$$

*is a sample from the pdf $p(x)$ (the inverse function of $p^{-1}(y)$).*

Two special cases that will turn out useful in the sequel are commented below. The connection between the GRoU and E-IoD methods can also be made apparent relying on Proposition 5.2, as shown below.

### Two Special Cases

Proposition 5.2 enables a straightforward connection between the E-IoD method, the fundamental theorem of simulation and the standard RoU technique. Choosing $h(y) = y$ (hence $\dot{h} = 1$), we have $U = Y$ and as a consequence $q(u) = q(y) = p^{-1}(y)$ and the region $\mathcal{A}_h$ is exactly $\mathcal{A}_0$. In this case $v' \sim p(x)$ ($p(x)$ is the inverse pdf of $q(u)$). Indeed, Eq. (5.49) becomes

$$x' = v', \tag{5.51}$$

that is exactly the fundamental theorem of simulation, since $\mathcal{A}_h \equiv \mathcal{A}_0$. Alternatively, if we choose $h(y) = \sqrt{2y}$, then $y \geq 0$, since $h^{-1}(u) = \frac{1}{2}u^2$, we have

$$x' = \frac{v'}{u'}, \tag{5.52}$$

that corresponds to the standard RoU method.

### GRoU and E-IoD Methods

Proposition 5.2 quite readily shows that the E-IoD method can be seen as a GRoU algorithm as well. Indeed, if we set $h(y) = g^{-1}(y)$, then we obtain $x' = v'/\dot{g}(u')$ that is exactly equivalent to a sample drawn using the GRoU technique, as specified by Theorem 5.2.

Indeed, one can interpret that the GRoU scheme as an extension of the classical IoD method. Whilst the standard IoD sampler draws from $p(x)$ by generating points $(x', y)$ uniformly distributed in $\mathcal{A}_0$, where $y$ is distributed according to $p^{-1}(y)$, a GRoU sampler draws from $p(x)$ by generating points $(u, v)$ uniformly distributed in $\mathcal{A}_g$, where $U = g^{-1}(Y)$ is a transformed r.v. and $Y \sim p^{-1}(y)$.

From the assumptions in the GRoU Theorem 5.2, the function $h^{-1} = g$ should be increasing. However, the argument leading to the E-IoD method in this section actually shows that this condition is not strictly needed. Indeed, a GRoU algorithm can also be devised using a strictly decreasing function $g$, as will be explicitly shown in Sect. 5.9. In such case, the variate generated by the GRoU sampler has the form $x = -\frac{v}{\dot{g}(u)}$, hence we can rewrite the variate, in general, as

$$x = \frac{v}{\left|\dot{g}(u)\right|},$$

which coincides exactly with Eq. (5.50).

### 5.6.2 *GRoU Sampling and the Transformed Rejection Method*

In this section, we show that the region $\mathcal{A}_g$ can be also obtained with a transformation of a random variable. Let us recall the region defined by the GRoU scheme in Eq. (5.17),

$$\mathcal{A}_g = \left\{ (v, u) \in \mathbb{R}^2 : 0 \leq u \leq g^{-1}\left[ p\left( \frac{v}{\dot{g}(u)} \right) \right] \right\}, \tag{5.53}$$

where we have set $c_A = 1$ for simplicity, and recall also that $p(x) \propto p_o(x)$ is bounded, and $g(u)$ is an *increasing* function with $g(0) = 0$. We divide the analysis by considering first a decreasing $p(x)$ and then an increasing $p(x)$. In both cases, in this section we consider the maximum of $p(x)$ is located at 0 (more general cases are tackled in Sect. 5.7).

#### Decreasing vs. Increasing $p(x)$

Let us assume $y = p_{dec}(x) \propto p_o(x)$ is decreasing with an unbounded support $\mathcal{D} = [0, +\infty)$. From the definition of $\mathcal{A}_g$, we have $u \leq g^{-1}\left[ p_{dec}\left( \frac{v}{\dot{g}(u)} \right) \right]$. Since $g$ is increasing (hence $g^{-1}$ is also increasing), we can write

$$g(u) \leq p_{dec}\left( \frac{v}{\dot{g}(u)} \right).$$

Moreover, if $p_{dec}(x)$ is decreasing then $p_{dec}^{-1}(y)$ is decreasing as well and we have

$$p_{dec}^{-1}(g(u)) \geq \frac{v}{\dot{g}(u)}. \tag{5.54}$$

Since $\dot{g}(u) \geq 0$ (because $g$ is increasing), from Eq. (5.54) we readily obtain

$$v \leq p_{dec}^{-1}(g(u))\dot{g}(u).$$

Since $p_{dec}(x)$ is defined in $[0, +\infty)$, and

$$\mathcal{A}_g = \left\{ (v, u) \in \mathbb{R}^2 : 0 \leq u \leq g^{-1} \left[ p_{dec} \left( \frac{v}{\dot{g}(u)} \right) \right] \right\},$$

then we need $\frac{v}{\dot{g}(u)}$ to be positive. Since $\dot{g} > 0$ by assumption, we also need that $v > 0$, thus we can finally write

$$0 \leq v \leq p_{dec}^{-1}(g(u))\dot{g}(u).$$

These trivial calculations have led us to express the set $\mathcal{A}_g$ as

$$\mathcal{A}_{g,p_{dec}} = \left\{ (v, u) \in \mathbb{R}^2 : 0 \leq v \leq p_{dec}^{-1} \left( g(u) \right) \dot{g}(u) \right\}, \tag{5.55}$$

where $p_{dec}^{-1}(y)$ is the inverse of the target density $p_{dec}(x)$. It is important to remark that the inequalities depend on the sign of the first derivative of $g$ (positive, in this case) and $p_{dec}$ (negative, in this case).

A similar argument holds when the target pdf $p(x) = p_{inc}(x) \propto p_o(x)$ is monotonically *increasing* with an unbounded support, $x \in \mathcal{D} = (-\infty, 0]$, i.e., $x = p_{inc}^{-1}(y) \leq 0$. In this case we can rewrite $\mathcal{A}_g$ as

$$\mathcal{A}_{g,p_{inc}} = \left\{ (v, u) \in \mathbb{R}^2 : p_{inc}^{-1} \left( g(u) \right) \dot{g}(u) \leq v \leq 0 \right\}, \tag{5.56}$$

where $p_{inc}^{-1} \left( g(u) \right) \dot{g}(u) \leq 0$. Note that in this case the inverse pdf is $-p_{inc}^{-1}(y) = |p_{inc}^{-1}(y)| \geq 0$.

## GRoU as a Transformation of a r.v.

Consider an increasing function $h(y)$, a random variable $Y$ with a decreasing pdf $p^{-1}(y)$, $y \in (0, \sup p(x)]$, and the transformed variable $U = h(Y)$ with (unnormalized) density $q(u) = p^{-1}(h^{-1}(u))\dot{h}^{-1}(u)$. The region below $q(u)$ is

$$\mathcal{A}_h = \left\{ (v, u) \in \mathbb{R}^2 : 0 \leq v \leq p^{-1}(h^{-1}(u))\dot{h}^{-1}(u) \right\}, \tag{5.57}$$

and we can easily note that Eq. (5.55) is equivalent to Eq. (5.57) when

$$y = g(u) = h^{-1}(u). \tag{5.58}$$

On the other hand, $Y$ has pdf $|p^{-1}(y)|$, where $p^{-1}(y)$ is increasing and negative, then we obtain

$$\mathcal{A}_h = \left\{(v, u) \in \mathbb{R}^2 : p^{-1}(h^{-1}(u))\dot{h}^{-1}(u) \leq v \leq 0\right\}, \tag{5.59}$$

exactly as in Eq. (5.56), if we set $g(u) = h^{-1}(u)$.

The cases of interest are those in which the region $\mathcal{A}_g$ and $\mathcal{A}_h$ are bounded. Specifically, in Sect. 3.7.1 we have discussed the properties that a transformation $h(y)$ has to fulfill in order to obtain a bounded region $\mathcal{A}_h$, while in Sect. 5.5.2 we have described the conditions that ensure a bounded set $\mathcal{A}_g$. These conditions coincide if we set $g(u) = h^{-1}(u)$ and the GRoU method only requires an additional assumption, namely $g(0) = 0$. Hence, we can state the following proposition.

**Proposition 5.3** *The region $\mathcal{A}_g$, described by the GRoU method, can be obtained as a transformation $h = g^{-1}$ of a random variable $Y$ distributed according to an inverse pdf $|p^{-1}(y)|$ where $p^{-1}(y)$ is monotonic (possibly with an asymptote at 0).*

This proposition states that the set $\mathcal{A}_g$, described by either Eq. (5.17) or Eq. (5.55) can be obtained by applying the transformed rejection method for unbounded pdfs to the inverse density $|p^{-1}(y)|$ (see Sect. 3.7.1). Figure 5.7b displays the region $\mathcal{A}_h$ (that coincides with $\mathcal{A}_g$ if $g = h^{-1}$) defined in Eq. (5.57). Figure 5.7c depicts the same region $\mathcal{A}_h$ rotated by 90°.

Clearly, Propositions 5.2 and 5.3 entail Proposition 5.1. So far, we have considered monotonic target pdfs $p_o(x) \propto p(x)$ with mode at 0. Similar considerations can be done for more general functions $p(x)$ as shown in Sect. 5.7.

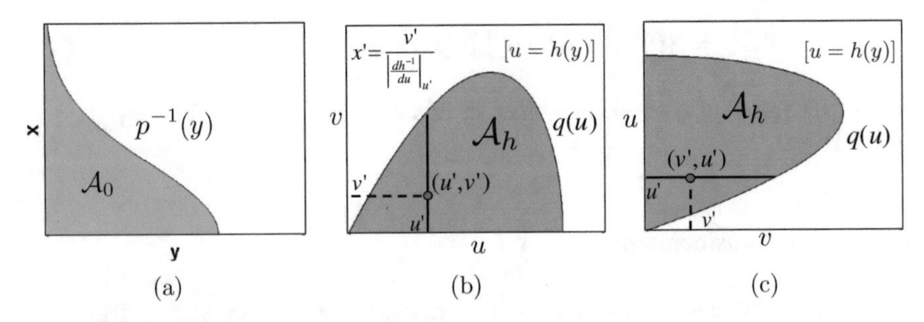


**Fig. 5.7** **(a)** Example of region $\mathcal{A}_0$ defined by the inverse density $p^{-1}(y)$. **(b)** The density $q(u) = \left|\frac{dh^{-1}}{du}\right| p^{-1}(h^{-1}(u))$ obtained transforming the r.v. $Y$, i.e., $U = h(Y)$. Generating uniformly the point $(u', v')$ in the region $\mathcal{A}_h$ we can obtain samples $x'$ from $p(x)$ using Eq. (5.50). **(c)** The region $\mathcal{A}_h$ rotated 90° in order to show it how appears when we apply the GRoU technique

### 5.6.3 Role of the Constant $c_A$

So far, in the previous analysis, for simplicity we have considered $c_A = 1$, in the definition $\mathcal{A}_g = \{(v, u) \in \mathbb{R}^2 : 0 \leq u \leq g^{-1}[c_A\, p(\frac{v}{\dot{g}(u)})]\}$. However, all the previous considerations remain valid for a generic positive value, $c_A > 0$. For the standard RoU method, we have already seen that the multiplication by the constant $c_A$ generates scaled versions of the region $\mathcal{A}$. This also occurs with the GRoU scheme and the corresponding region $\mathcal{A}_g$. Indeed, assume, for instance, the case of a decreasing $p(x)$, Eq. (5.55) becomes

$$\mathcal{A}_{g,p} = \left\{(v, u) \in \mathbb{R}^2 : 0 \leq v \leq p_{dec}^{-1}\left(\frac{g(u)}{c_A}\right)\dot{g}(u)\right\}. \tag{5.60}$$

We can also multiply both inequalities by a positive constant $1/c_A$, obtaining

$$\begin{aligned} \mathcal{A}_g &= \left\{(v, u) \in \mathbb{R}^2 : 0 \leq v \leq \frac{1}{c_A} p_{dec}^{-1}\left(\frac{g(u)}{c_A}\right)\dot{g}(u)\right\}, \\ &= \left\{(v, u) \in \mathbb{R}^2 : 0 \leq \frac{1}{c_A} v \leq p_{dec}^{-1}\left(\frac{g(u)}{c_A}\right)\frac{\dot{g}(u)}{c_A}\right\}, \end{aligned} \tag{5.61}$$

and $\mathcal{A}_g$ represents the area below $q(u) = p^{-1}\left(\frac{g(u)}{c_A}\right)\frac{\dot{g}(u)}{c_A}$, which is the (unnormalized) pdf of the r.v. $U = g^{-1}(c_A Y)$ where $Y$ is distributed according to the (non-normalized) density $p^{-1}(y)$. In this case, we have changed the measure $|\mathcal{A}_g|$ of the set $\mathcal{A}_g$. However, this change does not affect the marginal distributions of $V$ and $U$, where $(V, U) \sim \mathcal{U}(\mathcal{A}_g)$, as shown by the fundamental theorem of simulation (Sect. 2.4.3).[1] See also the related observation below.

## 5.7 How Does GRoU Work for Generic pdfs?

In Sect. 5.6, we have studied the connections among GRoU, E-IoD, and TRM considering monotonic target pdfs, either decreasing or increasing, always with a maximum at 0. These relationships clarify the underlying scheme behind the GRoU procedure. In this section, we investigate and discuss the connections between GRoU, E-IoD, and TDM for non-monotonic densities

(a) with only one mode, arbitrarily located (Sects. 5.7.2–5.7.3),
(b) and then with an arbitrary number of modes (Sect. 5.7.4).

For this purpose, firstly we need to clarify how the IoD can be extended for generic pdfs (possibly non-monotonic), as explained in the sequel.

[1] In the fundamental theorem of simulation of Sect. 2.4.3, the target pdf $p(x)$ is assumed unnormalized in general, so that multiplying $p(x)$ for a positive constant does not change the results of the theorem.

### 5.7.1 IoD for Arbitrary pdfs

Let us define the set of points

$$\mathcal{A}_{0|y} = \{x \in \mathcal{D} : \; p(x) > y\}, \tag{5.62}$$

for all $y \in \mathbb{R}^+$. Then we can define the *generalized* inverse pdf as

$$q_G^{-1}(y) \propto p_G^{-1}(y) = |\mathcal{A}_{0|y}|, \tag{5.63}$$

where $|\mathcal{A}_{0|y}|$ is the Lebesgue measure of $\mathcal{A}_{0|y}$. Then the IoD approach (and certain extended versions of Khintchine's theorem [4, 5, 9, 10, 26, 29]) can be summarized in this way: we can generate samples from $p(x)$ if

- we first draw $y'$ from $q_G^{-1}(y) \propto p_G^{-1}(y)$,
- and then draw $x'$ uniformly from $\mathcal{A}_{0|y'}$.

The resulting sample $x'$ is distributed according to $p(x)$. This approach is implicitly used in other Monte Carlo techniques as, for instance, the *slice sampling* algorithm [22, 28]. It is interesting to observe that the pdf $q_G^{-1}(y)$ has the following features:

(a) It is always monotonically non-increasing [8, 17].
(b) If the domain of $p_o(x) \propto p(x)$ is unbounded, it has a vertical asymptote at 0 and the minimum at $y = \sup_x p(x)$. The support set of $q_G^{-1}(y)$ is $(0, \sup_x p(x)]$.

Figure 5.8 shows an example of a bimodal pdf and the corresponding generalized inverse pdf $p_G^{-1}(y)$. Observe that, with the pdf $p(x)$ in Fig. 5.8a, the set $\mathcal{A}_{0|y}$ can be formed by two disjoint segments, as $S_1$ and $S_2$ in Fig. 5.8a, or a single one, depending on the value of $y$. Clearly, the length of the sets $S_1$ and $S_2$ depends on the four monotonic pieces $p_i(x)$, $i = 1, \ldots, 4$, which form $p(x)$. The IoD method consists of the following steps: (1) generate a sample $y'$ from $p_G^{-1}(y)$, in Fig. 5.8b and then,

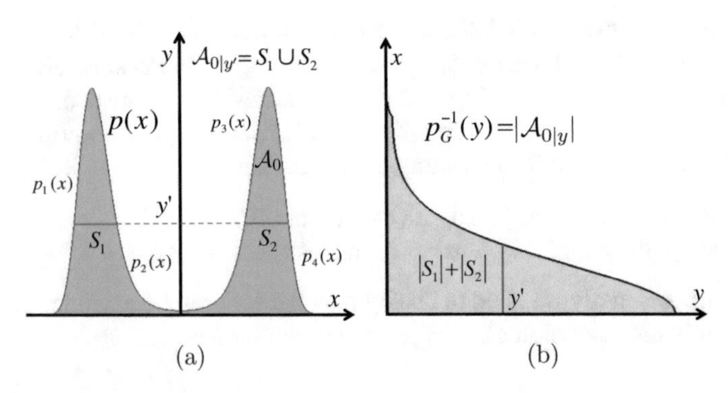


**Fig. 5.8** **(a)** A bimodal pdf $p(x)$. Monotonic parts of $p(x)$ are denoted as $p_i(x)$, $i = 1, \ldots, 4$. **(b)** The corresponding generalized inverse pdf $p_G^{-1}(y)$

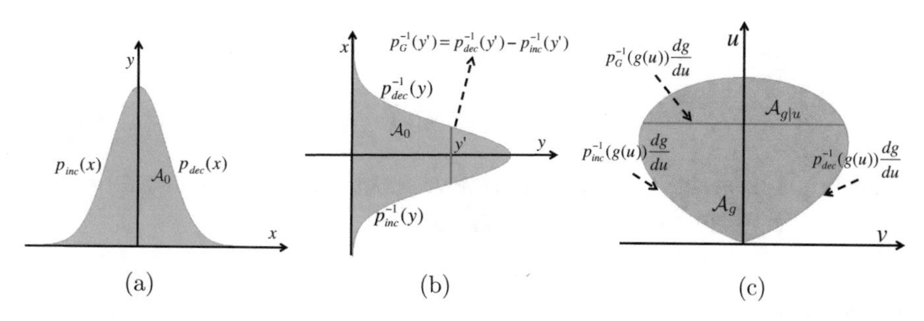

**Fig. 5.9** **(a)** A unimodal pdf $p(x)$ with the mode located at zero. **(b)** The region $\mathcal{A}_0$ rotated 90°. **(c)** The corresponding region $\mathcal{A}_g = \{(v, u) \in \mathbb{R}^2 : p_{inc}^{-1}(g(u))\, \dot{g}(u) \leq v \leq p_{dec}^{-1}(g(u))\, \dot{g}(u)\}$ obtained by the GRoU method using $g(u) = u^2/2$

considering, for instance, the case in Fig. 5.8a, (2) draw $x'$ uniformly distributed on $S_1 \cup S_2$.

### 5.7.2 GRoU for pdfs with a Single Mode at x = 0

In Sect. 5.6 we have already studied the region $\mathcal{A}_g$ when $p(x)$ is a monotonic function (increasing or decreasing) with a single mode at 0. If the target $p(x)$ is unimodal with its maximum at 0, we can divide the domain as $\mathcal{D} = \mathcal{D}_1 \cup \mathcal{D}_2$. Then for $x \in \mathcal{D}_1 = [0, +\infty)$ we obtain that $p(x) = p_{dec}(x)$ is decreasing, while for $x \in \mathcal{D}_2 = (-\infty, 0]$ it turns out that $p(x) = p_{inc}(x)$ is increasing. Hence, in this case, the region $\mathcal{A}_g$ can also be expressed as

$$\mathcal{A}_g = \{(v, u) \in \mathbb{R}^2 : p_{inc}^{-1}(g(u))\, \dot{g}(u) \leq v \leq p_{dec}^{-1}(g(u))\, \dot{g}(u)\}, \tag{5.64}$$

where we have put together Eq. (5.55) and Eq. (5.56).[2] Then, it can be interpreted that the GRoU method applies a transformation $U = g^{-1}(Y)$ over two random variables, $Y_1$ with pdf $p_{dec}^{-1}(y)$ and $Y_2$ with pdf $-p_{inc}^{-1}(y)$.

Figure 5.9a shows an example of a unimodal pdf (a standard Gaussian density) with mode located at zero. Figure 5.9b illustrates the same region $\mathcal{A}_0$ rotated 90°, i.e., switching the axes $x$ and $y$. In this case, the generalized inverse density associated to $p(x)$ is $p_G^{-1}(y) = p_{dec}^{-1}(y) - p_{inc}^{-1}(y)$. Since, in this case, $p(x)$ in this example is also symmetric, we have $p_G^{-1}(y) = 2p_{dec}^{-1}(y)$. Finally, Fig. 5.9c depicts the corresponding region $\mathcal{A}_g$, achieved with the special choice $g(u) = u^2/2$. Furthermore, let us observe that the Lebesgue measure of the subset $\mathcal{A}_{g|u} \subset \mathcal{A}_g$, defined as

$$\mathcal{A}_{g|u} \triangleq \{(v, z) \in \mathcal{A}_g : z = u\} \tag{5.65}$$

[2]Recall that the inequalities depend on the sign of the first derivative of $p(x)$ (i.e., whether $p(x)$ is increasing or decreasing).

for a fixed value of $u$, can be expressed as

$$|\mathcal{A}_{g|u}| = p_G^{-1}(g(u))\frac{dg}{du}. \tag{5.66}$$

On the right-hand side of the equation above we have the pdf of a transformed r.v. $U = g^{-1}(Y)$, where $Y$ is distributed as $p_G^{-1}(y)$.

### 5.7.3 GRoU for pdfs with a Single Mode at $x \neq 0$

In this section, we consider the application of the GRoU method to a unimodal density $p(x)$, with $x \in \mathbb{R}^+$ and the maximum located at $a \neq 0$. We can see an example of this kind of density in Fig. 5.10a. In Fig. 5.10b we depict the region $\mathcal{A}_0$ below $p(x)$ with the axis $x - y$ rotated by 90°. In this case the region $\mathcal{A}_0$ can be described as

$$\mathcal{A}_0 = \{(x, y) \in \mathbb{R}^2 : p_{inc}^{-1}(y) \le x \le p_{dec}^{-1}(y)\}. \tag{5.67}$$

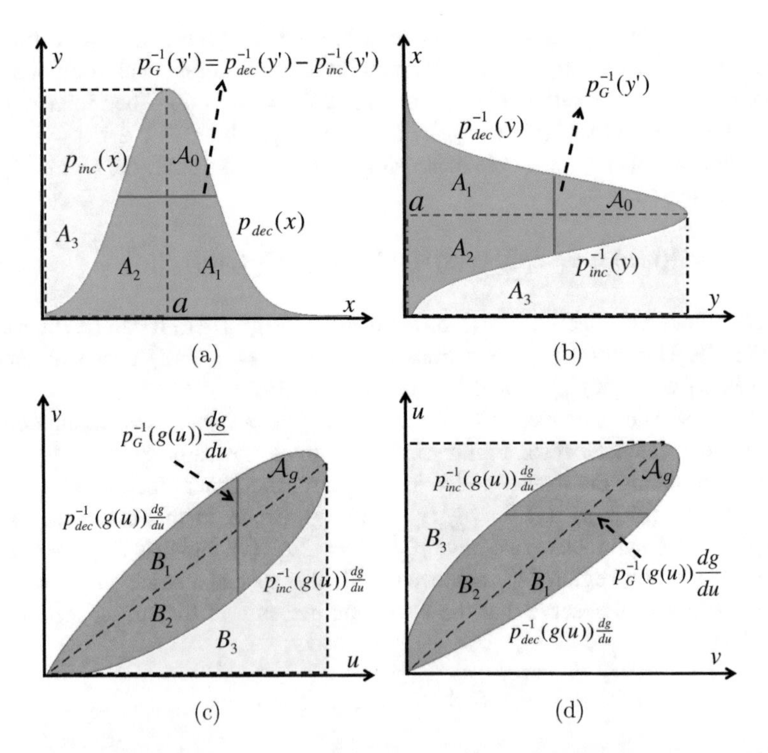


**Fig. 5.10** **(a)** An example of unimodal density $p(x)$. **(b)** The region $\mathcal{A}_0$ represented switching the axes $x - y$. **(c)** The region $\mathcal{A}_g$ obtained with the GRoU technique (using $g(u) = \frac{u^2}{2}$). **(d)** The same region $\mathcal{A}_g$ represented switching the axes $u - v$ in the previous picture (this is the typical representation of the GRoU regions)

Let us consider Fig. 5.10b. We define three different r.v.'s: $Y_1$ with pdf proportional to $p_{inc}^{-1}(y)$ (associated to the region $A_2$), $Y_2$ with pdf proportional to $p_{dec}^{-1}(y)$ (associated to the region $A_1 \cup A_2 \cup A_3$) and, finally, $Y_3$ distributed according to the generalized inverse density proportional to $p_G^{-1}(y) \propto p_{dec}^{-1}(y) - p_{inc}^{-1}(y)$ (associated to the region $A_1 \cup A_2$). Note that $\mathcal{A}_0$ consists only of $A_1$ and $A_2$, i.e., $\mathcal{A}_0 = A_1 \cup A_2$.

Next, we consider the transformed random variables $U_1 = g^{-1}(Y_1)$ and $U_2 = g^{-1}(Y_2)$, where $g^{-1}$ is an increasing function, and plot jointly the two pdfs $q_1(u) \propto p_{inc}^{-1}(g(u))\frac{dg}{du}$ and $q_2(u) \propto p_{dec}^{-1}(g(u))\frac{dg}{du}$, obtaining the regions $B_1$, $B_2$, and $B_3$ as represented in Fig. 5.10c. The region attained with the GRoU method is exactly $\mathcal{A}_g = B_1 \cup B_2$. Indeed we can write it explicitly as

$$\mathcal{A}_g = \left\{ (v, u) \in \mathbb{R}^2 : p_{inc}^{-1}(g(u))\frac{dg}{du} \leq v \leq p_{dec}^{-1}(g(u))\frac{dg}{du} \right\}. \tag{5.68}$$

Note that we can interpret that the boundary of $\mathcal{A}_g$ is obtained through a transformation of the contour of the region $\mathcal{A}_0$ [see Eqs. (5.67) and (5.68)]. Finally, recalling the subset $\mathcal{A}_{g|u} = \{(v, z) \in \mathcal{A}_g : z = u\}$, we note that in this case we also have $|\mathcal{A}_{g|u}| = p_G^{-1}(g(u))\frac{dg}{du}$, that is a transformation of the r.v. $Y_3 \sim p_G^{-1}(y) \propto p_{dec}^{-1}(y) - p_{inc}^{-1}(y)$.

*Remark 5.2* It is important to notice that the shape of the region $\mathcal{A}_g$ changes if the target pdf $p_o(x) \propto p(x)$ is shifted (i.e., applying the GRoU to $p(x - k)$ with $k$ constant). This property can be used to increase the acceptance rate in an RS scheme [32].

### 5.7.4 GRoU for Arbitrary pdfs

Consider finally a bounded target pdf $p_o(x) \propto p(x)$ with several modes. Assume that we find a partition of $\mathcal{D}$ consisting of $N$ disjoint sets $D_j$, $j = 1, \ldots, N$, i.e., $\mathcal{D} = \mathcal{D}_1 \cup \mathcal{D}_2 \cup \ldots \mathcal{D}_N$, such that $p(x)$ is monotonically increasing or decreasing, when restricted to a single subset. To be specific, we can define

$$p_j(x) = p(x) \quad \text{for} \quad x \in \mathcal{D}_j,$$

such that $p_j(x)$, $j = 1, \ldots, N$, are increasing or decreasing functions. Let us also assume that $p(x)$ is defined in $\mathcal{D} = \mathbb{R}$. Since $\int_{\mathcal{D}} p(x)dx < +\infty$, then the number $N$ of disjoint sets $D_j$ is necessarily even: the pieces $p_{2i-1}(x)$ are increasing functions whereas $p_{2i}(x)$ are decreasing functions, for $i = 1, \ldots, \frac{N}{2}$. Then, the region $\mathcal{A}_g$ generated by the GRoU method can be expressed as

$$\mathcal{A}_g = \mathcal{A}_{g,1} \cup \mathcal{A}_{g,2} \cup \ldots \cup \mathcal{A}_{g,\frac{N}{2}}, \tag{5.69}$$

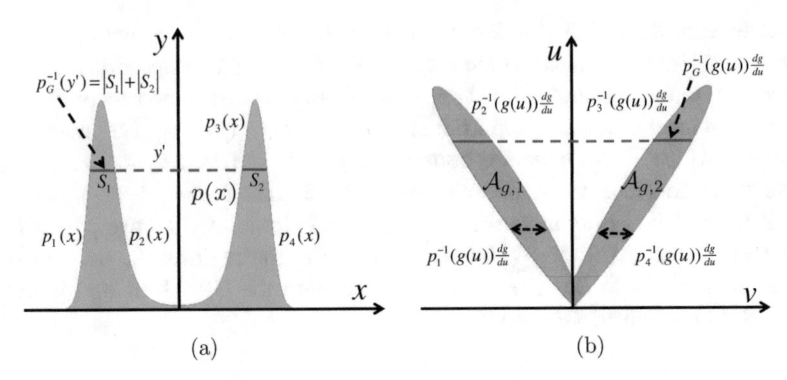


**Fig. 5.11** **(a)** A bimodal density $p_o(x) \propto p(x) = \exp\{-(x^2 - 4)^2/4\}$ formed by four monotonic pieces $p_i(x)$, $i = 1, \ldots, 4$. **(b)** The corresponding region $\mathcal{A}_g = \mathcal{A}_{g,1} \cup \mathcal{A}_{g,2}$ obtained by the GRoU method using $g(u) = \frac{u^2}{2}$

where

$$\mathcal{A}_{g,i} = \left\{(v, u) \in \mathbb{R}^2 : p_{2i-1}^{-1}\left(g(u)\right)\dot{g}(u) \leq v \leq p_{2i}^{-1}\left(g(u)\right)\dot{g}(u)\right\}, \tag{5.70}$$

for $i = 1, \ldots, \frac{N}{2}$. Figure 5.11a shows the bimodal pdf $p_o(x) \propto p(x) = \exp\{-(x^2 - 4)^2/4\}$. The corresponding region $\mathcal{A}_g$, obtained by the GRoU technique with $g(u) = \frac{1}{2}u^2$, is illustrated in Fig. 5.11b. For this example, we have $\mathcal{A}_{0|y} = S_1 \cup S_2$, so that $p_G^{-1}(y) = |S_1| + |S_2|$. Then, recalling the definition of the subset $\mathcal{A}_{g|u} = \{(v, z) \in \mathcal{A}_g : z = u\}$, we have again that $|\mathcal{A}_{g|u}| = p_G^{-1}(g(u))\frac{dg}{du}$, as depicted in Fig. 5.11b.

### 5.7.5 *Summary*

In Sects. 5.6 and 5.7, we have analyzed the procedure underlying the GRoU method for different types of target densities $p_o(x) \propto p(x)$. Specifically, increasing progressively the complexity of $p(x)$, we have considered: monotonic targets $p(x)$ with maximum at $x = 0$ in Sect. 5.6.2, unimodal $p(x)$ with mode at $x = 0$ in Sect. 5.7.2, unimodal $p(x)$ with mode at $x \neq 0$ in Sect. 5.7.3, and arbitrary bounded functions $p(x)$ in Sect. 5.7.4. Defining a partition of the domain, $\mathcal{D} = \mathcal{D}_1 \cup \mathcal{D}_2 \cup \ldots \mathcal{D}_N$, such that

$$p_j(x) = p(x), \quad \text{for} \quad x \in \mathcal{D}_j,$$

are increasing or decreasing functions, we can consider the inverse functions $p_j^{-1}(y)$, $j = 1, \ldots, N$. Then to the $j$th piece we can associate a r.v. $Y_j$ has density proportional to $p_j^{-1}(y)$. Therefore, given the considerations in the last two sections, we can state that:

- The boundary of the region $\mathcal{A}_g$ consists of $N$ pieces obtained as transformed r.v.'s, $U_j = g^{-1}(Y_j)$, where $Y_j$ with density proportional to $p_j^{-1}(y)$.

- The measure of $\mathcal{A}_g$ is related to the generalized inverse function $p_G^{-1}(y) = |\mathcal{A}_{0|y}|$ where

$$\mathcal{A}_{0|y} = \{x \in \mathcal{D} : \ p(x) > y\}.$$

Indeed, the measure of the subset $\mathcal{A}_{g,u} \subset \mathcal{A}_g$, defined as

$$\mathcal{A}_{g|u} = \{(v, z) \in \mathcal{A}_g : \ z = u\},$$

can be expressed as $|\mathcal{A}_{g|u}| = p_G^{-1}(g(u))\frac{dg}{du}$ (assuming $c_A = 1$), and thus we can write

$$\begin{aligned} |\mathcal{A}_g| &= \int_{\mathbb{R}+} p_G^{-1}(g(u))\frac{dg}{du}du \\ &= \int_{\mathbb{R}+} p_G^{-1}(y)dy \\ &= \int_{\mathcal{D}} p(x)dx, \end{aligned}$$

where the last equality holds because $p_G^{-1}(y)$ and $p(x)$ enclose the same area by definition. This is exactly the result given in Eq. (5.23) with $c_A = 1$.

## 5.8 Rectangular Region $\mathcal{A}_g$

One of the easiest cases in which the GRoU method can be used naturally to perform exact sampling arises when the region $A_g$ is rectangular. The analysis in the previous section is very useful to clarify which $g(u)$ produces a rectangular region $\mathcal{A}_g$ (see, e.g., Proposition 5.3). Indeed, we have seen that the GRoU method corresponds to a transformation of a r.v. $Y$ with pdf $p^{-1}(y)$, where we assume $p(x)$ is decreasing for simplicity, i.e., $U = g^{-1}(Y)$. The *inversion method* (described in Sect. 2.4.1) asserts that if the function $g^{-1}(y)$ is the cdf of $Y$, then the transformation produces a r.v. $U$ uniformly distributed in $[0, 1]$. Therefore if we choose

$$g^{-1}(y) = F_Y(y), \tag{5.71}$$

where $F_Y(y) = \int_{-\infty}^{y} p^{-1}(y)dy$ is proportional to the cdf of r.v. $Y$, then $\mathcal{A}_g$ is a rectangular region. Since $p^{-1}(y)$ is in general unnormalized, note that $F_Y(y) \to c_v$ with $y \to +\infty$, where

$$c_v = \int_{\mathcal{D}_Y} p^{-1}(y)dy = \int_{\mathcal{D}} p(x)dx.$$

Therefore, if $g^{-1}(y) = F_Y(y)$ then $U = F_Y(Y)$ is a uniform r.v. in $[0, c_v]$. In this case, $\mathcal{A}_g$ is a rectangle $0 \le u \le c_v$ and $0 \le v \le 1$ as we show below in Eq. (5.74). Indeed, when $g^{-1}(y) = F_Y(y)$, the region $\mathcal{A}_g$ is defined (for simplicity, we set $c_A = 1$) as

$$\mathcal{A}_g = \left\{(v, u) \in \mathbb{R}^2 : 0 \le u \le F_Y\left[p\left(\frac{v}{\dot{F}_Y^{-1}(u)}\right)\right]\right\}. \tag{5.72}$$

Since $\dot{F}_Y^{-1}(u) = \frac{1}{\dot{F}_Y(F_Y^{-1}(u))} = \frac{1}{p^{-1}(F_Y^{-1}(u))}$, we have

$$\mathcal{A}_g = \left\{(v, u) \in \mathbb{R}^2 : 0 \le u \le F_Y\left[p\left(vp^{-1}(F_Y^{-1}(u))\right)\right]\right\}, \tag{5.73}$$

and $x = vp^{-1}(F_Y^{-1}(u))$ is distributed as $p_o(x) \propto p(x)$ for the GRoU method. Since $0 \le F_Y(y) \le c_v$, the values of the variable $u$ are contained in $[0, c_v]$. The variable $v$ is contained in $[0, 1]$ independently of the values of $u$, because inverting the inequalities in Eq. (5.73) we obtain

$$0 \le v \le \frac{p^{-1}(F_Y^{-1}(u))}{p^{-1}(F_Y^{-1}(u))} = 1,$$

so that $\mathcal{A}_g$ is completely described by the inequalities

$$\mathcal{A}_g = \left\{(v, u) \in \mathbb{R}^2 :\ 0 \le v \le 1,\ \ 0 \le u \le c_v\right\}. \tag{5.74}$$

### 5.8.1 *Yet Another Connection Between IoD and GRoU*

In Sect. 5.6.1, we have introduced an extended version of the IoD technique that displays a direct connection with the GRoU methodology. In this section, we analyze the relationship between the GRoU scheme and the standard version of the IoD method described in Chap. 2 (for simplicity, we again assume $c_A = 1$). We have just seen above that the choice

$$Y = g(U) = F_Y^{-1}(U), \quad \text{where} \quad U \sim \mathcal{U}\left([0, c_v]\right),$$

yields the rectangular region $\mathcal{A}_g$ in Eq. (5.74) and the r.v.

$$X = Vp^{-1}(Y), \tag{5.75}$$

with $V \sim \mathcal{U}([0, 1])$, is distributed according to the target pdf $p_o(x) \propto p(x)$. However, Eq. (5.75) is exactly the same as Eq. (2.44) that corresponds to the standard IoD technique.

## 5.9 Relaxing Assumptions: GRoU with Decreasing $g(u)$

In this section, we show it is possible to relax some assumptions made in the statement of Theorem 5.2, which is the basis of the GRoU scheme. For instance, so far we have assumed that function $g$ is increasing. In this section, we discuss how to relax this assumption. Firstly, we highlight two observations:

- Consider the region $\mathcal{A}_g$ defined by the GRoU scheme with an increasing function $g$, i.e.,

$$\mathcal{A}_g = \left\{(v,u) \in \mathbb{R}^2 : 0 \leq u \leq g^{-1}\left[c_A\, p\left(\frac{v}{\dot{g}(u)}\right)\right]\right\},$$

  and let $(v,u)$ be uniformly distributed on it. The sample $x = -\frac{v}{\dot{g}(u)}$ is distributed according to the pdf $p_o(-x) \propto p(-x)$.
- Consider now the set, with an increasing $g$, defined as

$$\mathcal{A}'_g = \left\{(v,u) \in \mathbb{R}^2 : 0 \leq u \leq g^{-1}\left[c_A\, p\left(-\frac{v}{\dot{g}(u)}\right)\right]\right\}$$

  and let $(v,u)$ be uniformly distributed on it. The sample $x = -\frac{v}{\dot{g}(u)}$ is distributed according to $p_o(x) \propto p(x)$. The set $\mathcal{A}'_g$ is symmetric to $\mathcal{A}_g$ w.r.t. the axis $u$.

These considerations can be easily inferred from the proof of the GRoU Theorem 5.2.

In the same fashion, we can easily infer other two possible versions of the GRoU method when $g(u)$ is monotonically decreasing. For the sake of simplicity, consider also a decreasing target $p_o(x) \propto p(x)$ with $x \geq 0$. Let $g(u) : \mathbb{R}^- \to \mathbb{R}^+$ (i.e., $u \leq 0$) be strictly *decreasing*, $\dot{g} = \frac{dg}{du} < 0$. If $(v,u)$ is uniformly distributed in the set

$$\mathcal{A}_{g_{dec}} = \left\{(v,u) \in \mathbb{R}^2 : g^{-1}\left[c_A\, p\left(\frac{v}{\dot{g}(u)}\right)\right] \leq u \leq 0\right\}, \tag{5.76}$$

then $x = \frac{v}{\dot{g}(u)}$ is a sample from $p_o(x) \propto p(x)$. Moreover, if $(v,u)$ is uniformly distributed in the set

$$\mathcal{A}'_{g_{dec}} = \left\{(v,u) \in \mathbb{R}^2 : g^{-1}\left[c_A\, p\left(-\frac{v}{\dot{g}(u)}\right),\right] \leq u \leq 0\right\}, \tag{5.77}$$

then $x = -\frac{v}{\dot{g}(u)}$ has pdf $p_o(x)$. It is important to observe that $g^{-1}(y) : \mathbb{R}^+ \to \mathbb{R}^-$, i.e., $g^{-1}(y) \leq 0$. To clarify this point, we can consider again, as an example, the function $g(u) = \frac{u^2}{2}$ but now with $u \leq 0$ (since we need $g$ to be decreasing). In this case, the region $\mathcal{A}_{g_{dec}}$ has the same shape as the region $\mathcal{A}_g$ defined in the standard GRoU Theorem 5.2 using $g(u) = \frac{u^2}{2}$ with $u \geq 0$, but they are symmetric w.r.t. the origin of

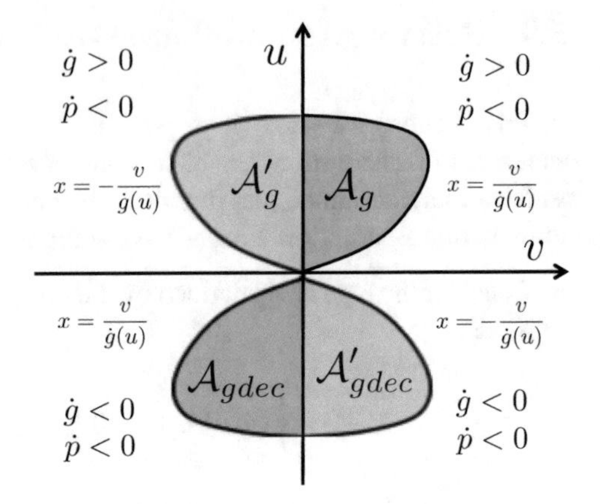


**Fig. 5.12** Summary of different cases, for a monotonically target $p_o(x) \propto p(x)$ with maximum at 0. **(a)** The different regions $\mathcal{A}_g$, $\mathcal{A}'_g$, $\mathcal{A}_{g_{dec}}$, and $\mathcal{A}'_{g_{dec}}$, defined in Eqs. (5.76) and (5.77), for a decreasing target $p_o(x) \propto p(x)$

the axes $(v, u) = (0, 0)$. Furthermore, the regions $\mathcal{A}'_{g_{dec}}$ and $\mathcal{A}_g$ are symmetric w.r.t. the axis $v$.

Figure 5.12 depicts an example of four possible regions $\mathcal{A}_g$, $\mathcal{A}'_g$, $\mathcal{A}_{g_{dec}}$ and $\mathcal{A}'_{g_{dec}}$ (for a monotonically decreasing target $p(x)$ with maximum at 0, i.e., $\mathcal{D} = \mathbb{R}^+$).

### 5.9.1 General Expression of a r.v. Transformation

Now, we are able to completely connect the GRoU method with a r.v. transformation. $\mathcal{A}'_{g_{dec}}$ in Eq. (5.77) can be rewritten as

$$\mathcal{A}'_{g_{dec}} = \left\{(v, u) \in \mathbb{R}^2 : 0 \leq v \leq -p^{-1}(g(u))\dot{g}(u)\right\}. \tag{5.78}$$

Recall that above $-\dot{g}(u) > 0$. Now, we can take together Eq. (5.55) and Eq. (5.78) above, to arrive at

$$\mathcal{A}_g = \left\{(v, u) \in \mathbb{R}^2 : 0 \leq v \leq p^{-1}(g(u))|\dot{g}(u)|\right\}, \tag{5.79}$$

where $\dot{g}(u)$ can be positive or negative. Equation (5.79) is clearly the expression of a transformation of a r.v. $Y$ with pdf proportional to $p^{-1}(y)$. Thus, considering $(v, u)$ uniformly distributed on $\mathcal{A}_g$ defined in Eq. (5.79), we obtain that

$$x = \frac{v}{|\dot{g}(u)|}, \tag{5.80}$$

is distributed as $p_o(x) \propto p(x)$.

## 5.10 Another View of GRoU

In Chap. 2, we have seen that the easiest and most basic sampling technique consists in finding a transformation that maps samples from one or more available random sources into samples distributed according to the target pdf $p_o(x) \propto p(x)$. We have also shown different examples and procedures to obtain this transformation. We have studied, for instance, transformations of the form

$$X = t(V, U), \qquad \text{where} \qquad (V, U) \sim \mathcal{U}(\mathcal{A}),$$

i.e., where the random vector $(V, U)$ is uniformly distributed on $\mathcal{A}$. In the previous chapters, we have considered "fixed" the domain $\mathcal{A}$ and have tried to find a suitable transformation $t$. An example is the *Box-Muller transformation* (see Chap. 2) that converts a uniform random vector $(V, U) \sim [0, 1] \times [0, 1]$ in a standard Gaussian r.v.

We can consider a different approach: assume the transformation $t(v, u)$ is fixed, and try to find the corresponding domain $\mathcal{A}$ of the uniform random vector $(V, U)$ ensuring that $X = t(u, v)$ is distributed as $p_o(x) \propto p(x)$. The solution is not straightforward. However, Jones' generalization [3, 18] of the GRoU method provides an answer. Indeed, the Jones' Theorem given in [18] asserts that the adequate region $\mathcal{A}$ is

$$\mathcal{A} = \Big\{(v, u) : 0 \leq u \leq \eta\big(t(v, u)\big)\Big\},$$

where

- the function $\eta(x)$ is such that

$$p_o(x) \propto \left[\frac{\partial r(x, z)}{\partial x}\right]_{z=\eta(x)},$$

- the function $r(x, z)$ is such that

$$f(x, z) = \frac{\partial r(x, z)}{\partial z},$$

- and $f(x, z)$ is the inverse function of $t(v, u)$ w.r.t. the variable $v$.

Therefore, given a fixed transformation, the GRoU method can be seen as a technique to find a suitable domain for two uniform r.v.'s $U$ and $V$, in such a way that the r.v. of interest $X$ can be sampled. Figure 5.13 below summarizes the steps to obtain the function $\eta(x)$ starting from $x = t(v, u)$. Further specific considerations about the Jones' generalization are provided in Appendix C.

Fig. 5.13 Relationships among functions in the Jones' generalization

$$x = t(v, u)$$
$$\downarrow$$
$$v = f(x, u)$$
$$\downarrow$$
$$r(x, u) = \int f(x, u)du$$
$$\downarrow$$
$$\delta(x, u) = \frac{\partial r(x, u)}{\partial x}$$
$$\downarrow$$
$$p(x) = \left. \frac{\partial r(x, u)}{\partial x} \right|_{u=\eta(x)} = \delta(x, \eta(x))$$

## 5.11 Summary

In this chapter, we have described in depth the features of the *ratio-of-uniforms* sampling techniques. The RoU and generalized RoU methods were introduced [19, 32] as bivariate transformations of the bidimensional region $\mathcal{A}_0$ below the target pdf $p_o(x) \propto p(x)$. This bivariate transformations follow from the equations $x = \frac{v}{g(u)}$ and $y = u$. These relationships describe all the points *within* a new transformed region $\mathcal{A}_g$.

If we are able to draw points $(v', u')$ uniformly from $\mathcal{A}_g$, we can also obtain samples from $p_o(x) \propto p(x)$, taking $x' = \frac{v'}{g(u')}$. The cases of interest occur when the region $\mathcal{A}_g$ is bounded and the GRoU method can be applied jointly with the rejection sampling principle, as discussed in Sects. 5.2 and 5.4 in detail. The test, for checking whether a point belongs to $\mathcal{A}_g$ or not, can be performed without knowing the boundary of $\mathcal{A}_g$ but only checking certain inequalities, properly defined by the GRoU methodology.

We have studied the conditions needed to guarantee that $\mathcal{A}_g$ is bounded, in Sect. 5.5.2. In Sects. 5.6, 5.7, and 5.8.1, we have described the connections of GRoU with other sampling methods. We have seen that GRoU techniques can be interpreted as transformations of r.v.'s $Y_i$ with pdfs proportional to the monotonic pieces $p_i^{-1}(y)$, $i = 1, .., N$, of the target density $p(x)$. These transformed densities describe disjoint parts of the boundary of the region $\mathcal{A}_g$ obtained using the GRoU scheme. We have also shown that, depending on the choice of the function $g$, the GRoU method coincides with other different techniques. For instance, Table 5.3 summarizes the relationship among GRoU, the different versions of the IoD method, and the fundamental theorem of simulation. The E-IoD technique has been described in Sect. 5.6.1.

**Table 5.3** GRoU and IoD methods

| Function $g$ | Related sampling method | Region $\mathcal{A}_g$ |
|---|---|---|
| $g(u) = u$ | Fund. theorem of simulation | $\mathcal{A}_g \equiv \mathcal{A}_0$ |
| $g(u) = F_Y^{-1}(u)$ | IoD | Rectangular $\mathcal{A}_g$ |
| Generic $g(u)$ | E-IoD | Generic |

In the last part of the chapter, from Sect. 5.9 to Sect. 5.10, we have presented several variants and extensions of the GRoU technique. The Jones' generalization (see also Appendix C) clarifies that GRoU algorithms can be seen sampling methods that choose (and keep fixed) a transformation of two uniform r.v.'s, $U_1$, $U_2$, and then construct a support domain for $U_1$ and $U_2$ in order to obtain samples from $p_o(x)$ (as remarked in Sect. 5.10).

## References

1. L. Barabesi, Optimized ratio-of-uniforms method for generating exponential power variates. Stat. Appl. **5**, 149–155 (1993)
2. L. Barabesi, *Random Variate Generation by Using the Ratio-of-Uniforms Method*. Serie Ricerca-Monografie 1 (Nuova Immagine, Siena, 1993)
3. G. Barbu, On computer generation of random variables by transformations of uniform varaibles. Soc. Sci. Math. R. S. Rom. Tome 26 **74**(2), 129–139 (1982)
4. M.C. Bryson, M.E. Johnson, Constructing and simulating multivariate distributions using Khintchine's theorem. J. Stat. Comput. Simul. **16**(2), 129–137 (1982)
5. Y.P. Chaubey, G.S. Mudholkar, M.C. Jones, Reciprocal symmetry, unimodality and Khintchine's theorem. Proc. R. Soc. A Math. Phys. Eng. Sci. **466**(2119), 2079–2096 (2010)
6. Y. Chung, S. Lee, The generalized ratio-of-uniform method. J. Appl. Math. Comput. **4**(2), 409–415 (1997)
7. J.H. Curtiss, On the distribution of the quotient of two chance variables. Ann. Math. Stat. **12**(4), 409–421 (1941)
8. P. Damien, S.G. Walker, Sampling truncated normal, beta, and gamma densities. J. Comput. Graph. Stat. **10**(2), 206–215 (2001)
9. B.M. de Silva, A class of multivariate symmetric stable distributions. J. Multivar. Anal. **8**(3), 335–345 (1978)
10. L. Devroye, Random variate generation for unimodal and monotone densities. Computing **32**, 43–68 (1984)
11. L. Devroye, *Non-uniform Random Variate Generation* (Springer, New York, 1986)
12. U. Dieter, Mathematical aspects of various methods for sampling from classical distributions, in *Proceedings of Winter Simulation Conference* (1989)
13. J.E. Gentle, *Random Number Generation and Monte Carlo Methods* (Springer, New York, 2004)
14. C. Groendyke, Ratio-of-uniforms Markov Chain Monte Carlo for Gaussian process models. Thesis in Statistics, Pennsylvania State University (2008)
15. W. Hörmann, A rejection technique for sampling from T-concave distributions. ACM Trans. Math. Softw. **21**(2), 182–193 (1995)
16. W. Hörmann, J. Leydold, G. Derflinger, *Automatic Nonuniform Random Variate Generation* (Springer, New York, 2003)

17. M.C. Jones, On Khintchine's theorem and its place in random variate generation. Am. Stat. **56**(4), 304–307 (2002)
18. M.C. Jones, A.D. Lunn, Transformations and random variate generation: generalised ratio-of-uniforms methods. J. Stat. Comput. Simul. **55**(1), 49–55 (1996)
19. A.J. Kinderman, J.F. Monahan, Computer generation of random variables using the ratio of uniform deviates. ACM Trans. Math. Softw. **3**(3), 257–260 (1977)
20. J. Leydold, Automatic sampling with the ratio-of-uniforms method. ACM Trans. Math. Softw. **26**(1), 78–98 (2000)
21. J. Leydold, Short universal generators via generalized ratio-of-uniforms method. Math. Comput. **72**, 1453–1471 (2003)
22. J.S. Liu, *Monte Carlo Strategies in Scientific Computing* (Springer, New York, 2004)
23. M.M. Marjanovic, Z. Kadelburg, Limits of composite functions. Teach. Math. **12**(1), 1–6 (2009)
24. G. Marsaglia, Ratios of normal variables and ratios of sums of uniform variables. Am. Stat. Assoc. **60**(309), 193–204 (1965)
25. J. Monahan, An algorithm for generating chi random variables. Trans. Math. Softw. **13**(2), 168–172 (1987)
26. R.A. Olshen, L.J. Savage, A generalized unimodality. J. Appl. Probab. **7**(1), 21–34 (1970)
27. C.J. Perez, J. Martín, C. Rojano, F.J. Girón, Efficient generation of random vectors by using the ratio-uniforms method with ellipsoidal envelopes. Stat. Comput. **18**(4), 209–217 (2008)
28. C.P. Robert, G. Casella, *Monte Carlo Statistical Methods* (Springer, New York, 2004)
29. L.A. Shepp, Symmetric random walk. Trans. Am. Math. Soc. **104**, 144–153 (1962)
30. S. Stefanescu, I. Vaduva, On computer generation of random vectors by transformations of uniformly distributed vectors. Computing **39**, 141–153 (1987)
31. I. Vaduva, Computer generation of random vectors based on transformations on uniform distributed vectors, in *Proceedings of Sixth Conference on Probability Theory, Brasov* (1982), pp. 589–598
32. J.C. Wakefield, A.E. Gelfand, A.F.M. Smith, Efficient generation of random variates via the ratio-of-uniforms method. Stat. Comput. **1**(2), 129–133 (1991)

# Chapter 6
# Independent Sampling for Multivariate Densities

**Abstract** In this chapter, we present several techniques for multivariate independent sampling. We recall some techniques introduced in the previous chapters and show how they can be adapted to a multidimensional setup. Additionally, we provide guidelines for their application in higher dimensional spaces. We also consider the problem of drawing uniformly from a measurable set embedded in $\mathbb{R}^n$. With this goal, an exhaustive description of transformations of random vectors is given, which extends the study of this approach in the previous chapters.

The problem of sampling a random vector can often be conveniently viewed as generating a (finite) sequence of statistically dependent scalar samples. Thus, in this chapter, we take a slight detour from the main course of the book and show different methods that yield dependent samples, including the use of stochastic processes. Furthermore, a collection of efficient samplers for specific multivariate distributions is described.

## 6.1 Introduction

In the previous chapters, we described a large collection of Monte Carlo methodologies for sampling from continuous probability density functions. However, the description was restricted to the univariate case. In this chapter, we look into different approaches for sampling from multivariate distributions. We recall several techniques presented in previous chapters, and show how they can be extended to be used in a multivariate setup. For each of them, we provide derivations and useful guidelines for their application in higher dimensional spaces. Furthermore, several other algorithms for the generation of specific multivariate distributions are presented.

The chapter is organized as follows. After briefly recalling the main notations to be used throughout the chapter in Sect. 6.2, we devote Sect. 6.3 to the description of different generic sampling procedures. In this section, we introduce miscellaneous techniques for representing statistical dependence (as the use of copula functions, for instance), and recall quickly other methods described in the previous chapters,


L. Martino et al., *Independent Random Sampling Methods*, Statistics and Computing, https://doi.org/10.1007/978-3-319-72634-2_6

as the rejection sampling (RS) and the ratio-of uniforms (RoU) algorithms. For both of them, we discuss their application in higher dimensional spaces.

Section 6.4 is devoted to the study of a large family of distributions, namely the *elliptically contoured distributions* [5, 15]. Specific methodologies can be applied for drawing from this family, as the polar methods and the vertical density representation (VDR). We revisit the VDR approach in Sect. 6.5. Both polar and VDR methods can be expressed as continuous mixture representations of the target density. In Sect. 6.6, we handle the generation of multivariate samples uniformly distributed in $n$-dimensional measurable sets embedded in $\mathbb{R}^n$, e.g., points uniformly distributed within a simplex or a hypersphere in $\mathbb{R}^n$. In the latter case, this problem involves the simulation from elliptically contoured distributions.

Some specific problems tackled in Sect. 6.6 can also be approached considering suitable transformations of random vectors in $\mathbb{R}^n$, as the transformation in polar coordinates [5, 15, 21]. In this case, the transformation involves $n$ variables and converts them into other suitable $n$ variables. The generic transformation case converting $m$ r.v's into $n$ r.v.'s ($m \neq n$) is discussed in Sect. 6.7. This section extends the considerations presented in Sect. 2.3.1 of Chap. 2 for the generic $m \neq n$ scenario. The case $m < n$ corresponds to the so-called *singular distributions* [5, Chap. 11], i.e., distributions that have all their probability mass on a subset of dimension $m$ in $\mathbb{R}^n$.

Section 6.8 provides a collection of sampling algorithms for generating samples from specific multivariate distributions, which deserve attention because of their implications in numerous multivariate statistical problems. In Sect. 6.9, we discuss the generation of random vectors with a specific dependence structure. Finally, some concluding considerations are made in Sect. 6.10.

## 6.2 Notation

The joint cumulative distribution of the point $[X_1, \dots, X_n]^\top \in \mathbb{R}^n$ is denoted as

$$F_{\mathbf{X}}(\mathbf{x}) = F_{\mathbf{X}}(x_1, \dots, x_n) = \text{Prob}(X_1 \leq x_1, \dots, X_n \leq x_n).$$

This cdf is related to the target density $p_o$ by the following integral,

$$F_{\mathbf{X}}(\mathbf{x}) = \int_{-\infty}^{x_1} \cdots \int_{-\infty}^{x_n} p_o(x_1, \dots, x_n) dx_1 \cdots dx_n = \int_{-\infty}^{\mathbf{x}} p_o(\mathbf{x}) d\mathbf{x},$$

so that we also have

$$p_o(x_1, \dots, x_n) = \frac{\partial^n F_{\mathbf{X}}(x_1, \dots, x_n)}{\partial x_1 \cdots \partial x_n}.$$

In the rest of this chapter, the $i$th marginal distribution will be denoted as

$$F_{X_i}(x_i) = F_{\mathbf{X}}(\infty, \ldots, \infty, x_i, \infty, \ldots, \infty) = \lim_{\mathbf{x}_{-i} \to +\infty} F_{\mathbf{X}}(\mathbf{x}),$$

where $\mathbf{x}_{-i} = [x_1, \ldots, x_{i-1}, x_{i+1}, \ldots, x_n]^\top$. The corresponding marginal pdfs are indicated as

$$f_i(x_i) = \int_{\mathbb{R}} \cdots \int_{\mathbb{R}} p_o(\mathbf{x}) d\mathbf{x}_{-i}, \qquad f_i(x_i) = \frac{dF_{X_i}(x_i)}{dx_i}, \tag{6.1}$$

for $i = 1, \ldots, n$. Finally, the conditional densities are defined as ratios of joint pdfs of subsets of components, e.g.,

$$\gamma_j(x_j | x_1, \ldots, x_{j-1}) = \frac{\int_{\mathbb{R}} \cdots \int_{\mathbb{R}} p_o(\mathbf{x}) dx_{j+1} \cdots dx_n}{\int_{\mathbb{R}} \cdots \int_{\mathbb{R}} p_o(\mathbf{x}) dx_j \cdots dx_n}. \tag{6.2}$$

## 6.3 Generic Procedures

In this section, we describe general approaches for multivariate sampling purposes. Most of them, as the rejection sampling principle and the ratio of uniforms technique, have been thoroughly discussed in the previous chapters for the scalar case. Here, we recall them and point out some relevant aspects, useful in multidimensional problems, for their application.

### *6.3.1 Chain Rule Decomposition*

A joint pdf $p_o(\mathbf{x}) \propto p(\mathbf{x})$ with $\mathbf{x} = [x_1, \ldots, x_n]^\top \in \mathbb{R}^n$ can be always expressed as a product of conditional densities, i.e.,

$$p_o(x_1, \ldots, x_n) = f_1(x_1)\gamma_2(x_2|x_1)\gamma_2(x_3|x_1, x_2) \ldots \gamma_n(x_n|x_1, \ldots, x_{n-1}),$$

where the marginal pdf $f_1(x_1)$ is defined in Eq. (6.1), and $\gamma_j(x_j|x_1, \ldots, x_{j-1})$, $j = 2, \ldots, n$, are defined in Eq. (6.2). The order of the variables can be arbitrarily chosen. If we are able to draw an independent sample from $f_1(x_1)$ and each conditional pdf $\gamma_i(x_i|x_1, \ldots, x_{i-1})$, then we can use the following procedure to generate samples from $p_o(\mathbf{x})$ [5, 9]:

1. Set $i = 1$.
2. Draw $x_i' \sim \gamma_i(x_i|x_1', \ldots, x_{i-1}')$ (denoting, for simplicity, $\gamma_1(x_1) = f_1(x_1)$).
3. If $i < n$, then set $i = i + 1$ and repeat from step 2. Otherwise, return $\mathbf{x}' = [x_1', \ldots, x_n']^\top$.

*Example 6.1* Consider a bivariate random vector $(X_1, X_2)$ uniformly distributed on a triangle of vertices $(0, 0)$, $(a, 0)$ and $(a, b)$. This region can be interpreted as the area below a linear pdf $\gamma_1(x_1) = \frac{b}{a}x_1$, with $0 \leq x_1 \leq b$. Then, from the Fundamental Theorem of Simulation (Sect. 2.4.3), we have that $\gamma_2(x_2|x_1)$ is a uniform pdf on $\left[0, \frac{b}{a}x_1\right]$. It is easy to draw from $f_1(x_1)$ using the inversion method, since $F_{X_1}(x_1) = \frac{b}{2a}x_1^2$ and $F_{X_1}^{-1}(u) = \sqrt{\frac{2a}{b}u}$. As a consequence, we can easily sample from $p_o(x_1, x_2)$ in three steps: (1) draw $u' \sim \mathcal{U}([0, 1])$, (2) set $x_1' = \sqrt{\frac{2a}{b}u'}$, and (3) then draw $x_2' \sim \mathcal{U}([0, \frac{b}{a}x_1'])$.

*Example 6.2* Consider a bivariate random vector $(X_1, X_2)$ uniformly distributed on a circle, i.e., the set $\{x_1, x_2 \in \mathbb{R} : x_1^2 + x_2^2 \leq 1\}$. A simple sampling method in this case is the following:

1. Draw $x_1' \sim \gamma_1(x_1) \propto \sqrt{1 - x_1^2}$, with $|x_1| \leq 1$.
2. Draw $x_2' \sim \mathcal{U}\left(\left[-\sqrt{1 - (x_1')^2}, \sqrt{1 - (x_1')^2}\right]\right)$.

Additional examples of application of this method are given in Sect. 6.8.

### 6.3.2 Dependence Generation

In order to generate samples from a random vector, it is often necessary to guarantee a certain dependence structure among the entries of the vector. In this section, we describe some general methods to produce multivariate samples which display a prescribed intra-vector dependence structure. Sequences of correlated samples can also be generated using stochastic processes, as shown in Sect. 6.9.

#### Copula Functions

Consider a random vector $\mathbf{X} = [X_1, \ldots, X_n]^\top \sim p_o(x_1, \ldots, x_n)$, with $X_i \sim f_i(x)$, where $f_i(x_i)$ is the marginal pdf of $i$th component. We assume the marginal pdfs $f_i(x_i)$ and the cdfs $F_{X_i}(x_i) = \int_{-\infty}^{x_i} f_i(z)dz$ are known. The *copula method* [21, 24] provides a convenient technique for describing dependence among the components of $\mathbf{X}$, while maintaining the marginal pdfs fixed. A copula function, $C : [0, 1]^N \to [0, 1]$, is defined as a cdf of $n$ dependent uniform r.v.'s $U_1, \ldots, U_N \sim \mathcal{U}([0, 1])$, i.e.,

$$C(u_1, \ldots, u_n) = \text{Prob}\{U_1 \leq u_1, \ldots, U_n \leq u_n\}. \tag{6.3}$$

The joint cdf of $\mathbf{X}$ can be written as

$$F_{\mathbf{X}}(x_1, \ldots, x_n) = C(F_{X_1}(x_1), \ldots, F_{X_N}(x_n)) \tag{6.4}$$

for a suitable choice of the function $C$ [21]. If the marginal cdfs $F_{X_i}$, $i = 1, \ldots, n$, are invertible, then we can generate a sample $\mathbf{x}' \sim p_o(x_1, \ldots, x_n)$ using the following procedure, which extends naturally the classical inversion method:

1. Draw $\mathbf{u}' = [u'_1, \ldots, u'_n]^\top \sim C(u_1, \ldots, u_n)$.
2. Set $\mathbf{x}' = [F_{X_1}^{-1}(u'_1), \ldots, F_{X_N}^{-1}(u'_n)]^\top$.

Note that, in general, it is not needed to have a closed form for the copula $C(u_1, \ldots, u_n)$ but only to be able to generate samples according to the distribution $C$. Different examples of specific copula functions are provided in Sect. 6.8.8. The copula approach follows by the simple observation below.

**Proposition 6.1 ([5, 24])** *Consider $n$ different random variables $X_i$ with marginal pdfs $f_i(x_i)$ and cdfs $F_{X_i}(x_i)$, $i = 1, \ldots, n$. Assume that the joint pdf can be expressed as*

$$p_o(x_1, \ldots, x_n) = \left[\prod_{i=1}^{n} f_i(x_i)\right] g(F_{X_1}(x_1), \ldots, F_{X_n}(x_n)), \tag{6.5}$$

*where $g(u_1, \ldots, u_n)$ is another joint pdf with uniform marginal densities. Then:*

- *The densities $f_i(x_i)$ are the marginal pdfs of the joint density $p_o(x_1, \ldots, x_n)$.*
- *Moreover, defining the uniform r.v.'s $U_i = F_{X_i}(X_i)$ (see Sect. 2.4.1), the random vector $[U_1, \ldots, U_n]^\top$ has joint pdf*

$$g(u_1, \ldots, u_n) = \frac{p_o(F_{X_1}^{-1}(u_1), \ldots, F_{X_n}^{-1}(u_n))}{\prod_{i=1}^{n} f_i(F_{X_i}^{-1}(u_i))}, \quad 0 \leq u_i \leq 1, \tag{6.6}$$

*and the corresponding cdf is*

$$C(u_1, \ldots, u_n) = F_{\mathbf{X}}(F_{X_1}^{-1}(u_1), \ldots, F_{X_n}^{-1}(u_n)), \tag{6.7}$$

*where $F_{\mathbf{X}}$ is the cdf of the random vector $[X_1, \ldots, X_n]^\top$, with pdf $p_o$.*

The previous proposition can be interpreted as a multivariate version of Theorem 2.1. If we are able to build a joint pdf $g(u_1, \ldots, u_n)$ with uniform marginal densities such that a prescribed dependence requirement is satisfied, then we can also construct another joint pdf with the desired marginals, preserving the dependence structures. Furthermore, a generation algorithm is automatically induced as shown above.

### Samples with a Specific Covariance Matrix

Consider a given $n \times n$ covariance matrix $\mathbf{\Sigma}$. The samples produced by the following generic procedure have exactly covariance matrix $\mathbf{\Sigma}$ [5]:

1. Obtain a matrix $\mathbf{A}$ such that $\boldsymbol{\Sigma} = \mathbf{A}\mathbf{A}^\top$ using, for instance, the Cholesky decomposition procedure [10].
2. Draw $n$ independent scalar samples $z'_1, \ldots, z'_n$ from a generic density $q(z)$ with zero mean, $E[Z] = \int_{\mathbb{R}} zq(z)dz = 0$, and unit variance, $E\left[(Z - E[Z])^2\right] = 1$.
3. Set $\mathbf{z}' = [z'_1, \ldots, z'_n]^\top$.
4. Return $\mathbf{z}' = \mathbf{A}\mathbf{x}'$.

This is the standard method, e.g., to produce a multidimensional Gaussian vector from the distribution $\mathcal{N}(\mathbf{0}, \boldsymbol{\Sigma})$ using just a (pseudo) random generator for the standard one-dimensional Gaussian distribution $\mathcal{N}(0, 1)$.

### Maximal Positive and Negative Dependence

Maximal positive or negative dependence between two r.v.'s $X_1$ and $X_2$ with cdfs $F_{X_1}$ and $F_{X_2}$, respectively, can be obtained using the following transformations [5, 27]

$$X_1 = F_{X_1}^{-1}(U), \qquad X_2 = F_{X_2}^{-1}(U), \qquad \text{namely} \quad X_1 = F_{X_1}^{-1}(F_{X_2}(X_2)),$$

and

$$X_1 = F_{X_1}^{-1}(U), \qquad X_2 = F_{X_2}^{-1}(1 - U), \qquad \text{namely} \quad X_1 = F_{X_1}^{-1}(1 - F_{X_2}(X_2)),$$

where $U \sim \mathcal{U}([0, 1])$. In the first case, the vector $[X_1, X_2]^\top$ has the maximum positive dependence between the two components with cumulative function $F_{\mathbf{X}}(x_1, x_2) = \min[F_{X_1}(x_1), F_{X_2}(x_2)]$. In the second case, the vector $[X_1, X_2]^\top$ has the maximum negative dependence between the two components with cumulative function $F_{\mathbf{X}}(x_1, x_2) = \max[0, F_{X_1}(x_1) + F_{X_2}(x_2) - 1]$. These statements are derived from the identities

$$\begin{aligned}
\text{Prob}\left\{F_{X_1}^{-1}(U) \leq x_1, F_{X_2}^{-1}(U) \leq x_2\right\} &= \text{Prob}\left\{U \leq F_{X_1}(x_1), U \leq F_{X_2}(x_2)\right\}, \\
&= \text{Prob}\left\{U \leq \min[F_{X_1}(x_1), F_{X_2}(x_2)]\right\},
\end{aligned}$$

and

$$\text{Prob}\{F_{X_1}^{-1}(U) \leq x_1, F_{X_2}^{-1}(1 - U) \leq x_2\} = \text{Prob}\left\{U \leq F_{X_1}(x_1), U \geq 1 - F_{X_2}(x_2)\right\}.$$

Recall that $P(U \leq b) = b$ and $P(1 - a \leq U \leq b) = max[0, b + a - 1]$ with $a, b \in [0, 1]$. Consider, for instance, the case of maximal dependence. The cdf is $F_{\mathbf{X}}(x_1, x_2) = \min[F_{X_1}(x_1), F_{X_2}(x_2)]$ and the corresponding pdf is non-zero only in the points of $\mathbb{R}^2$ belonging to the curve $x_1 = F_{X_1}^{-1}(F_{X_2}(x_2))$ (this is an example of *singular distribution* [5]; see Sect. 6.7.2). The marginal pdfs are clearly $f_1(x_1) = \frac{dF_{X_1}(x_1)}{dx_1}$ and $f_2(x_2) = \frac{dF_{X_2}(x_2)}{dx_2}$.

### 6.3.3 Rejection Sampling

The rejection sampling (RS) principle [35] has been thoroughly described in Chaps. 3 and 4. Let us now quickly review the RS approach in a multivariate setting. Given the pdf $p_o(\mathbf{x}) \propto p(\mathbf{x})$, $\mathbf{x} \in \mathbb{R}^n$, and a proposal density $\pi(\mathbf{x})$, if we know a constant $L$ such that $L\pi(\mathbf{x})$ is an envelope function for $p(\mathbf{x})$, i.e.,

$$L\pi(\mathbf{x}) \geq p(\mathbf{x}), \quad \forall \mathbf{x} \in \mathbb{R}^n, \tag{6.8}$$

for all $\mathbf{x} \in \mathbb{R}^n$, then we can first draw a sample from the proposal , $\mathbf{x}' \sim \pi(\mathbf{x})$, and then accept it with probability

$$p_A(\mathbf{x}') = \frac{p(\mathbf{x}')}{L\pi(\mathbf{x}')} \leq 1.$$

Otherwise, the proposed sample $\mathbf{x}'$ is discarded. The applicability and the performance of the RS method depend on the knowledge of a suitable constant $L$ that satisfies the inequality (6.8). In the sequel, we provide different inequalities that become useful to design accept–reject samplers for multidimensional target distributions.

#### Bounded Target with Bounded Support

If the target pdf, $p_o(\mathbf{x}) \propto p(\mathbf{x})$, is bounded and defined on a bounded domain, i.e.,

$$p(\mathbf{x}) \leq M, \quad \mathbf{x} \in \mathcal{D} \subseteq \mathbb{R}^n,$$

where $M < \infty$ is a constant and $\mathcal{D}$ is bounded. In this case, we can choose $L = M$ and $\pi(\mathbf{x}) = \frac{1}{|\mathcal{D}|}\mathbb{I}_{\mathcal{D}}(\mathbf{x})$ for applying a naive RS scheme. We simply need to be able to draw uniformly on $\mathcal{D}$. This is straightforward if $\mathcal{D}$ is a hyper-rectangle, i.e.,

$$\mathcal{D} = [a_1, b_1] \times [a_2, b_2] \times \ldots [a_n, b_n],$$

with $|a_i|, |b_i| < \infty$. In this scenario, if $p(\mathbf{x})$ is unimodal, the performance of the RS sampler can be easily improved splitting the rectangular domain $\mathcal{D}$ in different sub-rectangles (adaptively or not) and performing RS locally in a randomly selected sub-rectangle. This simple idea and possible adaptive variants have been already described in Sects. 4.3.2 and 4.5.2, for the univariate case. Other similar strategies can be found in literature, see, for instance, the *Ahrens method* for multivariate pdfs [13], [14, Chap. 11].

### Inequalities for Multidimensional RS

The previous ideas basically take advantage of the unimodality of $p(\mathbf{x})$. Now, let us define the concept of *ortho-unimodality* (and, as consequence, of *ortho-monotonicity*) [14, Chap. 11]. A pdf $p_o(\mathbf{x}) \propto p(\mathbf{x})$ is ortho-unimodal (at the origin) if all the (univariate) full-conditional densities are unimodal with mode at 0. Hence, we have $p(\mathbf{0}) \geq p(\mathbf{x})$. In the sequel, and unless otherwise stated, we are going to consider an ortho-unimodal target pdf defined in the unit hyper-cube, i.e., $\mathcal{D} = [0, 1]^n$. The definition of ortho-unimodality implies that for any $\mathbf{x} = [x_1, \dots, x_n]^\top$ and $\mathbf{z} = [z_1, \dots, z_n]^\top$ we have

$$p(\mathbf{z}) \geq p(\mathbf{x}), \qquad \text{for all } \mathbf{z} \in [0, \mathbf{x}] = [0, x_1] \times [0, x_2] \times \ldots \times [0, x_n],$$

namely for each $0 \leq z_i \leq x_i$, $i = 1, \dots, n$. Since the Lebesgue measure of $[0, \mathbf{x}]$ (defined above) is $\prod_{i=1}^n x_i$, if we denote $c_p = \int_{\mathcal{D}} p(\mathbf{x}) d\mathbf{x}$, then

$$p(\mathbf{x}) \prod_{i=1}^{n} x_i \leq c_p, \qquad \text{i.e.,} \qquad p(\mathbf{x}) \leq \frac{c_p}{\prod_{i=1}^n x_i}. \tag{6.9}$$

This inequality can be simply found in the univariate case, $x \in \mathbb{R}$, where we have $p(x) \leq \frac{c_p}{x}$ and $p(x)$ is monotonically decreasing in $[0, +\infty)$. Given the inequality (6.9) and since $p(\mathbf{0}) \geq p(\mathbf{x})$, we can combine them to obtain

$$p(\mathbf{x}) \leq f(\mathbf{x}) = \min\left[p(\mathbf{0}), \frac{c_p}{\prod_{i=1}^n x_i}\right], \qquad \forall \mathbf{x} \in \mathcal{D}, \tag{6.10}$$

which is appealing for RS. Indeed, we can choose a proposal pdf as $\pi(\mathbf{x}) \propto f(\mathbf{x})$, however there are different difficulties with this choice:

- In general, the constant $c_p = \int_{\mathcal{D}} p(\mathbf{x}) d\mathbf{x}$ is unknown and an upper bound $c' \geq c_p$ has to be found.
- The function $f(\mathbf{x})$ can be integrated only on a bounded domain (hence the assumption $\mathcal{D} = [0, 1]^n$).
- It is not straightforward to draw samples from $\pi(\mathbf{x}) \propto f(\mathbf{x})$. This density is called a *platymorphous* pdf [14, Chap. 11].

The last issue was solved in [6], where a generation method was proposed. The idea is to consider first the transformation $\mathbf{Y} = -\log(\mathbf{X})$ where $\mathbf{X} \sim \pi(\mathbf{x})$. Thus, the r.v. $\mathbf{Y}$ has the pdf

$$q(\mathbf{y}) \propto \min\left[c_p, p(\mathbf{0}) \exp\left(-\sum_{i=1}^{n} y_i\right)\right], \qquad y_i \geq 0,$$

where $\mathbf{y} = [y_1, \ldots, y_n]^\top$. For the sake of simplicity, we consider

$$q(\mathbf{y}) \propto \min\left[1, a\exp\left(-\sum_{i=1}^{n} y_i\right)\right], \qquad a \geq 1, y_i \geq 0,$$

so that a generation procedure can be outlined as follows:

1. Draw $z'$ from the univariate pdf $\phi(z) \propto z^{n-1}\min[1, ae^{-z}]$. This is possible via adaptive rejection sampling (see Chap. 4) since $\phi(z)$ is log-concave.
2. Generate the uniform spacings vector $\mathbf{d} = [d_1, \ldots, d_n]^\top$. The uniform spacings r.v.'s are obtained by taking the differences between uniform order statistics (see Sect. 2.3).
3. Return $\mathbf{y}' = [y'd_1, \ldots, y'd_n]^\top$.

As shown in [6] the sample $\mathbf{y}'$ is distributed according to $q(\mathbf{y})$, hence the transformation $\mathbf{x}' = \exp(-\mathbf{y}')$ yields the desired sample from $\pi(\mathbf{x}) \propto f(\mathbf{x})$. For an ortho-unimodal target pdf $p_o(\mathbf{x}) = p_o(x_1, \ldots, x_n) \propto p(\mathbf{x})$, other useful inequalities can be found. For instance, defining the univariate functions

$$\varphi_i(x) = p(0, \ldots 0, x, 0, \ldots, 0), \qquad i = 1, \ldots, n,$$

where $x$ is in the $i$th position, it is possible to prove that [14, Chap. 11]

$$p(\mathbf{x}) \leq \prod_{i=1}^{n} \varphi_i(x_i)^{1/n}, \tag{6.11}$$

$$p(\mathbf{x}) \leq \min\left[\varphi_1(x_1), \varphi_2(x_2), \ldots, \varphi_n(x_n)\right]. \tag{6.12}$$

These inequalities can be readily employed to design a rejection sampler as well.

### 6.3.4 RoU for Multivariate Densities

The ratio of uniforms (RoU) technique [22, 25, 36], thoroughly described in Chap. 5, can be easily extended for multivariate sampling purposes, as shown by the theorem below.

**Theorem 6.1** *Consider the target pdf $p_o(\mathbf{x}) \propto p(\mathbf{x})$ with $\mathbf{x} = (x_1, \ldots, x_n) \in \mathbb{R}^n$ and assume that the point $(v_1, \ldots, v_n, u) \in \mathbb{R}^{n+1}$ is a sample drawn uniformly from the set*

$$\mathcal{A}_{rn} = \left\{(v_1, \ldots, v_n, u) : 0 \leq u \leq \left[p\left(\frac{v_1}{u^r}, \ldots, \frac{v_n}{u^r}\right)\right]^{1/(rn+1)}\right\}, \tag{6.13}$$

*where $r \geq 0$. Then $\mathbf{x} = (x_1, \ldots, x_n)$, where $x_i = v_i/u^r$, is a sample from the distribution with density $p_o(\mathbf{x}) \propto p(\mathbf{x})$.*

*Proof* Assume that the r.v.'s $(V_1, \ldots, V_n, U)$ are distributed uniformly on $\mathcal{A}_{rn}$, and consider the direct and inverse transformations

$$\begin{cases} x_1 = \dfrac{v_1}{u^r} \\ \vdots \\ x_i = \dfrac{v_i}{u^r} \\ \vdots \\ y = u \end{cases} \longrightarrow \begin{cases} v_1 = x_1 y^r \\ \vdots \\ v_i = x_i y^r \\ \vdots \\ u = y \end{cases}. \tag{6.14}$$

Then, the joint pdf $q(\mathbf{x}, y)$ of the r.v.'s $(X_1, \ldots, X_n, Y)$ is

$$q(\mathbf{x}, y) = \frac{1}{|\mathcal{A}_{rn}|}|\mathbf{J}^{-1}| \quad \text{for all} \quad 0 \le y \le [p(x_1, \ldots, x_n)]^{1/(rn+1)}. \tag{6.15}$$

where $\mathbf{J}$ is the Jacobian matrix of the transformation and $|\mathbf{A}|$ represents the determinant of a matrix $\mathbf{A}$. Moreover, we can calculate easily the Jacobian of the inverse transformation, which yields

$$|\mathbf{J}^{-1}| = \det \begin{bmatrix} y^r & 0 & \ldots & 0 & x_1 r y^{r-1} \\ 0 & y^r & \ldots & 0 & x_2 r y^{r-1} \\ \vdots & \vdots & \ddots & \vdots & \vdots \\ 0 & 0 & \ldots & y^r & x_n r y^{r-1} \\ 0 & 0 & \ldots & 0 & 1 \end{bmatrix} = y^{nr}, \tag{6.16}$$

hence

$$q(\mathbf{x}, y) = \frac{1}{|\mathcal{A}_{rn}|} y^{rn} \quad \text{for all} \quad 0 \le y \le [p(x_1, \ldots, x_n)]^{1/(rn+1)}. \tag{6.17}$$

Finally, we integrate $q(\mathbf{x}, y)$ to obtain the marginal density $q(\mathbf{x})$,

$$\begin{aligned} \int_{-\infty}^{+\infty} q(\mathbf{x}, y) dy &= \int_0^{[p(\mathbf{x})]^{1/(rn+1)}} \frac{y^{rn}}{|\mathcal{A}_{rn}|} dy, \\ &= \frac{1}{|\mathcal{A}_{rn}|} \left[ \frac{y^{(rn+1)}}{rn+1} \right]_0^{[p(\mathbf{x})]^{1/(rn+1)}}, \\ &= \frac{p(\mathbf{x})}{(rn+1)|\mathcal{A}_{rn}|} = p_o(\mathbf{x}), \end{aligned}$$

where the first equality follows from Eq. (6.17). □

Hence, if we are able to draw points $(v_1, \ldots, v_n, u)$ uniformly from the set $\mathcal{A}_{rn}$ defined Eq. (6.13), then $\left(\frac{v_1}{u^r}, \ldots, \frac{v_n}{u^r}\right)$ is distributed as $p_o(\mathbf{x})$. The generation of samples uniformly on $\mathcal{A}$ can be performed by an RS procedure. Observe also that, for $r = 0$, the RoU theorem coincides with the fundamental theorem of simulation introduced in Sect. 2.4.3. The application of RoU in multidimensional spaces is often combined with the use of MCMC algorithms [12].

## 6.4 Elliptically Contoured Distributions

Elliptically contoured (also known as *elliptically symmetric*) distributions are distributions completely defined by their first and second moments, with pdfs whose contour lines are hyper-ellipses [5, 15]. This class includes, for instance, the Gaussian and sine-wave distributions [15] (other examples are given below). In one dimension, this class consists of all symmetric distributions. In dimension $n$, given $\mathbf{x} = [x_1, \ldots, x_n]^\top$, $\boldsymbol{\mu} = [\mu_1, \ldots, \mu_n]^\top$ and $\boldsymbol{\Sigma}$ an $n \times n$ positive definite matrix, an elliptically symmetric distribution is defined as

$$p_o(\mathbf{x}) \propto p(\mathbf{x}) = g\left((\mathbf{x} - \boldsymbol{\mu})^\top \boldsymbol{\Sigma}^{-1} (\mathbf{x} - \boldsymbol{\mu})\right), \qquad x \in \mathcal{D} \tag{6.18}$$

where $g(z) : \mathbb{R} \to \mathbb{R}^+$ is a one-dimensional positive function. We use $\mathcal{E}(\boldsymbol{\mu}, \boldsymbol{\Sigma}; g)$ to denote the cdf corresponding to the pdf in Eq. (6.18). The matrix $\boldsymbol{\Sigma}$ represents a scale parameter of the r.v. $\mathbf{X} \sim \mathcal{E}(\boldsymbol{\mu}, \boldsymbol{\Sigma}; g)$. Below we list some properties which become useful for the purpose of sampling [5, 15]:

- Given the linear transformation $\mathbf{Y} = \mathbf{B}\mathbf{X} + \boldsymbol{\eta}$ with $\mathbf{X} \sim \mathcal{E}(\boldsymbol{\mu}, \boldsymbol{\Sigma}; g)$, then it is possible to show that $\mathbf{Y}$ is distributed as $\mathcal{E}(\mathbf{B}\boldsymbol{\mu} + \boldsymbol{\eta}, \mathbf{B}\boldsymbol{\Sigma}\mathbf{B}^\top; g)$.
- The univariate r.v. $Z = (\mathbf{X} - \boldsymbol{\mu})^\top \boldsymbol{\Sigma}^{-1} (\mathbf{X} - \boldsymbol{\mu}) \in \mathbb{R}^+$ has pdf

$$q(z) \propto z^{\frac{n}{2}-1} g(z), \qquad z \in \mathbb{R}^+,$$

  when $\mathbf{X} \sim \mathcal{E}(\boldsymbol{\mu}, \boldsymbol{\Sigma}; g)$. The r.v. $R = \sqrt{Z}$ plays an important role in different sampling techniques and its density is

$$\varphi(r) \propto r^{n-1} g(r^2), \qquad r \in \mathbb{R}^+. \tag{6.19}$$

  These results can be derived directly by computing the determinant of the Jacobian matrix of the polar transformation [15] (see Sects. 6.6.2 and 6.7).
- All the marginal pdfs are themselves elliptically symmetric, with the same generator function $g(z)$. More specifically, consider a partition of the vectors $\mathbf{X} = [\mathbf{X}_1, \mathbf{X}_2]^\top$ and $\boldsymbol{\mu} = [\boldsymbol{\mu}_1, \boldsymbol{\mu}_2]^\top$ with dimension $1 \times k$ (for $\mathbf{X}_1$ and $\boldsymbol{\mu}_1$) and $1 \times (n - k)$ (for $\mathbf{X}_2$ and $\boldsymbol{\mu}_2$), respectively. Moreover, consider the corresponding

**Table 6.1** Examples of elliptically contoured pdfs

| $z = (\mathbf{x}-\boldsymbol{\mu})^\top \boldsymbol{\Sigma}^{-1}(\mathbf{x}-\boldsymbol{\mu}), \quad \mathbf{x} \in \mathbb{R}^n$ | | |
|---|---|---|
| Distribution | Function $g$ | Domain |
| Pearson type II | $g(z) = (1-z)^m$ | $z \leq 1, m > -1$ |
| Pearson type VII | $g(z) = (1+z)^{-m}$ | $z \in \mathbb{R}, m > \frac{n}{2}$ |
| Gaussian | $g(z) = \exp(-z)$ | $z \in \mathbb{R}$ |

partition of the scale matrix

$$\boldsymbol{\Sigma} = \begin{bmatrix} \boldsymbol{\Sigma}_{1,1} & \boldsymbol{\Sigma}_{1,2} \\ \boldsymbol{\Sigma}_{2,1} & \boldsymbol{\Sigma}_{2,2} \end{bmatrix},$$

with dimensions shown below

$$\begin{bmatrix} k \times k & n \times (n-k) \\ (n-k) \times n & (n-k) \times (n-k) \end{bmatrix}.$$

Then it can be proved that $\mathbf{X}_1 \sim \mathcal{E}(\boldsymbol{\mu}_1, \boldsymbol{\Sigma}_{1,1}; g)$ and $\mathbf{X}_2 \sim \mathcal{E}(\boldsymbol{\mu}_2, \boldsymbol{\Sigma}_{2,2}; g)$. Moreover, the conditional distributions of $\mathbf{X}_1$ given $\mathbf{X}_2 = \mathbf{x}_2$ and of $\mathbf{X}_2$ given $\mathbf{X}_1 = \mathbf{x}_1$ are also elliptically symmetric.

Table 6.1 provides some examples of elliptically contoured pdfs.

### *Radially (Spherically) Symmetric Distribution*

When $\boldsymbol{\mu} = \mathbf{0}$ and $\boldsymbol{\Sigma} = \mathbf{I}$ then $p_o(\mathbf{x})$ becomes

$$p_o(\mathbf{x}) \propto g\left(|\mathbf{x}|^2\right), \qquad |\mathbf{x}| = \sqrt{\mathbf{x}^\top \mathbf{x}} \in \mathbb{R}^+.$$

This kind of densities are also known as *radially symmetric* distributions [5, 33]. They are also defined in this way: given $\mathbf{X} \sim \mathcal{E}(\mathbf{0}, \mathbf{I}; g)$ and an orthogonal $n \times n$ matrix $\mathbf{P}$ (i.e., such that $\mathbf{P}\mathbf{P}^\top = \mathbf{I}$), then $\mathbf{P}\mathbf{X} \sim \mathcal{E}(\mathbf{0}, \mathbf{I}; g)$. Namely, this means that spherically symmetric distributions are invariant under rotations. Below we describe different strategies to draw samples from elliptically symmetric distribution.

## 6.4.1 *Polar Methods*

It is possible to draw samples from $p_o(\mathbf{x})$ using the following algorithm [5, 14]:

1. Draw a vector $\boldsymbol{\theta}' = [\theta_1', \ldots, \theta_n']$ uniformly *from the surface* of the unit $n$-dimensional hypersphere (see below and Sect. 6.6 for further details).

2. Draw $r' \sim \varphi(r) \propto r^{n-1} g(r^2)$.
3. Since $r'\boldsymbol{\theta}' \sim \mathcal{E}(\mathbf{0}, \mathbf{I}; g)$, compute an $n \times n$ matrix $\mathbf{A}$ such that $\boldsymbol{\Sigma} = \mathbf{A}\mathbf{A}^\top$ using, for instance, the Cholesky decomposition procedure [10], and then set

$$\mathbf{x}' = r'\mathbf{A}\boldsymbol{\theta}' + \boldsymbol{\mu}. \tag{6.20}$$

An alternative technique is given by the *Johnson-Ramberg method* [5]:

1. Draw a vector $\boldsymbol{b}' = [b'_1, \ldots, b'_n]$ uniformly *within* the unit $n$-dimensional hypersphere (e.g., using a rejection sampler).
2. Draw $z' \sim q(z) \propto z^n \frac{dg}{dz}$ $(z \geq 0)$.
3. Compute an $n \times n$ matrix $\mathbf{A}$ such that $\boldsymbol{\Sigma} = \mathbf{A}\mathbf{A}^\top$ using, for instance, the Cholesky decomposition procedure [10], and then set

$$\mathbf{x}' = z'\mathbf{A}\boldsymbol{b}' + \boldsymbol{\mu}. \tag{6.21}$$

Clearly, the method above is feasible only if we are able to draw from $q(z) \propto z^n \frac{dg}{dz}$. Other methods exists as well. For instance, see Sects. 6.5 and 6.8.

### Points Uniformly Distributed on a Unit Hypersphere

Consider $\mathbf{X} = [X_1, \ldots, X_n] \sim \mathcal{E}(\mathbf{0}, \mathbf{I}; g)$ then the random variable

$$\boldsymbol{\Theta} = \frac{\mathbf{X}}{\sqrt{X_1^2 + \ldots, X_n^2}} = \frac{\mathbf{X}}{|\mathbf{X}|},$$

is distributed uniformly on the surface of the unit hypersphere in $\mathbb{R}^n$ (see also Sect. 6.6). Since drawing uniformly from a hypersphere can be interpreted as choosing uniformly an angle in the space $\mathbb{R}^n$, this kind of pdfs defined on the surface of a unit hypersphere are often called *directional distributions* [5, 14, 15].

### Polar Methods as Continuous Mixtures

Consider for simplicity $\mathbf{X} \sim \mathcal{E}(\mathbf{0}, \mathbf{I}; g)$. The polar methods could also be interpreted as a continuous mixture (Sect. 2.3.5), i.e.,

$$p_o(\mathbf{x}) = \int_0^{+\infty} \phi(\mathbf{x}|r) f(r) dr, \quad r \in \mathbb{R}, \tag{6.22}$$

where either

- $f(r) = \varphi(r) \propto r^{n-1} g(r^2)$ (univariate) and $\phi(\mathbf{x}|r)$ is a uniform pdf defined on the surface of the hypersphere of radius $r$, or

- $f(r) = q(r) \propto r^n \frac{dg}{dr}$ (univariate) and $\phi(\mathbf{x}|r)$ is a uniform pdf within the hypersphere of radius $r$ (for the Johnson-Ramberg method).

Note that, in both cases, the univariate pdf $f(r)$ is defined with an unbounded support, i.e., $r \in [0, \infty)$. Below, we describe other continuous mixtures where the univariate pdf $f$ is defined on a bounded support.

## 6.5 Vertical Density Representation

In Sect. 2.4.2, we described a methodology for multidimensional sampling, termed vertical density representation (VDR) [31–33]. In the VDR approach, a multidimensional sampling problem is converted into another sampling problem where it is necessary to draw from a univariate density first and then uniformly from a suitably specified set. More specifically, assuming a bounded continuous target $p_o(\mathbf{x})$, the VDR can be seen as the continuous mixture

$$p_o(\mathbf{x}) = \int_0^M h(\mathbf{x}|z)q(z)dz, \quad z \in \mathbb{R}, \tag{6.23}$$

where:

- The univariate density $q(z)$ is the vertical pdf corresponding to $p_o(\mathbf{x})$, defined as

  $$q(z) \propto -z\frac{dA(z)}{dz}, \qquad 0 < z \leq M, \tag{6.24}$$

  where $A(z)$ is the Lebesgue measure of the set

  $$\mathcal{O}(z) = \{\mathbf{x} \in \mathcal{D} : p_o(\mathbf{x}) \geq z\}. \tag{6.25}$$

- The conditional pdf $h(\mathbf{x}|z)$ is uniform on the set

  $$\mathcal{C}(z) = \{\mathbf{x} \in \mathcal{D} : p_o(\mathbf{x}) = z\}. \tag{6.26}$$

  Note that $\mathcal{C}(z)$ is the boundary of $\mathcal{O}(z)$.
- Finally, $M = \max_{\mathbf{x}\in\mathcal{D}} p_o(\mathbf{x})$.

Once the vertical pdf is available, the VDR procedure involves two simple steps:

1. Draw a sample $z'$ from the univariate vertical density $q(z)$.
2. Draw a point $\mathbf{x}'$ from the uniform distribution on the set $\mathcal{C}(z')$.

Note the difference between Eqs. (6.22) and (6.23): the support of density $f$ in Eq. (6.22) is unbounded, while the vertical density $q$ has support in $(0, M]$. In the univariate case, the set $\mathcal{C}(z)$ consists of isolated points and the conditional pdf $h(x|z)$

collapses to a probability measure constructed as a convex combination of Dirac delta functions as shown in the first example below. See also Sect. 2.4.2, for further discussions.

*Example 6.3* Consider a univariate Gaussian density, i.e.,

$$p_o(x) = \frac{1}{\sqrt{2\pi}} \exp\left(-\frac{x^2}{2}\right), \quad x \in \mathbb{R}.$$

In this case, we have $A(z) = 2\sqrt{-2\log(\sqrt{2\pi}z)}$, hence the vertical density is

$$q(z) = \frac{2}{\sqrt{-2\log(\sqrt{2\pi}z)}}, \qquad 0 < z \leq \frac{1}{\sqrt{2\pi}}.$$

The conditional function $h(x|z)$ is formed by two delta functions, namely,

$$h(x|z) = \frac{1}{2}\delta(x - \phi(z)) + \frac{1}{2}\delta(x + \phi(z)),$$

with $\phi(z) = \sqrt{-2\log(\sqrt{2\pi}z)}$.

*Example 6.4* Consider a generic spherically symmetric density,

$$p_o(\mathbf{x}) \propto p(\mathbf{x}) = g\left(\mathbf{x}^\top \mathbf{x}\right), \quad \mathbf{x} \in \mathbb{R}^2,$$

where $g$ is a strictly decreasing and differentiable function. The set $\mathcal{O}(z) = \{\mathbf{x} \in \mathbb{R}^n : p(\mathbf{x}) \geq z\}$ can be rewritten as $\mathcal{O}(z) = \{\mathbf{x} \in \mathbb{R}^n : \mathbf{x}^\top \mathbf{x} \leq g^{-1}(z)\}$. Since $A(z) = |\mathcal{O}(z)|$, we can write

$$A(z) = \frac{2\pi^{\frac{n}{2}}}{n\Gamma(\frac{n}{2})} \left(g^{-1}(z)\right)^{\frac{n}{2}}.$$

where $\Gamma(\frac{n}{2})$ is the Gamma function. As a consequence, the vertical density is

$$q(z) = -\frac{\pi^{\frac{n}{2}}}{\Gamma(\frac{n}{2})} \left(g^{-1}(z)\right)^{\frac{n}{2}-1} \frac{dg^{-1}(z)}{dz}.$$

The contour of $\mathcal{O}(z)$ is the set $\mathcal{C}(z) = \{\mathbf{x} \in \mathbb{R}^n : \mathbf{x}^\top \mathbf{x} = g^{-1}(z)\}$, i.e., a hypersphere of radius $\sqrt{g^{-1}(z)}$.

*Example 6.5* Consider the class of multivariate exponential power distributions

$$p_o(\mathbf{x}) = c_m \exp\left(-(\mathbf{x}^\top \mathbf{x})^{m/2}\right), \quad \mathbf{x} \in \mathbb{R}^m, m \geq 1. \tag{6.27}$$

It is possible to show [20, 33] that the vertical density corresponding to $p_o(\mathbf{x})$ is a uniform distribution in $(0, c_m]$, i.e.,

$$q(z) = \frac{1}{c_m}, \quad 0 < z \leq c_m.$$

Hence sampling from $p_o(x)$ can be accomplished by drawing first $z' \sim \mathcal{U}((0, \frac{1}{c_m}])$ and then sampling uniformly from the set

$$\mathcal{C}(z) = \left\{ \mathbf{x} \in \mathbb{R}^m : \quad ||\mathbf{x}||^2 = -\left[ \log\left(\frac{z'}{c_m}\right)\right]^{2/m} \right\}.$$

### 6.5.1 *Inverse-of-Density Method*

The inverse-of-density (IoD) method [5, 16, 18], described in Sect. 2.4.4, is related to the VDR approach. Indeed, VDR is just one possibility to obtain a decomposition as in Eq. (6.23), where $h(\mathbf{x}|z)$ is uniform and $q(z)$ is a univariate pdf with bounded domain. If we recall the definition of $\mathcal{O}(z)$ in Eq. (6.25), and let $|\mathcal{O}(z)|$ denote its Lebesgue measure, then we can describe the IoD in two simple steps:

1. Draw $z'$ according to $p^{-1}(z) = |\mathcal{O}(z)|$,
2. Generate $\mathbf{x}'$ uniformly from $\mathcal{O}(z)$.

The obvious difference with the VDR technique is that, in step 2, sampling is carried out from $\mathcal{O}(z)$ and not just from its boundary $\mathcal{C}(z)$. The name of the "inverse-of-density" stems from the form of step 1.

## 6.6 Uniform Distributions in Dimension *n*

This section is focused on the problem of drawing samples uniformly distributed in a measurable $n$-dimensional set in $\mathbb{R}^n$. The case of $m$-dimensional sets embedded in the space $\mathbb{R}^n$, with $m < n$ will be handled in the next section (i.e., here we consider the case $m = n$). Considering a generic measurable set $\mathcal{A}$, obviously a simple rejection method can always be applied if we are able to obtain a hypercube $\mathcal{R}$ such that $\mathcal{R} \supseteq \mathcal{A}$ (see Fig. 6.1). Thus, we can draw uniformly from $\mathcal{R}$ and accept the points that belong to $\mathcal{A}$. However, in this section, we describe more specific methodologies which do not involve any rejection steps.

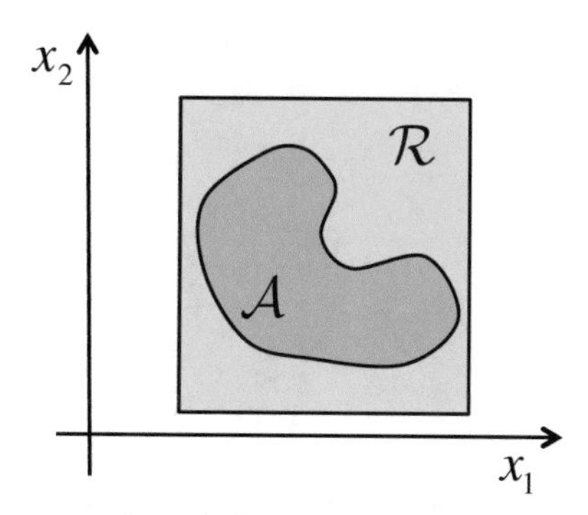


**Fig. 6.1** A generic bidimensional region $\mathcal{A}$ embedded in a rectangular region $\mathcal{R}$

### 6.6.1 Points Uniformly Distributed in a Simplex

A convex polytope in $\mathbb{R}^n$ generated by the vertices $\mathbf{v}_1, \ldots, \mathbf{v}_d$ is the set of points in $\mathbb{R}^n$ that can be obtained as a convex combination of the vectors $\mathbf{v}_1, \ldots, \mathbf{v}_d$. Namely, a vector $\mathbf{x}$ belongs to the convex polytope if it can be expressed as

$$\mathbf{x} = \sum_{i=1}^{d} a_i \mathbf{v}_i, \quad a_i \geq 0, \text{ and } \sum_{i=1}^{d} a_i = 1.$$

The set of vertices $\mathbf{v}_1, \ldots, \mathbf{v}_d$ is *minimal* when all $\mathbf{v}_i$'s are distinct, and none of the $\mathbf{v}_i$'s can be written as a *strictly* convex combination (i.e., at least one $a_i$ is different to 0 or 1) of the others [5, 21].

This combination above is strictly convex if there is at least one $a_i$ such that $0 < a_i < 1$ [5, 21].

A simplex $\mathcal{T}_n \subset \mathbb{R}^n$ is a convex polytope defined with a *minimal* set of $d = n + 1$ vertices. A simplex can be seen as a set $\mathcal{T}_n$ of points in the multidimensional space $\mathbb{R}^n$ that extends the bidimensional notion of triangle. Consider $n$ i.i.d. uniform r.v.'s in $[0, 1]$, denoted $U_1, \ldots, U_n$. Ordering these uniform variates in ascending order, we obtain the order statistics $U_{(1)}, \ldots, U_{(n)}$. The r.v.'s defined as

$$\begin{aligned} S_1 &= U_{(1)}, \\ S_2 &= U_{(2)} - U_{(1)}, \\ &\vdots \\ S_n &= U_{(n)} - U_{(n-1)}, \\ S_{n+1} &= 1 - U_{(n)}, \end{aligned} \tag{6.28}$$

are called *uniform spacings*. Note that $S_i \geq 0$ and $\sum_{i=1}^{n} S_i = 1$. It is possible to draw from the uniform distribution in the simplex $\mathcal{T}_n$ with the following two-step procedure:

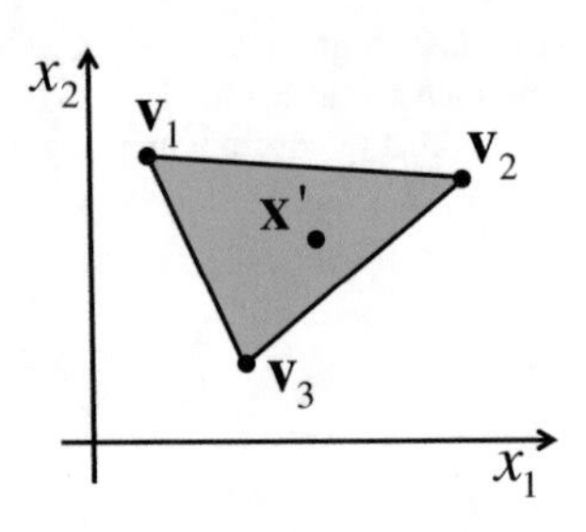


**Fig. 6.2** A generic triangle defined by the vertices $\mathbf{v}_1$, $\mathbf{v}_2$, and $\mathbf{v}_3$

1. Generate the uniform spacing r.v.'s $[s'_1, \ldots, s'_{n+1}]^\top$: to do this, draw $u'_1, \ldots, u'_n \sim \mathcal{U}([0, 1])$, order them to obtain $[u'_{(1)}, \ldots, u'_{(n)}]^\top$, and then set $s'_1 = u'_{(1)}$, $s'_i = u'_{(i)} - u'_{(i-1)}$, $i = 2, \ldots, n$ and $s_{n+1} = 1 - u'_{(n)}$ as described in Eq. (6.28).
2. Set $\mathbf{x}' = \sum_{i=1}^{n} s'_i \mathbf{v}_i$.

*Example 6.6* Consider a triangular set $\mathcal{T}_2$ in the plane $\mathbb{R}^2$ defined by the vertices $\mathbf{v}_1$, $\mathbf{v}_2$ and $\mathbf{v}_3$, as shown in Fig. 6.2. We can draw uniformly from a triangular region [30, 34], [5, p. 570] with the following steps:

1. Sample $u_1, u_2 \sim \mathcal{U}([0, 1])$.
2. If $u_1 > u_2$ then swap $u_1$ with $u_2$.
3. Set $\mathbf{x}' = \mathbf{v}_1 u_1 + \mathbf{v}_2(1 - u_2) + \mathbf{v}_3(u_2 - u_1)$.

The samples $\mathbf{x}'$ drawn with this convex combination are uniformly distributed within the triangle $\mathcal{T}_2$ with vertices $\mathbf{v}_1$, $\mathbf{v}_2$, and $\mathbf{v}_3$. Observe that the algorithm can be simplified setting directly

$$\begin{aligned} \mathbf{x}' = \mathbf{v}_1 \min[u_1, u_2] + \mathbf{v}_2(1 - \max[u_1, u_2]) + \\ + \mathbf{v}_3(\max[u_1, u_2] - \min[u_1, u_2]) \end{aligned} \tag{6.29}$$

at step 2. Different well-known distributions have been defined in a simplex support. Some examples are provided in Sect. 6.8.

### 6.6.2 *Sampling Uniformly Within a Hypersphere*

It is possible to draw from the uniform distribution in a (closed) hypersphere,

$$\mathcal{B}_r = \{\mathbf{x} \in \mathbb{R}^n : \quad |\mathbf{x}| = \sqrt{x_1^2 + \ldots, + x_n^2} \le r^2\}, \qquad r > 0,$$

using the simple algorithm below [5, 15, 21]:

1. Draw i.i.d. samples $z'_1, \ldots, z'_n \sim \mathcal{N}(0, 1)$, i.e., from a standard Gaussian distribution of mean 0 and variance 1.
2. Draw $v' \sim \mathcal{U}([0, r])$ and set $\rho' = v^{1/n}$.

3. Set

$$\mathbf{x}' = [x_1', \ldots, x_n']^\top = \left[\rho' \frac{z_1'}{|\mathbf{z}'|}, \ldots, \rho' \frac{z_n'}{|\mathbf{z}'|}\right]^\top,$$

where $|\mathbf{z}'| = \sqrt{z_1'^2 + \ldots, +z_n'^2}$.

Note that the sample $\rho'$ (the radius) is distributed according to the pdf

$$q(\rho) \propto \rho^{n-1} \qquad \text{with} \qquad \rho \in (0, r], \tag{6.30}$$

whereas a random uniform direction is chosen according to the vector $\left[\frac{z_1'}{|\mathbf{z}'|}, \ldots, \frac{z_n'}{|\mathbf{z}'|}\right]^\top$. See also Sect. 6.4.1, on directional distributions. The example below shows how to draw points uniformly within a circle using polar coordinates. As we have seen in Chap. 2 (and we recall in Sect. 6.7), we have to compute the determinant of the Jacobian matrix of the polar transformation, that is $|\mathbf{J}| = \rho$ for $n = 2$. See Appendix D on generic polar transformations.

*Example 6.7* In order to draw points uniformly distributed within a circle ($\mathbf{x} \in \mathbb{R}^2$) [23],

$$\mathcal{B}_r = \{[x_1, x_2]^\top : x_1^2 + x_2^2 \leq r^2\},$$

we have to

1. Draw an angle $\theta' \sim \mathcal{U}([0, 2\pi])$.
2. Draw $\rho' \sim q(\rho) \propto \rho$, with $0 < \rho \leq r$.
3. Set $x_1 = \rho' \cos(\theta')$ and $x_2 = \rho' \sin(\theta')$.

Observe that the radius $\rho$ is not uniformly distributed in $(0, r]$, but it is distributed as an increasing linear pdf. Figure 6.3a depicts 1000 points uniformly distributed in

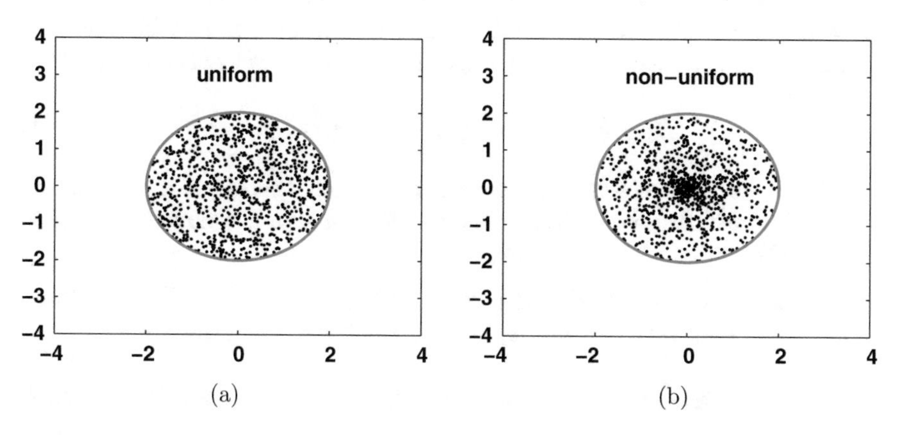


**Fig. 6.3** Points inside a circle of radius $r = 2$. **(a)** The points are uniformly distributed in the circle; $\theta' \sim \mathcal{U}([0, 2\pi])$ and $\rho' \sim q(\rho) \propto \rho$, with $0 < \rho \leq r$, in this case. **(b)** In this case, the points are non-uniformly distributed since the radius is (erroneously) chosen uniformly in $(0, r]$

a circle of radius $r = 2$. Choosing of $\rho$ uniformly in $(0, r]$ is a frequent mistake. Figure 6.3b shows the outputs in this case.

### 6.6.3 Points Uniformly Distributed Within a Hyperellipsoid

Consider now the set

$$\mathcal{E}_r = \{\mathbf{y} \in \mathbb{R}^n : \quad \mathbf{y}^\top \boldsymbol{\Sigma} \mathbf{y} \leq r^2\}, \qquad r > 0,$$

corresponding to a hyper-ellipsoid centered at the origin, where $\boldsymbol{\Sigma}$ is a positive definite $n \times n$ matrix that determines the "shape" of the set. It is possible to generate a sample $\mathbf{y}$ from the uniform distribution in $\mathcal{E}_r$ starting from a sample $\mathbf{x}$ from the uniform distribution in $\mathcal{B}_r$ of the same radius. In particular, let us consider the Cholesky decomposition of $\boldsymbol{\Sigma}^{-1}$, namely $\boldsymbol{\Sigma}^{-1} = \mathbf{A}\mathbf{A}^\top$ (which is well-defined if $\boldsymbol{\Sigma}$ is positive definite). If $\mathbf{x}$ is a uniform r.v. in $\mathcal{B}_r$, then $\mathbf{y} = \mathbf{A}\mathbf{x}$ is also a uniform random variable. Moreover,

$$\mathbf{y}^\top \boldsymbol{\Sigma} \mathbf{y} = (\mathbf{A}\mathbf{x})^\top \boldsymbol{\Sigma} (\mathbf{A}\mathbf{x}) = \mathbf{x}^\top \mathbf{A}^\top (\mathbf{A}\mathbf{A}^\top)^{-1} \mathbf{A}\mathbf{x} = \mathbf{x}^\top \mathbf{x} \leq r^2,$$

hence $\mathbf{y}$ is, indeed, a uniform r.v. in the hyperellipsoid $\mathcal{E}_r$.

## 6.7 Transformations of a Random Variable

In Chap. 2, we have described different transformations which map one (or many) realizations from an available random source into samples distributed according to a target distribution. Specifically, we have considered two random vectors $\mathbf{Z} = [Z_1, Z_2, \ldots, Z_m]^\top \in \mathbb{R}^m$ and $\mathbf{X} = [X_1, X_2, \ldots, X_n]^\top \in \mathbb{R}^n$ with joint pdfs $q(z_1, z_2, \ldots z_m)$ and $p_o(x_1, x_2, \ldots, x_n)$, respectively, and related by the transformation

$$\boldsymbol{\phi} = [\phi_1, \ldots, \phi_n]^\top : \mathbb{R}^m \to \mathbb{R}^n, \tag{6.31}$$

i.e., $\mathbf{X} = \boldsymbol{\phi}(\mathbf{Z})$ (where $\boldsymbol{\phi}$ is an injective differentiable function). We have already studied different cases depending on the values of $m$, $n$ or the monotonicity of $\boldsymbol{\phi}$. For example:

1. The case $m = n$ was discussed in Sect. 2.3.1 (and in Sect. 6.6, as well) for an invertible $\boldsymbol{\phi}$; the two joint pdfs, $p_o$ and $q$, are linked by the relationship

$$p_o(x_1, \ldots, x_n) = q\left(\phi_1^{-1}(x_1, \ldots, x_n), \ldots, \phi_n^{-1}(x_1, \ldots, x_n)\right) \left|\det \mathbf{J}^{-1}\right|, \tag{6.32}$$

where $|\mathbf{J}^{-1}|$ represents the determinant of the $n \times n$ Jacobian matrix of the inverse transformation $\boldsymbol{\phi}^{-1}$.

2. The case $m > n$ has been partially studied in Sect. 2.3.2 considering the specific value $n = 1$. Namely, we have considered a transformation $X = \phi(Z_1, \ldots, Z_m)$ invertible (at least) w.r.t. the variable $z_1$, so that the relationship

$$p_o(x) = \int_{\mathbb{R}} \cdots \int_{\mathbb{R}} q\left(\phi_1^{-1}(x, z_2, \ldots, z_m), z_2, \ldots, z_m\right) \left|\frac{\partial \phi_1^{-1}}{\partial x}\right| dz_2 \cdots dz_m, \tag{6.33}$$

holds.

In this section, we first extend the equality (6.33) to the case where $1 \leq n < m$. Then, we look into the case $n > m$, and finally we consider the problem of producing samples on differentiable manifolds, including some examples.

### 6.7.1 Many-to-Few Transformations (m > n)

Let us consider the system of equations

$$\begin{cases} X_1 = \phi_1(Z_1, Z_2, \ldots, Z_m), \\ \quad \vdots \\ X_n = \phi_n(Z_1, Z_2, \ldots, Z_m), \end{cases} \quad \text{with} \quad m > n, \tag{6.34}$$

where $[Z_1, Z_2, \ldots, Z_m]^\top$ has pdf $q(z_1, z_2, \ldots, z_m)$ and $m - n = k$. We extend this system adding $k$ new equations into (6.34) (obtaining an $m \times m$ system)

$$\begin{cases} X_1 &= \phi_1(Z_1, Z_2, \ldots, Z_m), \\ &\vdots \\ X_n &= \phi_n(Z_1, Z_2, \ldots, Z_m). \\ X_{n+1} &= Z_{n+1}, \\ &\vdots \\ X_{m=n+k} &= Z_{m=n+k}, \end{cases} \tag{6.35}$$

where $m - n$ auxiliary random variables

$$X_{n+1} = Z_{n+1}, \ldots, X_m = Z_m,$$

are chosen arbitrarily (i.e., the order of the $Z_i$'s is arbitrary, but the inverse transformation must exist).

We assume that the extended square ($m \times m$) system is invertible, i.e., $\det \mathbf{J} \neq 0$. Then, the inverse transformation exists, and it can be written as

$$\begin{cases} Z_1 &= \phi_1^{-1}(X_1, \ldots, X_n, X_{n+1}, \ldots, X_m), \\ &\vdots \\ Z_n &= \phi_n^{-1}(X_1, \ldots X_n, X_{n+1}, \ldots, X_m), \\ Z_{n+1} &= X_{n+1}, \\ &\vdots \\ Z_m &= X_m. \end{cases} \tag{6.36}$$

The Jacobian matrix of the inverse transformation is $\mathbf{J}^{-1}$. Thus, the joint pdf of $[X_1, \ldots, X_m]^\top$ can be obtained via (6.36) as

$$h(x_1, \ldots, x_m) = q\left(\phi_1^{-1}(x_1, \ldots, x_m), \ldots, \phi_n^{-1}(x_1, \ldots, x_m), x_{n+1}, \ldots, x_m\right) \left|\det \mathbf{J}^{-1}\right|,$$

and integrating out $x_{n+1}, \ldots, x_m$, we obtain the relationship

$$p_o(x_1, \ldots, x_n) = \int_{\mathbb{R}} \cdots \int_{\mathbb{R}} h(x_1, \ldots, x_m) dx_{n+1} \ldots dx_m. \tag{6.37}$$

Denoting $\mathbf{x} = [x_1, \ldots, x_n]^\top$, $\boldsymbol{\phi}^{-1} = [\phi_1^{-1}, \ldots, \phi_n^{-1}]^\top$ and $\mathbf{y} = [x_{n+1}, \ldots, x_m]^\top$, we can rewrite Eq. (6.37) in the compact form

$$p_o(\mathbf{x}) = \int_{\mathbb{R}^k} q(\boldsymbol{\phi}^{-1}(\mathbf{x}), \mathbf{y}) \left|\det \mathbf{J}^{-1}\right| d\mathbf{y}. \tag{6.38}$$

where $k = m - n$.

### 6.7.2 *Few-to-Many Transformations: Singular Distributions* ($m < n$)

Distributions that concentrate all their probability mass on a curve or a surface embedded in a higher dimensional space are called *singular distributions* [5]. Specifically, consider the transformation $\mathbf{X} = \boldsymbol{\phi}(\mathbf{Z})$, where $\mathbf{X} \in \mathbb{R}^n$, $\mathbf{Z} \in \mathbb{R}^m$, and $\boldsymbol{\phi} = [\phi_1, \ldots, \phi_n]^\top : \mathbb{R}^m \to \mathbb{R}^n$ with $n > m$, namely,

$$\begin{cases} X_1 = \phi_1(Z_1, Z_2, \ldots, Z_m), \\ \quad \vdots \\ X_n = \phi_n(Z_1, Z_2, \ldots, Z_m). \end{cases} \quad \text{with} \quad m < n, \tag{6.39}$$

where $\boldsymbol{\phi}$ is an injective differentiable function with full-rank Jacobian matrix $\mathbf{J}$, so that $\mathbf{J}^\top\mathbf{J}$ is invertible. Since $m < n$, the system of Eqs. (6.39) describes (in parametric form) a hypersurface of dimension $m$ embedded in $\mathbb{R}^n$,

$$\mathcal{H} = \{\mathbf{x} \in \mathbb{R}^n : \quad \mathbf{x} = \boldsymbol{\phi}(\mathbf{z}) \text{ for some } \mathbf{z} \in \mathbb{R}^m\} \subset \mathbb{R}^n,$$

where the $z_i$'s play the role of parameters. Thus, depending on the distribution of choice for the random vector $\mathbf{Z}$, the random vector $\mathbf{X}$ given by the transformation $\mathbf{X} = \boldsymbol{\phi}(\mathbf{Z})$ is distributed according to certain pdf on this $m$-dimensional hypersurface. Namely, the vector $\mathbf{X}$ has a pdf of the form

$$p_o(\mathbf{x}) = h(\mathbf{x})\mathbb{I}_{\mathcal{H}}(\mathbf{x}), \tag{6.40}$$

where

$$\mathbb{I}_{\mathcal{H}}(\mathbf{x}) = \begin{cases} 1 & \text{if} \quad x \in \mathcal{H}, \\ 0 & \text{if} \quad x \notin \mathcal{H}. \end{cases} \tag{6.41}$$

The function $h(\mathbf{x}) : \mathbb{R}^n \to \mathbb{R}$ is such that the (hyper) surface integral on $\mathcal{H}$ is

$$\begin{aligned} \int_{\mathcal{H}} h(\mathbf{x}) d\mathcal{H} &= \int_{\mathbb{R}^m} h(\boldsymbol{\phi}(\mathbf{z}))\sqrt{|\det \mathbf{J}^\top\mathbf{J}|}d\mathbf{z} \\ &= \int_{\mathbb{R}^m} p_o(\boldsymbol{\phi}(\mathbf{z}))\sqrt{|\det \mathbf{J}^\top\mathbf{J}|}d\mathbf{z} = 1. \end{aligned} \tag{6.42}$$

We have denoted with $d\mathcal{H}$ the infinitesimal hypersurface element and $\mathbf{J}$ is the $n \times m$ Jacobian matrix of the transformation $\boldsymbol{\phi}$, namely

$$\mathbf{J}(\mathbf{z}) = \begin{bmatrix} \frac{\partial \phi_1}{\partial z_1} & \frac{\partial \phi_1}{\partial z_2} & \cdots & \frac{\partial \phi_1}{\partial z_m} \\ \frac{\partial \phi_2}{\partial z_1} & \frac{\partial \phi_2}{\partial z_2} & \cdots & \frac{\partial \phi_2}{\partial z_m} \\ \vdots & \vdots & \vdots & \vdots \\ \frac{\partial \phi_n}{\partial z_1} & \frac{\partial \phi_n}{\partial z_2} & \cdots & \frac{\partial \phi_n}{\partial z_m} \end{bmatrix}. \tag{6.43}$$

For the sake of simplicity, we skip the dependence on $\mathbf{z}$ hereafter and write $\mathbf{J}$ instead of $\mathbf{J}(\mathbf{z})$. If $m = n$ then $\sqrt{|\det \mathbf{J}^\top\mathbf{J}|} = |\mathbf{J}|$ and we obtain the classical change-of-variables formula. From Eq. (6.42), we can write

$$\text{Prob}\{\mathbf{X} \in [\boldsymbol{\phi}(\mathbf{a}), \boldsymbol{\phi}(\mathbf{b})]\} = \text{Prob}\{\mathbf{Z} \in [\mathbf{a}, \mathbf{b}]\} = \int_{\mathbf{a}}^{\mathbf{b}} p_o(\boldsymbol{\phi}(\mathbf{z}))\sqrt{|\det \mathbf{J}^\top\mathbf{J}|}d\mathbf{z},$$

**Table 6.2** Special cases of $\sqrt{|\det \mathbf{J}^\top \mathbf{J}|}$ with a generic parametrization

| $m$ | $n$ | $\sqrt{\lvert\det \mathbf{J}^\top \mathbf{J}\rvert}$ |
|---|---|---|
| 1 | $\forall n \in \mathbb{N}$ | $\sqrt{\sum_{i=1}^{n} \left(\frac{\partial \phi_i}{\partial z}\right)^2}$ |
| 2 | $\forall n \in \mathbb{N}$ | $\sqrt{\left\lvert\left[\sum_{i=1}^{n} \left(\frac{\partial \phi_i}{\partial z_1}\right)^2\right]\left[\sum_{i=1}^{n} \left(\frac{\partial \phi_i}{\partial z_2}\right)^2\right] - \left[\sum_{i=1}^{n} \frac{\partial \phi_i}{\partial z_1}\frac{\partial \phi_i}{\partial z_2}\right]^2\right\rvert}$ |

where $[\mathbf{a}, \mathbf{b}] = [a_1, b_1] \times \ldots \times [a_m, b_m]$. Namely, the density of $\mathbf{Z}$ is

$$q(\mathbf{z}) = p_o(\boldsymbol{\phi}(\mathbf{z}))\sqrt{|\det \mathbf{J}^\top \mathbf{J}|}. \tag{6.44}$$

However, Eq. (6.44) is most useful when the goal is to produce samples from the r.v. $\mathbf{X} = \boldsymbol{\phi}(\mathbf{Z})$ on $\mathcal{H}$ with a prescribed distribution, $p_o(\mathbf{x})$. In this class of problems, what we need is to identify the pdf $q(\mathbf{z})$ that actually yields the desired form of $p_o(\mathbf{x})$, and that is precisely given in Eq. (6.44). Table 6.2 shows the analytic form of $\sqrt{|\det \mathbf{J}^\top \mathbf{J}|}$ for two special cases $m = 1$ and $m = 2$, for all $n \in \mathbb{N}$.

Let us denote as $\mathbf{z} = \boldsymbol{\phi}^{-1}(\mathbf{x}) : \mathcal{H} \to \mathbb{R}^m$ the inverse function of the transformation in Eq. (6.39). Thus, we also have

$$\begin{aligned} p_o(\mathbf{x}) &= q(\boldsymbol{\phi}^{-1}(\mathbf{x}))\sqrt{|\det \mathbf{J}^\top \mathbf{J}|^{-1}_{\boldsymbol{\phi}^{-1}(\mathbf{x})}}\,\mathbb{I}_{\mathcal{H}}(\mathbf{x}), \\ &= \frac{q(\boldsymbol{\phi}^{-1}(\mathbf{x}))}{\sqrt{|\det \mathbf{J}^\top \mathbf{J}|_{\boldsymbol{\phi}^{-1}(\mathbf{x})}}}\mathbb{I}_{\mathcal{H}}(\mathbf{x}), \end{aligned} \tag{6.45}$$

where $|\det \mathbf{J}^\top \mathbf{J}|^{-1}$ is evaluated at $\mathbf{z} = \boldsymbol{\phi}^{-1}(\mathbf{x})$ (and we have used the property $\det \mathbf{A}^{-1} = (\det \mathbf{A})^{-1}$, of a generic invertible square matrix $\mathbf{A}$). Note that $\boldsymbol{\phi}^{-1}$ is well-defined because $\boldsymbol{\phi}$ is injective. Thus, if we restrict its image to the manifold $\mathcal{H}$, $\boldsymbol{\phi} : \mathbb{R}^m \to \mathcal{H}$, it becomes bijective.

*Example 6.8* Consider the curve $\mathcal{H}$ in $\mathbb{R}^2$ with the following parametric form

$$\begin{cases} X_1 = r\cos(Z), \\ X_2 = r\sin(Z), \end{cases} \quad \text{with} \quad Z \in [0, 2\pi), r > 0, \tag{6.46}$$

that describes a circle with radius $r$. If $Z$ is uniformly distributed in $[0, 2\pi)$, i.e.,

$$q(z) = \frac{1}{2\pi}\mathbb{I}_{[0,2\pi)}(z),$$

and since

$$\sqrt{|\det \mathbf{J}^\top \mathbf{J}|} = r, \qquad \sqrt{|\det \mathbf{J}^\top \mathbf{J}|^{-1}} = \frac{1}{r},$$

then, from Eqs. (6.44) and (6.45), we have

$$p_o(x_1, x_2) = \frac{1}{2\pi r} \mathbb{I}_{\mathcal{H}}(x_1, x_2),$$

where $\mathcal{H}$ is the circle described in the system above. Actually, $[X_1, X_2]^\top$ is uniformly distributed on $\mathcal{H}$. Furthermore, note that $\int_0^{2\pi} \sqrt{|\det \mathbf{J}^\top \mathbf{J}|} dz = 2\pi r$ is the length of the circumference.

### 6.7.3 *Sampling a Uniform Distribution on a Differentiable Manifold*

Given the arbitrary parametrization $\mathbf{Z} \in [\mathbf{a}, \mathbf{b}] = [a_1, b_1] \times \ldots \times [a_m, b_m]$, we assume that

$$|\mathcal{H}_{\mathbf{a},\mathbf{b}}| = \int_{[\mathbf{a},\mathbf{b}]} \sqrt{|\det \mathbf{J}^\top \mathbf{J}|} d\mathbf{z} < \infty,$$

where $|\mathcal{H}_{\mathbf{a},\mathbf{b}}|$ denotes the Lebesgue measure of the manifold

$$\mathcal{H}_{\mathbf{a},\mathbf{b}} = \{\mathbf{x} \in \mathbb{R}^n : \mathbf{x} = \boldsymbol{\phi}(\mathbf{z}) \text{ for some } \mathbf{z} \in [\mathbf{a}, \mathbf{b}]\}.$$

The uniform density on $\mathcal{H}_{\mathbf{a},\mathbf{b}}$ can be written as

$$p_o(\mathbf{x}) = \frac{1}{|\mathcal{H}_{\mathbf{a},\mathbf{b}}|} \mathbb{I}_{\mathcal{H}_{\mathbf{a},\mathbf{b}}}(\mathbf{x}), \tag{6.47}$$

where, clearly, $|\mathcal{H}_{\mathbf{a},\mathbf{b}}|$ is a constant. Replacing Eq. (6.47) into Eq. (6.44), it is straightforward to see that we can obtain a pdf $p_o(\mathbf{x})$ of this type if we choose

$$q(\mathbf{z}) \propto \sqrt{|\det \mathbf{J}^\top \mathbf{J}|}, \tag{6.48}$$

as the density of the random parameter $\mathbf{Z}$. Therefore, in order to draw points uniformly on $\mathcal{H}_{\mathbf{a},\mathbf{b}}$ we can take the steps below:

1. Draw $\mathbf{z}'$ from $q(\mathbf{z}) \propto \sqrt{|\det \mathbf{J}^\top \mathbf{J}|}$.
2. Set $\mathbf{x}' = \boldsymbol{\phi}(\mathbf{z}')$.

*Example 6.9* Consider a curve $\mathcal{H}$ in $\mathbb{R}^2$ (with finite length) described by the function $x_2 = \psi(x_1)$, where $\psi$ is invertible and $x_1 \in \mathcal{D}_1 \subseteq \mathbb{R}$ (and $x_2 = \psi(x_1) \in \mathcal{D}_2 \subseteq \mathbb{R}$). The curve is expressed using the Cartesian parametrization. In order to draw uniformly from the curve, we can:

1. Draw $x_1' \sim g_1(x_1) \propto \sqrt{1 + \left(\frac{d\psi(x_1)}{dx_1}\right)^2}$,
2. Set $x_2' = \psi(x_1')$.
3. Return the point $(x_1', x_2')$.

The point $(x_1', x_2')$ belongs to the curve described by $x_2 = \psi(x_1)$ in $\mathbb{R}^2$ and is uniformly distributed on it, i.e.,

$$p_o(x_1, x_2) = \frac{1}{|\mathcal{H}|}\mathbb{I}_{\mathcal{H}}(x_1, x_2),$$

where

$$\begin{aligned}|\mathcal{H}| = \int_{\mathcal{H}} d\mathcal{H} &= \int_{\mathcal{D}_1} \sqrt{1 + \left(\frac{d\psi}{dx_1}\right)^2} dx_1, \\ &= \int_{\mathcal{D}_2} \sqrt{1 + \left(\frac{d\psi^{-1}}{dx_2}\right)^2} dx_2.\end{aligned}$$

Indeed, in this case $x_1$ plays the role of a parameter in the system

$$\begin{aligned} x_1 &= z, \\ x_2 &= \psi(z), \end{aligned} \tag{6.49}$$

i.e., $\phi_1(z) = z$, $\phi_2(z) = \psi(z)$ and, as a consequence, $q(x_1) = g(x_1)$. The marginal pdf of $X_1$ is

$$q(x_1) = g_1(x_1) \propto \sqrt{1 + \left(\frac{d\psi}{dx_1}\right)^2},$$

by construction. Thus, using Eq. (6.45), we have

$$\begin{aligned} p_o(x_1, x_2) &= p_o(x_1, \psi(x_1))\mathbb{I}_{\mathcal{H}}(x_1, x_2), \\ &= g_1(x_1)\frac{1}{\sqrt{1 + \left(\frac{d\psi}{dx_1}\right)^2}}\mathbb{I}_{\mathcal{H}}(x_1, x_2), \\ &\propto \mathbb{I}_{\mathcal{H}}(x_1, x_2). \end{aligned}$$

The marginal pdf of $X_2$ is clearly given by the expression of a transformation of a r.v., i.e.,

$$g_2(x_2) = g_1(\psi^{-1}(x_2)) \left| \frac{d\psi^{-1}}{dx_2} \right|. \tag{6.50}$$

Hence, replacing $g_1(x_1)$ in the formula above, we have

$$g_2(x_2) \propto \sqrt{1 + \left( \frac{d\psi}{dx_1}\bigg|_{\phi^{-1}(x_2)} \right)^2} \left| \frac{d\psi^{-1}}{dx_2} \right|.$$

Finally, recalling that $\frac{d\psi}{dx_1}\Big|_{\psi^{-1}(x_2)} = \left(\frac{d\psi^{-1}}{dx_2}\right)^{-1}$, then we can write

$$g_2(x_2) \propto \sqrt{\frac{\left(\frac{d\psi^{-1}}{dx_2}\right)^2 + 1}{\left(\frac{d\psi^{-1}}{dx_2}\right)^2}} \left| \frac{d\psi^{-1}}{dx_2} \right| = \sqrt{1 + \left( \frac{d\psi^{-1}}{dx_2} \right)^2}.$$

The expression above suggests that clearly we can also draw points uniformly from the curve $x_2 = \psi(x_1)$, considering $x_1 = \psi^{-1}(x_2)$ and taking $x_2$ as a parameter.

*Example 6.10* Consider the ellipse described by the equation

$$x_1^2 + \frac{1}{9}x_2^2 = 1.$$

It can be parameterized as $x_1 = \cos(z)$, $x_2 = 3\sin(z)$, with $z \in [0, 2\pi]$. Hence, to draw uniformly from this ellipse we can:

1. Draw $z'$ from

$$q(z) \propto \sqrt{\left(\frac{dx_1}{dz}\right)^2 + \left(\frac{dx_2}{dz}\right)^2} = \sqrt{\sin^2(z) + 9\cos^2(z)}.$$

2. Then set $x_1' = \cos(z')$, $x_2' = 3\sin(z')$.

Figure 6.4a depicts 200 points generated by the previous algorithm.

*Example 6.11* Consider now the following spiral in parametric form

$$x_1 = z\cos(2\pi z), \quad x_2 = z\sin(2\pi z), \quad z \in [0, 2\pi].$$

In this case the pdf $q(z)$, after some simple manipulations, can be expressed as

$$q(z) \propto \sqrt{\left(\frac{dx_1}{dz}\right)^2 + \left(\frac{dx_2}{dz}\right)^2} = \sqrt{4\pi^2 z^2 + 1}. \tag{6.51}$$

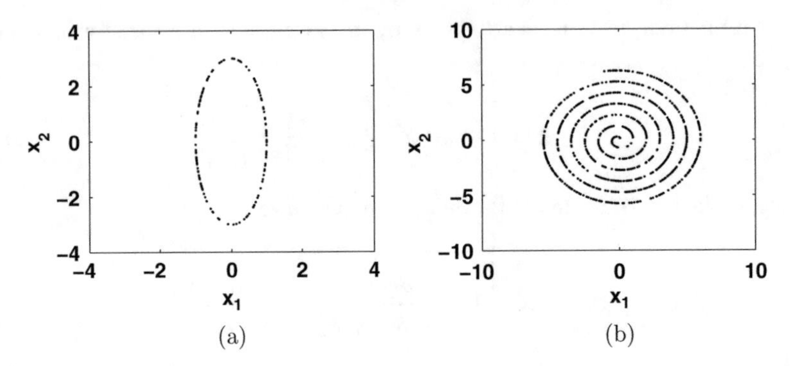


**Fig. 6.4** **(a)** Points (200) uniformly distributed on an ellipse. **(b)** Points (2000) uniformly distributed on an spiral

We can use the previous procedure to yield points uniformly on the spiral. Figure 6.4b shows 2000 samples.

## 6.8 Sampling Techniques for Specific Distributions

This section provides a collection of sampling algorithms for specific multivariate distributions which appear frequently in various applications [4, 9, 15].

### *6.8.1 Multivariate Gaussian Distribution*

Consider a mean vector $\boldsymbol{\mu} = [\mu_1, \ldots, \mu_n] \in \mathbb{R}^n$ and an $n \times n$ covariance matrix $\boldsymbol{\Sigma}$. The multivariate Gaussian density has the form

$$p_o(\mathbf{x}) = (2\pi)^{-\frac{n}{2}} |\det \boldsymbol{\Sigma}|^{-\frac{1}{2}} \exp\left(-\frac{1}{2}(\mathbf{x} - \mu)^\top \boldsymbol{\Sigma}^{-1}(\mathbf{x} - \mu)\right).$$

The following procedure generates a sample $\mathbf{x}'$ from a generic multivariate Gaussian distribution, i.e., $\mathbf{x}' \sim \mathcal{N}(\boldsymbol{\mu}, \boldsymbol{\Sigma})$:

1. Compute a factor $\mathbf{A}$ such that $\boldsymbol{\Sigma} = \mathbf{A}\mathbf{A}^\top$.
2. Draw $v_k \sim \mathcal{N}(0, 1)$, for $k = 1, \ldots, n$.
3. Set $\mathbf{x}' = \boldsymbol{\mu} + \mathbf{A}\mathbf{v}$ where $\mathbf{v} = [v_1, \ldots, v_n]^\top$.

The computation of the matrix $\mathbf{A}$ can be carried out in different manners. The Cholesky factorization and the eigendecomposition are the most typical procedures [10]:

- *Cholesky decomposition:* any positive definite matrix $\mathbf{\Sigma}$ can be factorized as $\mathbf{\Sigma} = \mathbf{L}\mathbf{L}^\top$, where $\mathbf{L}$ is a lower triangular matrix. Hence, we can set $\mathbf{A} = \mathbf{L}$.
- *Spectral decomposition:* any symmetric matrix $\mathbf{\Sigma}$ can be decomposed as $\mathbf{\Sigma} = \mathbf{U}\mathbf{\Lambda}\mathbf{U}^\top$, where $\mathbf{U}$ is a unitary matrix (i.e., $\mathbf{U}\mathbf{U}^\top = \mathbf{I}$) whose columns are eigenvectors, and $\mathbf{\Lambda}$ is a diagonal matrix containing the eigenvalues. In this case, we set $\mathbf{A} = \mathbf{U}\mathbf{\Lambda}^{\frac{1}{2}}$.

### 6.8.2 Multivariate Student's t-Distribution

A multivariate Student's $t$ density with $\nu$ degrees of freedom, location parameter $\boldsymbol{\mu}$, and scale matrix $\mathbf{\Sigma}$ is defined as

$$p_o(\mathbf{x}) \propto \left(1 + \frac{(\mathbf{x} - \mu)^\top \mathbf{\Sigma}^{-1}(\mathbf{x} - \mu)}{\nu}\right)^{-\frac{\nu+n}{2}}.$$

A r.v. $X$ distributed as a univariate Student's $t$ pdf, $p_o(x)$ with $x \in \mathbb{R}$, $\mu = 0$ and $\nu > 0$, can be expressed as the ratio between a Gaussian r.v. $Z$ and the square root of a chi-square r.v. $Y$ with $\nu$ degrees of freedom, more specifically, $X = \sqrt{\nu}\frac{Z}{\sqrt{Y}}$. Hence, a sampling technique for generating a multivariate Student's $t$-Distribution can follow the steps below:

1. Compute the matrix $\mathbf{A}$ such that $\mathbf{\Sigma} = \mathbf{A}\mathbf{A}^\top$.
2. Draw $v'_k \sim \mathcal{N}(0, 1)$, for $k = 1, \ldots, n$ and define $\mathbf{v}' = [v'_1, \ldots, v'_n]^\top$.
3. Draw $z' \sim q(x) \propto x^{\frac{\mu}{2}-1}e^{-x}$.
4. Set $\mathbf{x}' = \boldsymbol{\mu} + \frac{\sqrt{\nu}}{z'}\mathbf{A}\mathbf{v}'$.

### 6.8.3 Wishart Distribution

Consider a positive definite $n \times n$ matrix $\mathbf{X} = \{x_{i,j}\}$ with $i, j = 1, \ldots, n$. The Wishart distribution is defined on the space of positive definite matrices in $\mathbb{R}^{n\times n}$. The corresponding pdf is

$$p_o(\mathbf{X}) \propto |\det \mathbf{X}|^{\frac{\nu-n-1}{2}} \exp\left(-\frac{1}{2}\mathrm{trace}(\mathbf{\Sigma}^{-1}\mathbf{X})\right), \tag{6.52}$$

where $\nu \geq n$ is the number of degrees of freedom and $\mathbf{\Sigma}$ is an $n \times n$ covariance matrix. The Wishart distribution is often interpreted as a multivariate extension of the Gamma distribution. It is frequently used in Bayesian statistics because it is the conjugate prior for the precision matrix $\mathbf{\Sigma}^{-1}$ of an $n$-dimensional Gaussian distribution $\mathcal{N}(\mathbf{0}, \mathbf{\Sigma})$.

When $\nu$ is an integer, the Wishart distribution represents the sums of squares (and cross-products) of $n$ draws from a multivariate Gaussian distribution. Specifically, given $L$ random vectors $\mathbf{Y}_i \sim \mathcal{N}(\mathbf{0}, \mathbf{\Sigma})$, $i = 1, \ldots, L$, of dimension $n$, the matrix

$$\mathbf{X}' = \sum_{i=1}^{L} \mathbf{Y}_i^\top \mathbf{Y}_i,$$

has a Wishart density with $\nu = L$ degrees of freedom and $n \times n$ scale matrix $\mathbf{\Sigma}$. Then, trivially we can derive the following sampling method:

1. Draw $L$ multivariate Gaussian samples $\mathbf{y}_i = [y_{i,1}, \ldots, y_{i,n}]^\top \sim \mathcal{N}(\mathbf{0}, \mathbf{\Sigma})$, $i = 1, \ldots, L$.
2. Set $\mathbf{x}' = \sum_{i=1}^{L} \mathbf{y}_i^\top \mathbf{y}_i$.

The previous procedure is inefficient, especially for large $\nu = L$. An alternative algorithm is based on the so-called *Bartlett decomposition* [5, 15]:

1. Compute an $n \times n$ lower triangular matrix $\mathbf{A}$ via the Cholesky decomposition, such that $\mathbf{\Sigma} = \mathbf{A}\mathbf{A}^\top$.
2. Draw independently the $\frac{n(n+1)}{2}$ samples, $c'_{ii} \sim q(c) \propto c^{\nu-i+1} e^{-\frac{c^2}{2}}$ and $c'_{ij} \sim \mathcal{N}(0, 1)$, with $i = 1, \ldots, n$ and $j = 1, \ldots, i-1$. Then construct the lower triangular matrix

$$\mathbf{C} = \begin{bmatrix} c'_{11} & 0 & \cdots & 0 \\ c'_{21} & c'_{22} & \cdots & 0 \\ \vdots & \vdots & \vdots & \vdots \\ c'_{n1} & c'_{n2} & \cdots & c'_{nn} \end{bmatrix}, \tag{6.53}$$

3. Return $\mathbf{X}' = \mathbf{A}\mathbf{C}\mathbf{C}^\top\mathbf{A}^\top$.

### 6.8.4 Inverse Wishart Distribution

Consider a positive definite $n \times n$ matrix $\mathbf{V} = \{v_{i,j}\}$ with $i, j = 1, \ldots, n$. The *inverse* Wishart distribution is defined as

$$p_o(\mathbf{V}) \propto |\det \mathbf{V}|^{-\frac{\nu+n+1}{2}} \exp\left(-\frac{1}{2}\text{trace}(\mathbf{\Phi}\mathbf{V}^{-1})\right), \quad \nu \geq n, \tag{6.54}$$

where $\mathbf{\Phi}$ is an $n \times n$ (precision) matrix. The inverse Wishart pdf is used as conjugate prior over the covariance matrix $\mathbf{\Sigma}$ of a multivariate Gaussian likelihood function. In the univariate case $n = 1$, the inverse Wishart distribution becomes the *inverse-Gamma distribution*.

A possible procedure to generate a sample matrix $\mathbf{V}'$ with inverse Wishart distribution is to draw a matrix $\mathbf{W}$ with Wishart density, with parameters $\nu$ and $\boldsymbol{\Sigma} = \boldsymbol{\Phi}^{-1}$, and then set $\mathbf{V}' = \mathbf{W}^{-1}$.

### 6.8.5 Multivariate Gamma Samples

In many applications, there is a need of generating non-negative random vectors and Gamma random variables are often considered. Let us define an incidence matrix $\mathbf{T}$ of dimensions $m \times n$, consisting of 0 and 1 entries. A simple way to generate multidimensional Gamma variates is the following [4, 28]:

1. Draw $v_j \sim \mathcal{G}(\alpha_1, 1)$, with $j = 1, \ldots, m$, i.e., with pdf $q(v) \sim x^{\alpha_1 - 1} e^{-x}$.
2. Draw $z_k \sim \mathcal{G}(\alpha_2, 1)$, with $k = 1, \ldots, n$, i.e., with pdf $h(z) \sim x^{\alpha_2 - 1} e^{-x}$.
3. Let $\mathbf{v}' = [v_1, \ldots, v_m]^\top$ and $\mathbf{z}' = [z_1, \ldots, z_n]^\top$, then return

$$\mathbf{g}' = \mathbf{v}' + \mathbf{T}\mathbf{z}'.$$

Note that every element in the vector $\mathbf{g}'$ is a sum of independent Gamma r.v.'s, and the marginal pdf of each component is also a Gamma pdf. Clearly, different choices of the matrix $\mathbf{T}$ yield different correlation structures among the components of $\mathbf{g}'$ [28].

### 6.8.6 Dirichlet Distribution

Let $a_1, \ldots, a_{n+1}$ be positive scalar values (called concentration parameters). The corresponding *Dirichlet distribution* has pdf

$$p_o(x_1, \ldots, x_n) \propto x_1^{a_1 - 1} x_2^{a_2 - 1} \cdots x_n^{a_n - 1} (1 - x_1 - x_2 \cdots - x_n)^{a_{n+1} - 1},$$

defined over the simplex described by the inequalities $x_i > 0$, for all $i = 1, \ldots, n$, and $\sum_{i=1}^n x_i < 1$. In Bayesian statistics the Dirichlet distribution is often used as a prior over probability mass functions $\{p_i\}_{i=1}^{n+1}$ (defined as $p_i = x_i$, $\forall i$, and $p_{n+1} = 1 - x_1 - x_2 \cdots - x_n$) [5, 7, 8, 27]. It can be seen as a multivariate generalization of the Beta distribution. A simple procedure to draw from a Dirichlet distribution involves the generation of Gamma samples:

1. Draw $z_k' \sim q(z_k) \propto z_k^{a_k - 1} e^{-z_k}$, i.e., from a Gamma distribution with parameter $a_k$ and unit scale parameter, $\mathcal{G}(a_k, 1)$, for $k = 1, \ldots, n+1$.
2. Set $s = \sum_{k=1}^{n+1} z_k'$.
3. Return $x_i' = \frac{z_i'}{s}$ for $i = 1, \ldots, n$.

### *6.8.7 Cook-Johnson's Family*

The so-called *Cook-Johnson uniform distribution* [2, 15] has cdf

$$F_{\mathbf{X}}(x_1, \ldots, x_n) = \left( \sum_{i=1}^{n} x_i^{-\frac{1}{a}} + 1 - n \right)^{-a}, \quad a > 0,\ 0 \leq x_i \leq 1, \tag{6.55}$$

and density

$$p_o(x_1, \ldots, x_n) \propto \prod_{i=1}^{n} x_i^{-\frac{1}{a}-1} \left( \sum_{i=1}^{n} x_i^{-\frac{1}{a}} + 1 - n \right)^{-a-n}, \quad a > 0,\ 0 \leq x_i \leq 1,$$

This distribution is invariant under permutations of the components $x_i$'s and it has marginal uniform pdfs so that it can be interpreted as a multivariate generalization of the uniform distribution or as a copula function. A sampling procedure for this distribution is given below:

1. Draw $z_i \sim \exp(-z)$, for $i = 1, \ldots, n$.
2. Draw $v \sim \mathcal{G}(a, 1)$.
3. Set $x_i = \left(1 + \frac{z_i}{v}\right)^{-a}$, for $i = 1, \ldots, n$.

If we denote the Cook-Johnson cdf as

$$C(u_1, \ldots, u_n) = F_{\mathbf{X}}(x_1, \ldots, x_n),$$

and use it as a copula function, different well-known multivariate distributions can be built as $C(F_{X_1}(x_1), \ldots, F_{X_n}(x_n))$ where $F_{X_i}(x_i)$ is the marginal cdf of the $i$th component $x_i$ of the new distribution. A straightforward sampling method in this case consists of the steps below:

1. Draw $z_i \sim \exp(-z)$, for $i = 1, \ldots, n$.
2. Draw $v \sim \mathcal{G}(a, 1)$.
3. Set $u_i = \left(1 + \frac{z_i}{v}\right)^{-a}$, for $i = 1, \ldots, n$.
4. Set $x_i = F_{X_i}^{-1}(u_i)$, for $i = 1, \ldots, n$.

Examples of distributions that can be written (and sampled) using the Cook-Johnson copula include the following [2, 15]:

- The multivariate Burr pdf

$$p_o(x_1, \ldots, x_n) \propto \left[ \prod_{i=1}^{n} d_i c_i x_i^{c_i - 1} \right] \left( 1 + \sum_{i=1}^{n} d_i x_i^{c_j} \right)^{-a-n}, \quad a, c_i, d_i > 0,\ x_i > 0,$$

where the $i$th marginal cdf and its inverse, are

$$F_{X_i}(x_i) = 1 - \left(1 + d_i x_i^{c_i}\right)^{-a}, \quad F_{X_i}^{-1}(u_i) = \left(d_i^{-1}\left[\left(1 - u_i^{-\frac{1}{a}}\right) - 1\right]\right)^{\frac{1}{c_i}}.$$

- The multivariate Pareto pdf

$$p_o(x_1, \ldots, x_n) \propto \left[\prod_{i=1}^{n} \theta_i\right]^{-1} \left(1 + \sum_{i=1}^{n} \theta_i^{-1} x_i + 1 - n\right)^{-a-n}, \quad x_i > \theta_i > 0,$$

where its marginal cdf, and its inverse, are

$$F_{X_i}(x_i) = 1 - \left(\frac{\theta_i}{x_i}\right)^{a}, \quad F_{X_i}^{-1}(u_i) = \theta_i \left(1 - u_i\right)^{-\frac{1}{a}}.$$

- Gumbel's multivariate logistic distribution, which can be described by the joint cdf

$$F_{\mathbf{X}}(x_1, \ldots, x_n) = \left(1 + \sum_{i=1}^{n} e^{-x_i}\right)^{-a}, \qquad a > 0, x_i > 0, \quad i = 1, \ldots, n.$$

and the corresponding pdf

$$p_o(x_1, \ldots, x_n) \propto \left[\prod_{i=1}^{n} e^{-x_i}\right] \left(1 + \sum_{i=1}^{n} e^{-x_i}\right)^{-a-n}.$$

The $i$th marginal cdf and its inverse are

$$F_{X_i}(x_i) = \left[1 + \exp(-x_i)\right]^{-a}, \quad F_{X_i}^{-1}(u_i) = -\log(u_i^{-\frac{1}{a}} - 1), \quad i = 1, \ldots, n.$$

## 6.8.8 Some Relevant Bivariate Distributions

The densities considered in this section have attracted some attention in the literature [4, 9, 15, 24]. Note that several families of bivariate pdfs can be constructed by way of the copula function approach (see Sect. 6.3.2).

### Morgenstern's Distribution

The density

$$p_o(x_1, x_2) \propto 1 + \alpha(2x_1 - 1)(2x_2 - 1), \qquad 0 \le x_1, x_2 \le 1, \quad |\alpha| \le 1,$$

has been studied in [5, 15]. The marginal distributions are both uniform in $[0, 1]$ and the dependence properties are controlled by the parameter $\alpha$ (hence, its cdf can be used as a copula function). The linear correlation coefficient is $-\frac{\alpha}{3}$. The method based on the chain rule decomposition, described in Sect. 6.3.1, can be used in this case. Namely, we can use the formula

$$p_o(x_1, x_2) = q(x_2|x_1)f_1(x_1),$$

and the method consists of drawing $x'_1$ from $f_1(x_1)$ and then $x'_2 \sim q(x_2|x'_1)$. We already know that the marginal density $f_1(x_1)$ is uniform in $[0, 1]$. Additionally, $q(x_2|x_1)$ is a linear function (given $x_1$) that can be sampled, e.g., by the inversion method. For instance, since $q(x_2|x_1)$ is a trapezoid density, a possible sampling algorithm is the following:

1. Draw $x'_1 \sim \mathcal{U}([0, 1])$.
2. Draw $u' \sim \mathcal{U}([0, 1])$ and $v' \sim \mathcal{U}([0, 1])$.
3. Set

$$x'_2 = \begin{cases} \min\left[u', -\frac{v'}{\alpha(2x'_1-1)}\right], & \text{if } x'_1 < \frac{1}{2}, \\ \max\left[u', 1 - \frac{v'}{\alpha(2x'_1-1)}\right], & \text{if } x'_1 \geq \frac{1}{2}. \end{cases} \tag{6.56}$$

4. Return $[x'_1, x'_2]^\top$.

### Ali-Mikhail-Haq's Distribution

Consider the following bivariate cumulative distribution [4, 15]

$$F_{\mathbf{X}}(x_1, x_2) = \frac{x_1 x_2}{1 - \alpha(1 - x_1)(1 - x_2)}, \qquad 0 \leq x_1, x_2 \leq 1,$$

with $|\alpha| \leq 1$. The marginal pdfs $f_i(x_i)$ are both uniform in $[0, 1]$, hence this cdf is a copula function as well. The chain rule decomposition method can be applied again. Indeed, one conditional cumulative function is

$$F_{X_2|X_1}(x_2|x_1) = \frac{x_2[1 - \alpha(1 - x_2)]}{(1 - \alpha(1 - x_1)(1 - x_2))^2}, \tag{6.57}$$

and the equation $F_{X_2|X_1}(x_2|x_1) = k$ with $k \in [0, 1]$ has one solution in the interval $[0, 1]$. This solution can be computed analytically, so that the inversion method can be applied.

### Gumbel's Bivariate Exponential Distribution

Consider the density [4, 15]

$$p_o(x_1,x_2) = [(1+\theta x_1)(1+\theta x_2)-\theta]\exp(-x_1-x_2-\theta x_1x_2), \quad x_1,x_2>0, \qquad (6.58)$$

with $0 \leq \theta \leq 1$. Both marginal pdfs are exponential densities, i.e., $f_i(x_i) = \exp(-x_i)$, with $i = 1,2$. The two components are independent if $\theta = 0$. It is also possible to show that the maximum linear correlation between the two components is obtained for $\theta = 1$, reaching the value $-0.43$ (i.e., weak negative dependence). The conditional pdf of $x_2$ given $x_1$ is

$$q(x_2|x_1) = \frac{p_o(x_1,x_2)}{f_1(x_1)} = [(1+\theta x_1)(1+\theta x_2)-\theta]\exp(-x_2-\theta x_1x_2),$$

that can be rewritten (after some manipulations) as

$$q(x_2|x_1) = w\,[\lambda\exp(-\lambda x_2)] + (1-w)\left[\lambda^2 x_2\exp(-\lambda x_2)\right], \qquad (6.59)$$

where $\lambda = 1+\theta x_1$ and $w = \frac{\lambda-\theta}{\lambda}$. Equation (6.59) represents a mixture, with weight $w$, of an exponential pdf with parameter $\lambda$, and a Gamma pdf (we denote the corresponding distribution as $\mathcal{G}(2,\lambda)$). Hence, since $p_o(x_1,x_2) = q(x_2|x_1)f_1(x_1)$, we can draw from the pdf in Eq. (6.58) by sampling $x_1' \sim \exp(-x_i)$ first, using the inversion method, and then draw a sample $x_2' \sim q(x_2|x_1')$. Samples from the Gamma distribution $\mathcal{G}(2,\lambda)$ can be generated as a sum of two independent exponential r.v.'s with parameter $\lambda$ [5].

### Wrapped Cauchy Distribution

Consider the pdf [15]

$$p_o(r,\theta) \propto r\exp\left(-\frac{r^2}{2}\right)\frac{(1-a^2)}{2\pi(1+a^2-2a\cos\theta)}, \quad \text{with} \quad |a|\leq 1.$$

If $r$ and $\theta$ are interpreted as polar coordinates for a complex r.v., then the alternative density

$$p_o(x_1,x_2) \propto \exp\left(-\frac{x_1^2+x_2^2}{2}\right)\frac{(1-a^2)}{2\pi(1+a^2-2ax_1\left(x_1^2+x_2^2\right)^{-1/2})},$$

where $\cos\theta = \frac{x_1}{\sqrt{x_1^2+x_2^2}}$ is fully equivalent, with $x_1$, $x_2$ being the Cartesian coordinates for the complex r.v.

This distribution results from the “wrapping” of the Cauchy distribution around the unit circle. It is often used in *directional statistics*: a branch of statistics that studies directions (unit vectors in $\mathbb{R}^n$) and rotations in $\mathbb{R}^n$, for instance. More generally, directional statistics deals with observed data on compact Riemannian manifolds. A possible generation procedure is given below:

1. Draw $r' \sim q(r) \propto r \exp\left(-\frac{r^2}{2}\right)$, which is a Rayleigh pdf. It is possible to draw from a Rayleigh pdf by the inversion method: draw a sample $v' \sim \mathcal{U}([0, 1])$ and then set $r' = \sqrt{-2 \log v'}$.
2. Draw $u' \sim \mathcal{U}([0, 1])$.
3. Set $\theta' = \tan(\pi u' - \frac{1}{2}) \mod (2\pi)$.
4. Return the pair$(r', \theta')$.

### Von Mises Distribution

Another distribution often used in directional statistics is the von Mises distribution [9, 15]. The corresponding pdf can be written as

$$p_o(\theta) \propto \exp(k \cos \theta), \qquad 0 \leq \theta < 2\pi,$$

where $k > 0$ is a constant value. A rejection sampling method can be applied using a uniform proposal pdf (since $\exp(k) \geq \exp(k \cos \theta)$):

1. Draw $u_1, u_2 \sim \mathcal{U}([0, 1])$.
2. If $u_2 \leq \frac{\exp(k \cos 2\pi u_1)}{\exp(k)} = \exp(k(\cos 2\pi u_1 - 1))$ then set $\theta' = 2\pi u_1$. Otherwise, reject $u_1$ and repeat from step 1.

## 6.9 Generation of Stochastic Processes

In this section, we describe different sampling techniques for generating random vectors that constitute a finite representation of a continuous-time random processes [9, 14, 17, 21]. Each finite realization of a stochastic process contains a sequence of dependent random variables with a specific structure. In general, we consider random processes $\{X_t\}$ with $t \in \mathbb{R}$ (continuous-time), with the exception of Sect. 6.9.6 where the time variable is discrete. The r.v.’s $X_t$ forming the stochastic process can take continuous values ($X_t \in \mathbb{R}$) or only discrete values ($X_t \in \mathbb{N}$; see Sects. 6.9.5 and 6.9.1).

### *6.9.1 Markov Jump Processes*

A *Markov jump process* (MJP) is a Markov process with a continuous-time index and a countable, i.e., discrete, state space. To be specific, a MJP is a continuous-time stochastic process, $\{X_t\}$, with $t \in \mathbb{R}$, $X_t \in \mathcal{X} \subseteq \mathbb{N}$, and the Markov property [1, 17, 21]

$$p(x_{t+\tau}|x_s, s \leq t) = p(x_{t+\tau}|x_t), \qquad \forall s, \tau, t \in \mathbb{R}.$$

There also exist several stochastic processes with a continuous state space, i.e., $X_t \in \mathcal{X} \subseteq \mathbb{R}$, and where the Markov property is satisfied (e.g., the *Wiener processes*). In this section, we focus on the case $\mathcal{X} \subseteq \mathbb{N}$. A time-homogenous MJP is usually defined using the so-called $Q$-matrix,

$$\mathbf{Q} = \{q_{i,j}\} = \begin{bmatrix} -q_1 & q_{1,2} & q_{1,3} & \cdots \\ q_{2,1} & -q_2 & q_{2,3} & \cdots \\ q_{3,1} & q_{3,2} & -q_3 & \cdots \\ \vdots & \vdots & \vdots & \ddots \end{bmatrix},$$

where

$$\begin{aligned} q_{i,j} &= \lim_{\tau \to 0} \frac{p(x_{t+\tau} = j|x_t = i)}{\tau}, & i \neq j, \quad i,j \in \mathcal{X}, \\ q_{i,i} &= -q_i = -\lim_{\tau \to 0} \frac{1 - p(x_{t+\tau} = i|x_t = i)}{\tau}, & i \in \mathcal{X}. \end{aligned} \tag{6.60}$$

These coefficients $q_{i,j}$ are finite and positive, i.e., $0 \leq q_{i,j} < \infty$, and such that

$$q_i = \sum_{j \neq i} q_{i,j}.$$

Hence, the sum of the coefficients in a row of $\mathbf{Q}$ is zero. The process is called homogenous if the entries of $\mathbf{Q}$, $q_{i,j}$, are constant, i.e., independent from the time index $t$. The matrix $\mathbf{Q}$ induces a transition matrix $\mathbf{R} = \{R_{i,j}\}$. Indeed, the probability of jumping from the state $i$ to $j$ is

$$R_{i,j} = \frac{q_{i,j}}{q_i}, \qquad \text{with } i \neq j, \tag{6.61}$$

by definition. Moreover, the time $\Delta(x_t = i)$ that $X_t$ dwells at the $i$th state has an exponential distribution $\text{Exp}(q_i)$, i.e.,

$$\Delta(x_t = i) \sim q_i e^{-q_i \tau}, \quad \tau \geq 0. \tag{6.62}$$

Denoting as $Z_n$, with $n \in \mathbb{N}$, the values of the process $X_t$ at the jump times $s_n$ (see Fig. 6.5), the resulting process $\{Z_n\}$ is a Markov chain with transition matrix $\mathbf{R}$. Furthermore, in some cases there exists a *stationary distribution*,

$$\lim_{t \to +\infty} p(x_t = y|x_0) = \phi(y), \qquad \text{with} \qquad \sum_{y \in \mathcal{X}} \phi(y) = 1,$$

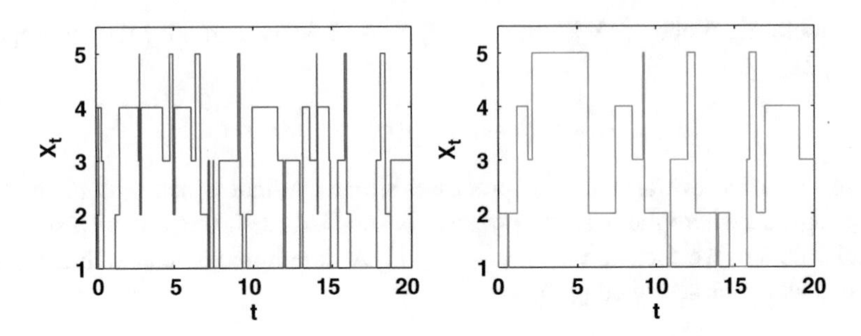


**Fig. 6.5** Two independent realizations of a Markov jump process with $X_t \in \{1, 2, 3, 4, 5\}$

that is also solution of the equation $\boldsymbol{\phi}^\top \mathbf{Q} = \mathbf{0}$, where $\boldsymbol{\phi} = [\phi(0), \phi(1), \phi(2), \ldots]^\top$. Denote as $s_1, s_2, \ldots$ the jump times and, for the sake of simplicity, denote the corresponding dwelling times as $\delta_1, \delta_2, \ldots$ A MJP can be generated in this way:

1. Start with $n = 0$, $t = 0$, and $s_0 = 0$. Choose $x_0$ and set $y_0 = x_0$.
2. Draw $\delta_{n+1} \sim q_{y_n} e^{-q_{y_n}\tau}$, $\tau \geq 0$.
3. Set $x_t = y_n$ for $s_n \leq t \leq s_n + \delta_{n+1}$.
4. Set $s_{n+1} = s_n + \delta_{n+1}$.
5. Draw $y_{n+1}$ with probability according to the $y_n$th row of the transition matrix $\mathbf{R}$.

Figure 6.5 illustrates two independent realizations of a MJP with

$$\mathbf{Q} = \begin{bmatrix} -(a_1 + a_2) & a_1 & a_2 & 0 & 0 \\ b_1 & -(b_1 + a_2) & 0 & a_2 & 0 \\ b_2 & 0 & -(b_2 + a_1) & 0 & a_1 \\ 0 & 0 & -b_1 & b_1 & 0 \\ 0 & b_2 & 0 & 0 & -b_2 \end{bmatrix},$$

where $a_1 = a_2 = b_1 = b_2 = 1$ and $X_t \in \{1, 2, 3, 4, 5\}$.

### 6.9.2 Gaussian Processes

Consider a stochastic process $\{X_t\}_{t=-\infty}^{+\infty} \in \mathbb{R}$ such that every finite-dimensional vector

$$[Z_1, \ldots, Z_n]^\top = [X_{t_1}, \ldots, X_{t_n}]^\top,$$

where $t_1 < t_2 < \ldots < t_n$ are real time indices, follows a multivariate Gaussian distribution. In this case, the stochastic process $\{X_t\}$ is called a Gaussian process (GP) [26]. Note that both $t$ and $X_t$ are continuous (real) variables. A GP is completely defined by its mean and covariance functions, i.e.,

$$\mu(t) = E[X_t], \qquad t \in \mathbb{R},$$

and

$$K(t,s) = \text{Cov}(X_t, X_s) \qquad t,s \in \mathbb{R}.$$

The covariance function is often also called the *kernel* of the GP. It is possible to generate easily a finite realization of a GP at the given time steps $t_1, \dots, t_n$, by way of the procedure below:

1. Given $t_1, \dots, t_n$, compute the mean $\tilde{\boldsymbol{\mu}} = [\tilde{\mu}_1, \dots, \tilde{\mu}_n]^\top = [\mu_{t_1}, \dots, \mu_{t_n}]^\top$ and the covariance matrix $\boldsymbol{\Sigma}(j,k) = K(t_j, t_k)$, with $j,k \in \{1,\dots,n\}$, where $\boldsymbol{\Sigma}(j,k)$ denotes the entry in row $j$ and column $k$ of the matrix $\boldsymbol{\Sigma}$.
2. Compute the Cholesky decomposition $\boldsymbol{\Sigma} = \mathbf{A}\mathbf{A}^\top$.
3. Draw $v_i \sim N(0,1)$, for $i = 1, \dots, n$, to obtain $\mathbf{v} = [v_1, \dots, v_n]^\top$.
4. Calculate $\mathbf{z} = \tilde{\boldsymbol{\mu}} + \mathbf{A}\mathbf{v}$, where $\mathbf{z} = [z_1, \dots, z_n]^\top = [x_{t_1}, \dots, x_{t_n}]^\top$ is a realization of the GP.

Some common choices of $\mu(t)$ and $K(t,s)$ are presented in Table 6.3.

Figure 6.6 shows 20 different realizations $[X_{t_1}, \dots, X_{t_n}]^\top$ of a GP with

$$\mu(t) = 0, \qquad K(t,s) = \exp\left(-\frac{(t-s)^2}{2}\right), \tag{6.63}$$

and the time grid $t_1 = -5, t_2 = -4.9, t_3 = -4.8, \dots, t_n = 5$.

**Table 6.3** Special cases of Gaussian processes

| Type | Mean $\mu(t)$ | Kernel $K(t,s)$ |
|---|---|---|
| Gaussian kernel | 0 | $\exp\left(-\frac{(t-s)^2}{2\sigma_p^2}\right)$ |
| Wiener process | 0 | $\min\{t,s\}$ |
| Standard Brownian bridge | 0 | $\min\{t,s\} - st$ |
| Ornstein-Uhlenbeck process | $e^{-\theta t}\mu_0 + \nu(1 - e^{-\theta t})$ | $\frac{\sigma^2}{2\theta} e^{-\theta(t+s)} \left(e^{2\theta \min\{t,s\}} - 1\right)$ |

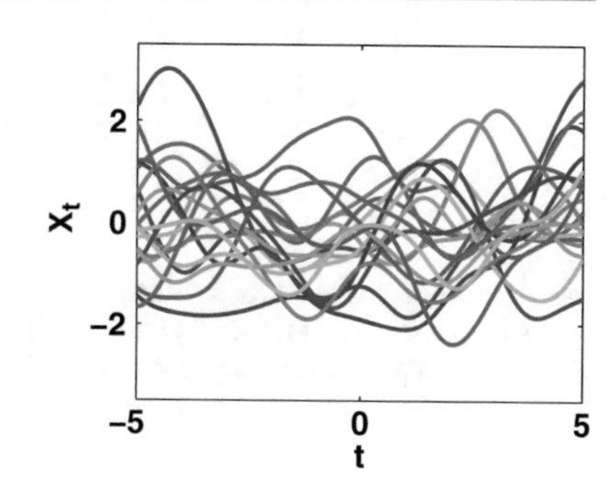


**Fig. 6.6** Twenty independent realizations of a GP with a Gaussian kernel ($\sigma_p = 2$) and zero mean, evaluated between $t_1 = -5$ and $t_n = 5$ with step $10^{-1}$

**Interpolation: Noise-Free Observations**

Consider a set of fixed points $(s_j, y_j) \in \mathbb{R}^2$, with $j = 1, \ldots, J$ (where $J$ is a constant value). It is possible to draw GP functions passing through these points using simple properties of multivariate Gaussian pdfs [26]. Considering $n$ abscissas $t_1, \ldots, t_n$, we are interested in drawing the vector $[Z_1, \ldots, Z_n]^\top = [X_{t_1}, \ldots, X_{t_n}]^\top$ conditional on $[X_{s_1} = y_1, \ldots, X_{s_J} = y_J]^\top$, where $\{X_t\}$ is a GP. We denote

$$\mathbf{A}(i,k) = K(s_i, s_k), \quad \mathbf{B}(r,m) = K(t_r, t_m), \quad \mathbf{C}(r,k) = K(t_r, s_k),$$

where $i, k \in \{1, \ldots, J\}$ and $r, m \in \{1, \ldots, n\}$, hence the matrix $\mathbf{A}$ has dimension $J \times J$, $\mathbf{B}$ has dimension $n \times n$, and $\mathbf{C}$ is an $n \times J$ matrix. Then, it is easy to show that [26]

$$q(\mathbf{z}|\mathbf{t}, \mathbf{s}, \mathbf{y}) \sim \mathcal{N}(\mathbf{z}; \boldsymbol{\mu}, \boldsymbol{\Sigma}), \tag{6.64}$$

where $\mathbf{t} = [t_1, \ldots, t_n]^\top$, $\mathbf{s} = [s_1, \ldots, s_J]^\top$, $\mathbf{y} = [y_1, \ldots, y_J]^\top$, and

$$\boldsymbol{\mu} = \mathbf{C}\mathbf{A}^{-1}\mathbf{y}, \qquad \boldsymbol{\Sigma} = \mathbf{B} - \mathbf{C}\mathbf{A}^{-1}\mathbf{C}^\top. \tag{6.65}$$

Note that $\boldsymbol{\mu}$ is an $n \times 1$ vector and $\boldsymbol{\Sigma}$ has dimension $n \times n$. Therefore, the generation procedure consists of the following step:

1. Compute $\boldsymbol{\mu} = \mathbf{C}\mathbf{A}^{-1}\mathbf{y}$ and $\boldsymbol{\Sigma} = \mathbf{B} - \mathbf{C}\mathbf{A}^{-1}\mathbf{C}^\top$.
2. Draw $[z'_1, \ldots, z'_n]^\top \sim \mathcal{N}(\mathbf{z}; \boldsymbol{\mu}, \boldsymbol{\Sigma})$.
3. Set $[x'_{t_1}, \ldots, x'_{t_n}]^\top = [z'_1, \ldots, z'_n]^\top$.

Figure 6.7a illustrates 20 independent realizations of a GP with a Gaussian kernel given $(-10, 3)$, $(-2, 1)$, $(4, -7)$ and $(10, 5)$, i.e., $[X_{-10} = 3, X_{-2} = 1, X_4 = -7, X_{10} = 5]^\top$.

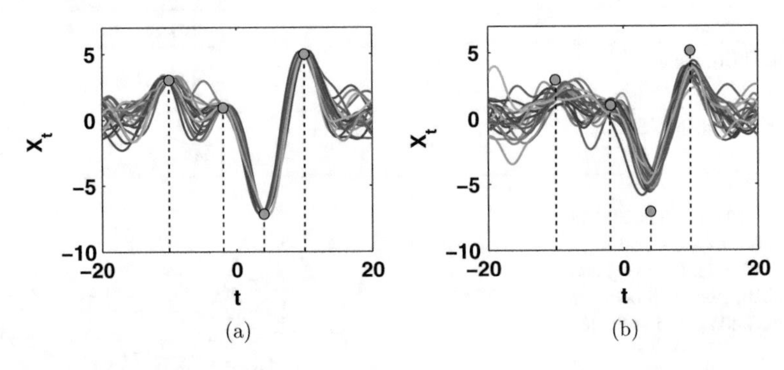


**Fig. 6.7** **(a)** Independent realizations of a GP with a Gaussian kernel ($\sigma_p = 2$) given four observed points $(-10, 3)$, $(-2, 1)$, $(4, -7)$ and $(10, 5)$, evaluated between $t_1 = -20$ and $t_n = 20$ with step $10^{-1}$. **(b)** Independent realizations of the GP considering the observed values $[Y_{-10} = 3, Y_{-2} = 1, Y_4 = -7, Y_{10} = 5]^\top$ contaminated by Gaussian noise with $\sigma = 0.5$

### Regression

In the case that each element of the vector of observations $\mathbf{y} = [y_1, \ldots, y_J]^\top$ is contaminated by independent Gaussian noise with zero mean and variance $\sigma^2$, we have for instance

$$\mu = \mathbf{C}(\mathbf{A} + \sigma^2\mathbf{I})^{-1}\mathbf{y} = \mathbf{C}\widetilde{\mathbf{A}}^{-1}\mathbf{y}, \tag{6.66}$$

where $\widetilde{\mathbf{A}} = \mathbf{A} + \sigma^2\mathbf{I}$ and $\mathbf{I}$ is the unit diagonal matrix of dimension $n \times n$. The sampling procedure described above for the interpolation case is still valid for the regression scenario replacing $\mathbf{A}$ with $\widetilde{\mathbf{A}}$. Some random samples from a GP posterior pdf in the regression case are shown in Fig. 6.7b with $\sigma = 0.5$.

## 6.9.3 *Wiener Processes*

A Wiener process (WP) is a stochastic process $\{W_t\}$, $t \in \mathbb{R}$, with the following specific properties [1, 17, 21]:

- The r.v. defined by the increments $R_{t,s} = W_t - W_s$ with $t > s$, for any $t, s \in \mathbb{R}$, is independent from the past path $\{W_\tau\}_{\tau \leq s}$.
- The r.v. $R_{t,s}$ has a Gaussian distribution, specifically

$$R_{t,s} = W_t - W_s \sim \mathcal{N}(0, t - s).$$

- The trajectory $\{W_t\}$ forms a *continuous* path [21, Chap. 5].

The Wiener process is often interpreted as the continuous counterpart of a discrete random walk (see Fig. 6.8 for some examples). A WP is a special case of Gaussian process with $\mu(t) = 0$ and $K(t, s) = \min[t, s]$. It also satisfies the Markov property $p(w_{t+\tau}|w_s, s \leq t) = p(w_{t+\tau}|w_t)$, where

$$p(w_{t+\tau}|w_t) = \frac{1}{\sqrt{2\pi\tau}} \exp\left(-\frac{(w_{t+\tau} - w_t)^2}{2\tau}\right).$$

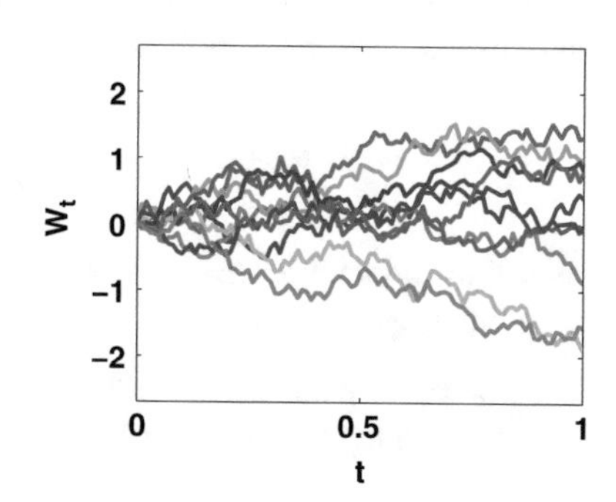


**Fig. 6.8** Ten independent realizations of a Wiener process

As a consequence, we have already shown different procedures for generating a discrete subset of a WP, i.e., considering $t_1 < t_2 \dots < t_n$ and $[W_{t_1}, \dots, W_{t_n}]$. However, another simple algorithm can be provided using the specific properties of a WP. It is common to consider $t_k > t_0 = 0$, for $k = 1, \dots, n$ and $W_0 = 0$. Then the sampling algorithm is the following:

1. Set $w_0 = 0$ at $t_0 = 0$ $(0 < t_1 < t_2 \dots < t_n)$.
2. Draw $n$ independent Gaussian samples $v_i$, i.e., $v_i \sim N(0, 1)$ for $i = 1, \dots, n$.
3. Set $w_{t_k} = \sum_{i=1}^{k} \sqrt{t_i - t_{i-1}} v_i$, $k = 1, \dots, n$.

The algorithm is exact; it returns a discrete subset of the true continuos random path.

### 6.9.4 Brownian Motion

A WP is also referred to as *standard Brownian motion*. More in general, it is also common to consider the stochastic process $\{B_t\}$ that satisfies

$$B_t = \nu t + \sigma W_t, \qquad t \in \mathbb{R}^+,$$

as a *Brownian motion* [21], where $\nu$, $\sigma$ are constants and $W_t$ is a WP. If $\nu = 0$ we recover the standard Brownian motion (i.e., a WP). Since we are able to generate a WP then we can easily generate a Brownian motion:

1. Draw samples from a WP at $0 < t_1 < t_2 \dots < t_n$, i.e., $w_{t_1}, \dots, w_{t_n}$.
2. Set $b_{t_k} = \nu t_k + \sigma w_{t_k}$, for $k = 1, \dots, n$.

#### Geometric Brownian Motion

A geometric Brownian motion [1, 17, 21] is defined by the stochastic differential equation (SDE),

$$dX_t = \nu X_t dt + \sigma X_t dW_t, \qquad t \in [0, T],$$

where $\nu$, $\sigma$ are constants and $W_t$ is a WP. In this case, the SDE can be solved analytically, having the strong solution

$$X_t = X_0 \exp\left(\left(\nu - \frac{\sigma^2}{2}\right) t + \sigma W_t\right), \qquad t \in [0, T].$$

Hence, the geometric Brownian motion can be easily simulated at $0 < t_1 < t_2 \dots < t_n$, in this way:

1. Draw $x_0$ from some prior pdf.
2. Draw $n$ independent Gaussian samples $v_i$, i.e., $v_i \sim N(0, 1)$ for $i = 1, \dots, n$.

3. Set

$$x_{t_k} = x_0 \exp\left(\left(\nu - \frac{\sigma^2}{2}\right) t_k + \sigma \sum_{i=1}^{k} \sqrt{t_i - t_{i-1}} v_i\right), \qquad k = 1, \ldots, n.$$

## Standard Brownian Bridge

The standard Brownian bridge, $\{X_t\}$, [1, 17, 21] is a stochastic process defined in the interval $t$ă $\in [0, 1]$. Specifically, the Brownian bridge is a WP with $t \in [0, 1]$ such that $X_0 = X_1 = 0$. It possible to show that

$$X_t = (1 - t) W_{\frac{t}{1-t}}, \qquad t \in [0, 1],$$

and

$$X_t = W_t - t W_1, \qquad t \in [0, 1],$$

where $W_t$ is a Wiener process and $X_t$ is a standard Brownian bridge. Figure 6.9 provides some examples of standard Brownian bridge path. Moreover, the standard Brownian bridge can be seen as a GP with $\mu(t) = 0$ and $K(t, s) = \min\{t, s\} - st$. Thus, given $t_0 = 0 < t_1 < t_2 \ldots, < t_n = 1$, it can be generated as a GP but we can also use alternative sampling algorithms, as the procedure below:

1. Generate a WP at $t_0 = 0 < t_1 < t_2 \ldots, < t_n = 1$, i.e., $w_{t_0}, w_{t_1} \ldots, w_1 = w_{t_n}$.
2. Set $x_{t_k} = w_{t_k} - t_k w_1$, with $k = 0, \ldots, n$.

## General Brownian Bridge

The Brownian bridge can be generalized considering a random process $\{X_t\}$ with $t \in [t_0, t_n]$ with the boundary conditions $X_{t_0} = a$ and $X_{t_n} = b$ [21]. This process is

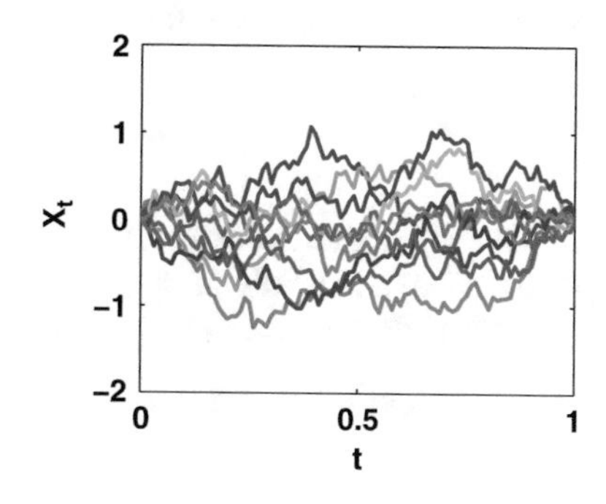


**Fig. 6.9** Ten independent realizations of a standard Brownian bridge

still a GP with mean function

$$\mu(t) = a + \frac{(b-a)(t-t_0)}{(t_n - t_0)},$$

and kernel

$$K(t,s) = \min\{t - t_0, s - t_0\} - \frac{(s-t_0)(t-t_0)}{(t_n - t_0)}.$$

This general version of the Brownian bridge is often used to interpolate a discrete approximation of a WP. Namely, given $W_{t_0} = a$ and $W_{t_n} = b$, additional points of the WP at $t_1, \ldots, t_{n-1} \in [t_0, t_n]$ can be drawn using the Brownian bridge, following the procedure below:

1. Set $k = 0$.
2. Draw $v_k \sim \mathcal{N}([0, 1])$.
3. Set

$$W_{t_k} = W_{t_{k-1}} + \frac{t_k - t_{k-1}}{t_n - t_{k-1}}(b - W_{t_{k-1}}) + \sqrt{\frac{(t_n - t_k)(t_k - t_{k-1})}{t_n - t_{k-1}}}v_k.$$

4. If $k < n - 1$ then set $k = k + 1$ and repeat from step 2. Otherwise, stop.

### 6.9.5 *Poisson Processes*

Poisson processes [3, 19, 29] are often applied to model spatio-temporal distributions of random points. More specifically, it is a continuous-time process that counts the number of events associated to a phenomenon and the time at which they occur. Let us consider a collection of random vectors $\{\mathbf{x}^{(i)}\} \in \mathcal{D} \subset \mathbb{R}^n$, with $i = 1, 2, \ldots$ Moreover, we define a counting r.v. $N(\mathcal{D})$, that provides the number of random points $\mathbf{x}^{(i)}$ within $\mathcal{D}$. We assume that $N(\mathcal{D})$ satisfies the following properties:

- $N(\mathcal{D}) \sim \text{Poisson}(\mu_{\mathcal{D}})$, for any subset $\mathcal{D} \subset \mathbb{R}^n$ and $\mu_{\mathcal{D}} > 0$, i.e., with pdf

$$N(\mathcal{D}) \sim \frac{e^{-\mu_{\mathcal{D}}}\mu_{\mathcal{D}}^n}{n!}, \quad n = 0, 1, 2, \ldots$$

  Namely, in one realization we have $N(\mathcal{D}) = n'$ with $n' \in \mathbb{N}$.
- Given a collection of disjoint sets $\mathcal{D}_1, \mathcal{D}_2, \ldots, \mathcal{D}_m$, the corresponding r.v.'s $N(\mathcal{D}_1), N(\mathcal{D}_2), \ldots, N(\mathcal{D}_m)$ are independent.

In the cases of interest, the mean $\mu_{\mathcal{D}}$ can be expressed as

$$\mu_{\mathcal{D}} = \int_{\mathcal{D}} \lambda(\mathbf{x})d\mathbf{x}, \tag{6.67}$$

where the function $\lambda(\mathbf{x})$ is often called *intensity* or *rate function*. The two previous properties induce an important feature of a Poisson process: given $N_{\mathcal{D}} = m$, the $m$ random vectors $\{\mathbf{x}^{(i)}\}_{i=1}^{m}$ are distributed according to the density $q(\mathbf{x}) = \frac{\lambda(\mathbf{x})}{\mu_{\mathcal{D}}}$ [3, 21], i.e.,

$$\mathbf{x}^{(i)} \sim q(\mathbf{x}) = \frac{1}{\mu_{\mathcal{D}}}\lambda(\mathbf{x}), \qquad i = 1, 2, \ldots, m.$$

Hence, a Poisson process can be generated with the following two steps:

1. Generate a number of points, $m'$, according to a Poisson probability mass function, i.e., $m' \sim \frac{e^{-\mu_{\mathcal{D}}}\mu_{\mathcal{D}}^{m}}{m!}$, $m = 0, 1, 2, \ldots$, where $\mu_{\mathcal{D}} = \int_{\mathcal{D}} \lambda(\mathbf{x})d\mathbf{x}$.
2. Then, draw $\mathbf{x}^{(i)} \sim \frac{1}{\mu_{\mathcal{D}}}\lambda(\mathbf{x})$, $i = 1, 2, \ldots, m'$. The collection $\{\mathbf{x}^{(i)}\}_{i=1}^{m'}$ represents the realization of the Poisson process.

### Homogenous Poisson Processes

Assume that the subset $\mathcal{D} \subset \mathbb{R}^n$ is finite. In this case, if we consider a constant rate $\lambda(\mathbf{x}) = \lambda$, then

$$\mu_{\mathcal{D}} = \int_{\mathcal{D}} \lambda(\mathbf{x})d\mathbf{x} = |\mathcal{D}|\lambda, \tag{6.68}$$

and the density $q(\mathbf{x}) = \mathbb{I}_{\mathcal{D}}(\mathbf{x})$ is a uniform pdf in $\mathcal{D}$. In Fig. 6.10, we provide four independent realizations with $\mathcal{D} = [0, 1] \times [0, 1]$ (so that $|\mathcal{D}| = 1$) and $\lambda = 50$. Since in this case $\mu_{\mathcal{D}} = \lambda = 50$, the procedure used to generate the Poisson process consists in: (a) drawing $m' \sim \frac{e^{-\lambda}\lambda^m}{m!}$ with $m \in \mathbb{N}$ and then (b) drawing $\mathbf{x}' = [x_1, x_2]^\top \sim \mathcal{U}(\mathcal{D})$.

### One-Dimensional Poisson Processes

Following the notation used for a MJP, let us denote the points of the Poisson process as $s_1, s_2, \ldots$ which, in this case, can indicate jumps or arrival times. We denote the holding or inter-arrival times as $\delta_i = s_i - s_{i-1}$. Considering a homogeneous Poisson process with constant $\lambda \in \mathbb{R}^+$, these holding times are exponentially-distributed [3, 21], i.e., $\delta_i \sim \lambda e^{-\lambda\tau}$ [see Eq. (6.62)]. Therefore, the procedure to generate a realization of a Poisson process in the interval $[0, T]$ can be outlined as follows:

1. Set $n = 0$, $s_0 = 0$.
2. Draw $u' \sim \mathcal{U}([0, 1])$ and set $\delta_{n+1} = -\frac{1}{\lambda}\log(u')$.
3. Set $s_{n+1} = s_n + \delta_{n+1}$.
4. If $s_{n+1} \leq T$ set $n = n + 1$ and go back to step 2. stop. Otherwise, if $s_{n+1} > T$, then return $\{s_1, \ldots, s_n\}$ as points of the Poisson process and stop.

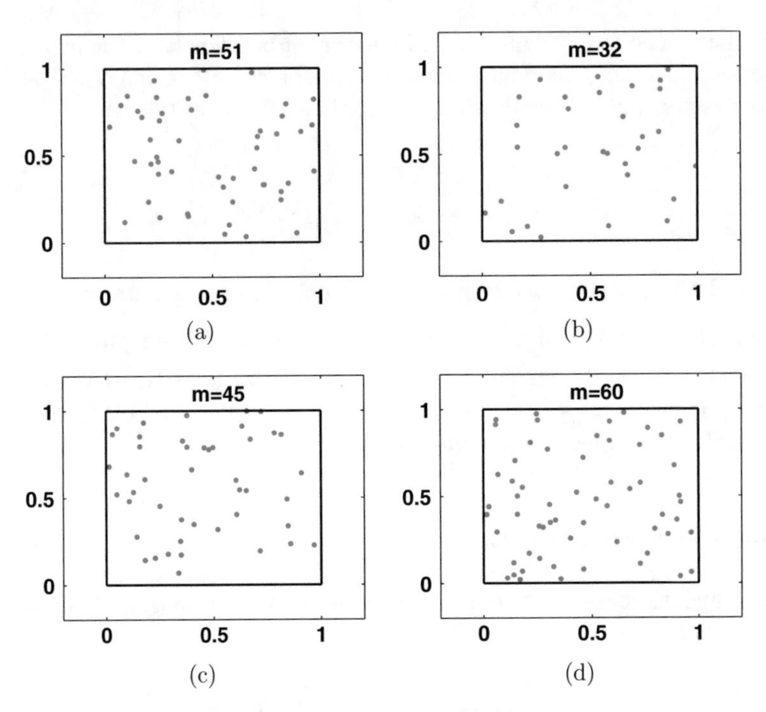


**Fig. 6.10** Four independent realizations of a homogeneous Poisson process with $\lambda = 50$ and $\mathcal{D} = [0, 1] \times [0, 1]$. The number of points $m'$ is drawn according to pmf $\frac{e^{-\lambda}\lambda^m}{m!}$, with $m \in \mathbb{N}$, and more specifically (**a**) $m' = 51$ points; (**b**) $m' = 32$ points; (**c**) $m' = 45$ points; (**d**) $m' = 60$ points

Furthermore, the counting r. v. $N_t = N([0, t])$, that represents the number of events $s_i$ within the interval $[0, t]$, forms a *Poisson counting process*: this is a MJP on $\mathcal{X} = \mathbb{N}$, $N_0 = 0$, $q_{i,j} = \lambda$ for $j = i + 1$, $i \in \mathbb{N}$, and $q_{i,j} = 0$ for $j \neq i + 1$. If $\lambda$ is time-dependent, i.e., $\lambda(t)$, the Poisson process is non-homogeneous. In this case, assuming a known upper bound

$$\Lambda \geq \lambda(t), \qquad t \in [0, T],$$

the sampling procedure becomes:

1. Set $n = 0$, $s_0 = 0$.
2. Draw $u' \sim \mathcal{U}([0, 1])$ and set $\delta = -\frac{1}{\Lambda}\log(u')$.
3. Set $s' = s_n + \delta$.
4. If $s' > T$ then stop.
5. Otherwise if $s' \leq T$:

   (a) Draw $v' \sim \mathcal{U}([0, 1])$.
   (b) If $v' \leq \frac{1}{\Lambda}\lambda(s')$, set $s_{n+1} = s'$, $n = n + 1$ and repeat from step 2.
   (c) Otherwise, if $v' > \frac{1}{\Lambda}\lambda(s')$, reject $s'$ and repeat from step 2.

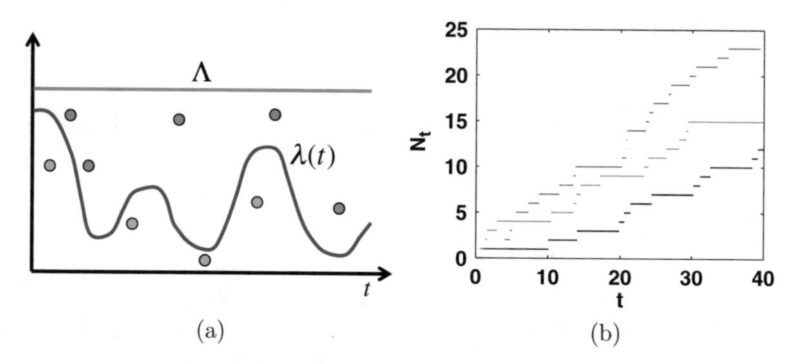


**Fig. 6.11** **(a)** Example of rate function $\lambda(t)$ and the upper bound $\Lambda \geq \lambda(t)$. A time instant $s'$ is accepted if the point $(s', v'\Lambda)$ with $v' \sim \mathcal{U}([0, 1])$ is below the graph of $\lambda(t)$. **(b)** Three independent realizations of a non-homogeneous Poisson process $N_t$ with $\lambda(t) = \sin(t)^2$

The underlying idea of the algorithm above is to generate a bidimensional homogenous Poisson process on $[0, T] \times [0, \Lambda]$ and then accept only the points that lie below the graph of $\lambda(t)$ (see Fig. 6.11a). Indeed, if all the points were accepted, we could obtain a homogenous Poisson process with rate $\Lambda$. Instead, accepting the points with probability $\frac{1}{\Lambda}\lambda(t)$ we get the corresponding non-homogenous process with rate $\lambda(t)$. Figure 6.11b illustrates three independent realizations of a non-homogeneous Poisson process with $\lambda(t) = \sin(t)^2$.

### 6.9.6 Dirichlet Processes: "Rich Get Richer"

Dirichlet processes are stochastic processes whose realizations can be interpreted as random discrete probability measures [7, 8, 11]. Hence, a Dirichlet process (DP) can also be seen as a "distribution over distributions." In the same way that the Dirichlet distribution is the conjugate prior for categorial distributions, DPs are used in non-parametric Bayesian inference as priors for (infinite) discrete models [7, 11]. A DP is completely specified by a *base density* $q_0(x)$ (which is the expected value of the process, as clarified later) and a positive real number $\alpha$ (usually named *concentration parameter*) [7, 11]. Without loss of generality, we assume $q_0(x)$ is a continuous pdf.

The *Chinese restaurant process* [8] is a procedure that yields a finite approximation of a DP realization. It generates a sequence of r.v.'s $\{X_k\}_{k\in\mathbb{N}}$ of the form

$$X_k = \begin{cases} X_i, \ i \in \{1, \ldots, k-1\}, & \text{with prob. } \frac{1}{k-1+\alpha} \\ X' \sim q_0(x), & \text{with prob. } \frac{\alpha}{k-1+\alpha}. \end{cases} \tag{6.69}$$

Drawing $x_1 \sim q_0(x)$, the equation above provides a recursive generation procedure (for $k \geq 2$):

1. Draw an index $j' \in \{1, \dots, k\}$ according to the normalized weights

$$\bar{w}_i = \frac{1}{k-1+\alpha}, \quad \text{for } i = 1, \dots, k-1, \text{ and } \bar{w}_k = \frac{\alpha}{k-1+\alpha}.$$

2. If $j' = k$, draw $x' \sim q_0(x)$ and set $x_k = x'$.
3. Otherwise, i.e., $j' \neq k$, set $x_k = x_{j'}$.
4. Set $k = k + 1$ and repeat from step 1.

Observe that, in general, at the $k$th iteration, several values $x_j$ are repeated. Let us denote as $m \leq k$ the number of unique values in the vector $[x_1, \dots, x_k]^\top$. Furthermore, we indicate as $[x_1^*, \dots, x_m^*]^\top$ the vector of unique values in $[x_1, \dots, x_k]^\top$, and with $\#x_j^*$ the number of elements in $[x_1, \dots, x_k]^\top$ equal to $x_j^*$, for $j = 1, \dots, m$. Then, the algorithm above can also be summarized as

$$x_k = \begin{cases} x_j^*, & \text{with prob. } \frac{\#x_j^*}{k-1+\alpha}, \\ x' \sim q_0(x), & \text{with prob. } \frac{\alpha}{k-1+\alpha}. \end{cases} \tag{6.70}$$

Hence, DPs tend to repeat the previously generated samples in a "rich get richer" fashion: the probability $\frac{\#x_j^*}{k-1+\alpha}$ of repeating the value $x_j^*$ becomes greater as the number $\#x_j^*$ grows. Note also that the probability of adding a new value $\frac{\alpha}{k-1+\alpha}$ decreases with $k$. Thus, a single (infinite) DP realization $\{x_k\}_{k=1}^{\infty}$ can also be expressed as a random probability measure made up of a countably infinite number of point masses, i.e.,

$$q(x) = \sum_{j=1}^{\infty} \gamma_j \delta(x - x_j^*), \qquad x_j^* \sim q_0(x), \tag{6.71}$$

where $\delta(x - x_j^*) = 1$ if $x = x_j^*$, and $\delta(x - x_j^*) = 0$ otherwise. The weights can be computed following the so-called stick-breaking procedure [7, 8, 11] as

$$\gamma_m = \beta_m' \prod_{j=1}^{m-1} (1 - \beta_j'), \quad \text{with} \quad \beta_j' \sim \alpha(1-\beta)^{\alpha-1}. \tag{6.72}$$

Namely, $\beta_j'$ are independent samples from a Beta distribution $\text{BETA}(1, \alpha)$. Note that $\beta_j' \in (0, 1)$ and $\sum_{j=1}^{\infty} \gamma_j = 1$, by construction. The probability $q(x)$ in Eq. (6.71) represents the stationary distribution of the stochastic process generated as (6.69). Hence, another way to generate one representation of the random measure $q(x)$ is the following procedure (starting with $i = 1$):

1. Draw $x_i^* \sim q_0(x)$,
2. Draw $\beta_i' \sim \alpha(1-\beta)^{\alpha-1}$.
3. Set $\gamma_i = \beta_i' \prod_{j=1}^{i-1} (1 - \beta_j')$.
4. Set $i = i + 1$ and repeat from step 1.

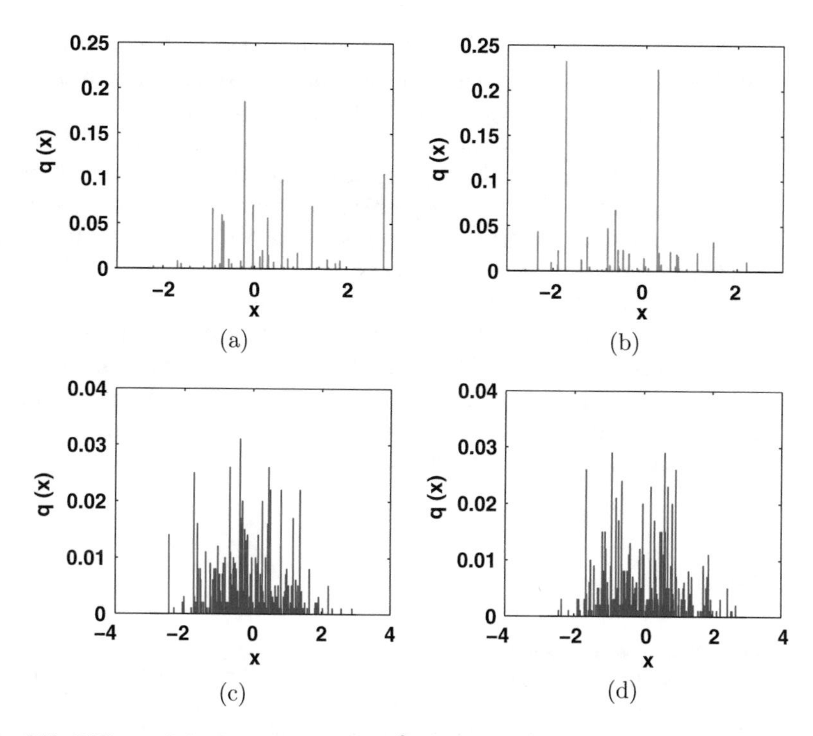


**Fig. 6.12** Different finite DP realizations ($10^3$ samples) using a standard Gaussian base pdf ($q_0(x) = \mathcal{N}(x; 0, 1)$) and with a concentration parameter $\alpha = 10$ in figures **(a)** and **(b)**, whereas with $\alpha = 100$ in figures **(c)** and **(d)**

It can be shown that the expected value of a realization $q(x)$ in Eq. (6.71) coincides with the base pdf $q_0(x)$.

Observe also that the parameter $\alpha$ defines how strong the concentration of $q(x)$ in Eq. (6.71) is: if $\alpha = 0$, the samples $x_j^*$ are all concentrated in the single value $x_1 \sim q_0(x)$, whereas when we increase $\alpha$, $q(x)$ tends to place significant mass on more "atoms," getting closer and closer to the base pdf $q_0(x)$, i.e., $q(x) \to q_0(x)$ as $\alpha \to +\infty$. Figure 6.12 shows four independent finite DP realizations formed by 1000 samples with $\alpha = 10$ (Fig. 6.12a, b), and with $\alpha = 100$ (Fig. 6.12c, d). We have considered $q_0(x) = \mathcal{N}(x; 0, 1)$ as base density.

## 6.10 Summary

In this chapter, we have described different approaches for sampling from multivariate distributions. First, we have recalled several techniques already introduced in previous chapters, providing derivations and useful guidelines for their application in higher dimensional spaces. For instance, in Sect. 6.3 we have recalled the rejection sampling and the ratio-of-uniforms techniques. In the same section, various methods for representing statistical dependence have been introduced.

In Sect. 6.4, we have described sampling techniques for a large family of distributions, namely, the class of elliptically contoured distributions. In Sect. 6.5, we have extended the vertical density representation for the multidimensional case. In Sect. 6.6, we have tackled the problem of generating samples uniformly distributed in $n$-dimensional measurable sets embedded in $\mathbb{R}^n$, e.g., points uniformly distributed within a simplex or a hypersphere in $\mathbb{R}^n$.

Generic transformations of random variables, converting $m$ r.v's into $n$ r.v.'s ($m \neq n$) have been discussed in Sect. 6.7. Furthermore, a collection of sampling algorithms for generating samples from specific multivariate distributions have been provided in Sect. 6.8. Finally, we have described the generation of random vectors with a specific dependence structure in Sect. 6.9. This includes sampling from some of the most common classes of stochastic processes.

## References


1. E. Çinlar, *Introduction to Stochastic Processes* (Prentice Hall, Englewood Cliffs, 1975)
2. R.D. Cook, M.E. Johnson, A family of distributions for modelling non-elliptically symmetric multivariate data. J. R. Stat. Soc. Ser. B Methodol. **43**(2), 210–218 (1981)
3. D.R. Cox, V.I. Isham, *Point Processes* (Chapman and Hall, Boca Raton, 1980)
4. J. Dagpunar, *Principles of Random Variate Generation* (Clarendon Press, Oxford/New York, 1988)
5. L. Devroye, *Non-uniform Random Variate Generation* (Springer, New York, 1986)
6. L. Devroye, Random variate generation for multivariate unimodal densities. ACM Trans. Model. Comput. Simul. **7**(4), 447–477 (1997)
7. T.S. Ferguson, A Bayesian analysis of some nonparametric problems. Ann. Stat. **1**, 209–230 (1973)
8. B.A. Frigyik, A. Kapila, M.R. Gupta, Introduction to the Dirichlet distribution and related processes. UWEE Technical Report Number UWEETR-2010-0006, University of Washington (1992)
9. J.E. Gentle, *Random Number Generation and Monte Carlo Methods* (Springer, New York, 2004)
10. G.H. Golub, C.F. Van Loan, *Matrix Computations*, 3rd edn. (Johns Hopkins University Press, Baltimore, 1996)
11. D. Görür, Non parametric Bayesian discrete latent variable models for unsupervised learning. Doctoral Thesis, Max Planck Institute for Biological Cybernetic (2007)
12. C. Groendyke, Ratio-of-uniforms Markov Chain Monte Carlo for Gaussian process models. Thesis in Statistics, Pennsylvania State University (2008)
13. W. Hörmann, A note on the performance of the Ahrens algorithm. Computing **69**, 83–89 (2002)
14. W. Hörmann, J. Leydold, G. Derflinger, *Automatic Nonuniform Random Variate Generation* (Springer, New York, 2003)
15. M.E. Johnson, *Multivariate Statistical Simulation*. Wiley Series in Probability and Statistics (Wiley, New York, 1987)
16. M.C. Jones, On Khintchine's theorem and its place in random variate generation. Am. Stat. **56**(4), 304–307 (2002)
17. S. Karlin, H.M. Taylor, *A First Course on Stochastic Processes* (Academic, New York, 1975)

18. A.Y. Khintchine, On unimodal distributions. Izvestiya NauchnoIssledovatelskogo Instituta Matematiki i Mekhaniki **2**, 1–7 (1938)
19. J.F.C. Kingman, *Poisson Processes* (Clarendon Press, Oxford, 1992)
20. T.J. Kozubowski, On the vertical density of the multivariate exponential power distribution. Statistics **36**, 219–221 (2002)
21. D.P. Kroese, T. Taimre, Z.I. Botev, *Handbook of Monte Carlo Methods*. Wiley Series in Probability and Statistics (Wiley, New York, 2011)
22. J. Leydold, Automatic sampling with the ratio-of-uniforms method. ACM Trans. Math. Softw. **26**(1), 78–98 (2000)
23. D. Luengo, L. Martino, Efficient random variable generation: ratio of uniforms and polar rejection sampling. IET Electron. Lett. **48**(6), 326–327 (2012)
24. J.F. Mai, M. Scherer, *Simulating Copulas: Stochastic Models, Sampling Algorithms and Applications*. Series in Quantitative Finance, vol. 4 (Imperial College Press, London, 2012)
25. L. Martino, J. Míguez, A rejection sampling scheme for posterior probability distributions via the ratio-of-uniforms method, in *European Signal Processing Conference (EUSIPCO)*, Aalborg (2010)
26. C.E. Rasmussen, C.K.I. Williams, *Gaussian Processes for Machine Learning* (The MIT Press, Cambridge, 2005)
27. C.P. Robert, G. Casella, *Monte Carlo Statistical Methods* (Springer, New York, 2004)
28. G. Ronning, A simple scheme for generating multivariate Gamma distributions with non-negative covariance matrix. Techometrics **19**, 179–183 (1977)
29. D.L. Snyder, M.I. Miller, *Random Point Processes in Time and Space* (Springer, New York, 1991)
30. W.E. Stein, M.F. Keblis, A new method to simulate the triangular distribution. Math. Comput. Model. **49**(5), 1143–1147 (2009)
31. M.D. Troutt, A theorem on the density ordinate and an alternative interpretation of the Box-Muller method. Statistics **22**, 463–466 (1991)
32. M.D. Troutt, Vertical density representation and a further remark on the Box-Muller method. Statistics **24**, 81–83 (1993)
33. M.D. Troutt, W.K. Pang, S.H. Hou, *Vertical Density Representation and Its Applications* (World Scientific, Singapore, 2004)
34. A.W. Van Kemp, Patchwork rejection algorithms. J. Comput. Appl. Math. **31**(1), 127–131 (1990)
35. J. von Neumann, Various techniques in connection with random digits, in *Monte Carlo Methods*, ed. by A.S. Householder, G.E. Forsythe, H.H. Germond. National Bureau of Standards Applied Mathematics Series (U.S. Government Printing Office, Washington, DC, 1951), pp. 36–38
36. J.C. Wakefield, A.E. Gelfand, A.F.M. Smith, Efficient generation of random variates via the ratio-of-uniforms method. Stat. Comput. **1**(2), 129–133 (1991)

# Chapter 7
# Asymptotically Independent Samplers

**Abstract** Markov Chain Monte Carlo (MCMC) methods are possibly the most popular tools for random sampling nowadays. They generate "chains" (sequences) of samples from a target distribution that can be selected with few constraints. However, as highlighted by the term "chain," the draws output by the MCMC method are statistically dependent (and often highly correlated), which makes such algorithms not directly comparable with the methods in the rest of this monograph. In this chapter, we describe two families of non-standard MCMC techniques that enjoy the property of producing samples that become asymptotically independent as a parameter grows to infinity or the number of random draws in the algorithm is increased. The methods of the first family are based on generating a pool of candidate samples at each iteration of the chain, instead of only one as in conventional procedures. The techniques in the second family rely on an adaptive, non-parametric approximation of the target density, which is improved as new samples are generated. We describe the general methodology for the two families, and provide some specific algorithms as examples.


## 7.1 Introduction

Although in the previous chapters we have described numerous sampling techniques for generating independent samples from different families of distributions, in many applications there is a need of more general, off-the-shelf sampling techniques. The class of Markov Chain Monte Carlo (MCMC) algorithms responds to this need: they can be applied to sample from virtually any kind of target density, without any previous theoretical study. The only requirement is to be able to evaluate, point-wise, a function proportional to the target density. Hence the range of application is vast.

The idea behind MCMC methods is to generate a Markov chain whose invariant density is exactly the target pdf [4, 12, 13, 31]. Their main advantage is that they can be easily applied to draw from virtually any kind of density. However, since a Markov chain is constructed, the generated samples are correlated and the variance of the resulting Monte Carlo estimators can be much higher than when




L. Martino et al., *Independent Random Sampling Methods*, Statistics and Computing, https://doi.org/10.1007/978-3-319-72634-2_7

using independent samples [4, 12, 23]. Furthermore, from a practical point of view, this correlation also implies that the generated chain can remain trapped in a local mode for an arbitrarily large numbers of iterations [4, 12, 31]. This slows down the convergence of the chain towards the invariant target distribution. Generally, MCMC methods require a sufficiently large number of iterations until the chain attains its invariant distribution. This is usually referred to as a "burn-in" period and the corresponding samples should be discarded [4, 12]. The problem is that, in general, it is difficult to estimate the length of this "burn-in" period. One ideal solution for reducing it is to minimize the correlation between samples.

This chapter is devoted to describe two classes of MCMC algorithms, formed by techniques that can yield asymptotically independent samples at the expense of an increase in their computational cost. In the first part of the chapter, we provide two examples of MCMC methods based on proposing $N$ different candidates at each iteration. One sample among them is chosen according to some suitable weights, and then it is tested in order to accept it as a new element of the chain or not. If $N$ grows the performance of this kind of techniques improves, becoming closer and closer to the performance of an exact sampler. Clearly, the use of a larger number of candidates requires more evaluations of the target pdf per iteration. Namely, the increase of the number of tries $N$ also implies an increase in the computational cost. These techniques can be easily used for sampling from multivariate target distributions (although, in this chapter, we use a scalar notation for the sake of simplicity).

In the second part of the chapter, we describe another kind of MCMC technique (for sampling from univariate target distributions), called *Independent Doubly Adaptive Rejection Metropolis Sampling* (IA$^2$RMS). In this case, we consider the use of an adaptive non-parametric proposal pdf within a Metropolis-Hastings (MH) algorithm [9, 27, 28]. The construction of the proposal is based on interpolation procedures given a set of support points similar to the adaptive rejection samplers described in Chap. 4. The number of support points increases with the iterations in a suitable way, providing a better proposal pdf (closer to the target) but also ensuring that the total number of points does not diverge. The proposal function built via interpolation becomes closer to the target as more nodes are used and hence the performance is improved (i.e., we have smaller variance in the Monte Carlo estimators). However, at the same time, the computational effort required for drawing samples from the proposal pdf grows. It is possible to show that the algorithm produces an ergodic chain (despite the use of an adaptive proposal), and the sequence of proposal pdfs converges to the target pdf, so that we obtain asymptotically independent samples.

The chapter is structured as follows. We recall the basics of MH algorithms together with some theoretical background in Sect. 7.2. Two MCMC schemes based on multiple candidates are described in Sect. 7.3 and the IA$^2$RMS is introduced in Sect. 7.4. Finally, we conclude with a summary in Sect. 7.5.

## 7.2 Metropolis-Hastings (MH) Methods

One of the most popular and widely applied MCMC algorithms, jointly with the Gibbs sampler, is the Metropolis-Hastings (MH) method [4, 9, 12, 13, 27, 28, 31]. Essentially, MCMC algorithms (and so the MH method as well) generate a Markov chain, $x_1, x_2, \ldots, x_t, \ldots$, whose stationary distribution is the target $p_o$, by using samples drawn from a simpler proposal density $\pi$. In the following, we focus on the MH method, hence the rest of the chapter is devoted to describe extensions of this standard algorithm.

In order to apply the MH method, the only requirement is to be able to evaluate point-wise a function proportional to the target, i.e., $p(x) \propto p_o(x)$. In general, the proposal density $\pi$ can be dependent on the previous state of the chain and we denote it as $\pi(x|x_{t-1})$. If $x_{t-1}$ plays the role of a location parameter in $\pi(x|x_{t-1})$, then this proposal pdf is often known as *random walk proposal*. A simpler choice can also be employed, considering an *independent proposal* pdf $\pi(x)$, where the adjective "independent" refers to the independence of the proposal pdf from the previous state of the chain $x_{t-1}$. Below, we describe the MH algorithm in detail.

### 7.2.1 *The Algorithm*

The MH algorithm is a simple, well-known, and widely used MCMC technique. This chapter presents several extensions of the standard MH scheme. For this reason, here we briefly recall the MH method and its main theoretical properties. Let us set $t = 1$ and an arbitrary initial state for the chain, $x_0$. Let us also consider a target density, $p_o(x) \propto p(x)$ with $x \in \mathcal{D} \subseteq \mathbb{R}$,[1] and a proposal density $\pi(x|x_{t-1})$, where $x_{t-1}$ denotes the state of the chain at the $(t-1)$th iteration ($t = 1, 2, \ldots$). The MH algorithm consists of the following steps [4, 16, 31]:

1. Choose an initial state $x_0$.
2. For $t = 1, \ldots, T$ :

   (a) Draw a sample $z' \sim \pi(x|x_{t-1})$.
   (b) Accept the new state, $x_t = z'$, with probability

$$\alpha(x_{t-1}, z') = \min\left[1, \frac{p(z')\pi(x_{t-1}|z')}{p(x_{t-1})\pi(z'|x_{t-1})}\right]. \tag{7.1}$$

   Otherwise (i.e., with probability $1 - \alpha_t$), set $x_t = x_{t-1}$.

3. Return the sequence of states $\{x_1, x_2, \ldots, x_t, \ldots, x_T\}$.

[1]For the sake of simplicity, in this chapter we consider only a scalar variable $x$, although both the MH algorithm and most of the methodologies described here can be directly applied or extended to higher dimensional spaces.

It can be easily proved, under some mild regularity conditions, that the pdf of the current state $x_t$, when $t$ grows, converges to the target density, $p_o(x)$ (recall that $p(x) \propto p_o(x)$). In the following, we show that the MH algorithm satisfies the so-called detailed balance condition which is sufficient to guarantee that the output chain is ergodic and has $p_o$ as stationary distribution [4, 12, 31].

### 7.2.2 *Invariant Distribution of the MH Algorithm*

Let us denote as $K(x_t|x_{t-1})$ the transition pdf (or *kernel*) that determines the move from the state $x_{t-1}$ to the state $x_t$. A generic MCMC technique has $p_o(x)$ as an invariant (or stationary) distribution [31] if its kernel satisfies

$$\int_{\mathcal{D}} K(x_t|x_{t-1})p_o(x_{t-1})dx_{t-1} = p_o(x_t). \tag{7.2}$$

A sufficient condition which implies the equation above is the *detailed balance condition* [31],

$$p_o(x_{t-1})K(x_t|x_{t-1}) = p_o(x_t)K(x_{t-1}|x_t). \tag{7.3}$$

If the condition above is fulfilled, $p_o$ is invariant w.r.t. $K$ and the chain is also *reversible* [4, 31]. In the following, we show that the MH technique yields a reversible chain, with invariant pdf $p_o(x) \propto p(x)$, by proving that it fulfills the detailed balance condition. Note that we only have to consider the case $x_t \neq x_{t-1}$, since the case $x_t = x_{t-1}$ is trivial (the kernel is a delta in this case). For $x_t \neq x_{t-1}$, the kernel of the MH algorithm is

$$K(x_t|x_{t-1}) = \pi(x_t|x_{t-1})\alpha(x_{t-1}, x_t), \qquad x_t \neq x_{t-1},$$

so that, recalling also that $p_o(x) \propto p(x)$,

$$\begin{aligned} p(x_{t-1})K(x_t|x_{t-1}) &= p(x_{t-1})\pi(x_t|x_{t-1})\alpha(x_{t-1}, x_t), \\ &= p(x_{t-1})\pi(x_t|x_{t-1}) \min\left[1, \frac{p(x_t)\pi(x_{t-1}|x_t)}{p(x_{t-1})\pi(x_t|x_{t-1})}\right], \\ &= \min\left[p(x_{t-1})\pi(x_t|x_{t-1}), p(x_t)\pi(x_{t-1}|x_t)\right], \end{aligned} \tag{7.4}$$

where we have replaced the expression of $\alpha(x_{t-1}, x_t)$ in Eq. (7.1). Finally, we can observe that (7.4) is symmetric w.r.t. the variables $x_{t-1}$ and $x_t$ (i.e., they can be interchanged without varying the expression), then we can write $p(x_{t-1})K(x_t|x_{t-1}) = p(x_t)K(x_{t-1}|x_t)$, which is precisely the detailed balance condition.

### 7.2.3 Acceptance Rate in MH-Type Methods

In every MH-type algorithm, a tentative sample is drawn from a proposal distribution and then a test is carried out to determine whether the state of the chain should "jump" to the new proposed value or not. This test depends on the acceptance probability $\alpha$. If the jumps are not accepted (with probability $1 - \alpha$), the chain remains in the same state as before, exactly as in the standard MH method. In all cases, the acceptance probability $\alpha$ is designed in order to fulfill the detailed balance condition.[2]

For general MH-type techniques, we can define the acceptance rate as

$$a_R = \int_{\mathcal{D}^2} \alpha(x, z)\pi(z|x)p_o(x)dzdx \tag{7.5}$$

$$\approx \frac{1}{T}\sum_{t=1}^{T} \alpha(x_{t-1}, z_t), \tag{7.6}$$

where the latter expression is a Monte Carlo approximation of the integral in Eq. (7.5), $x_{t-1}$ represents the state of an MH chain at the $(t-1)$th iteration, and $z_t$ is the proposed sample at the $t$th iteration,[3] i.e., $z_t \sim \pi(z|x_{t-1})$. Clearly, $0 \leq a_R \leq 1$.

Given a target $p_o(x)$ and choosing the class of the proposal functions to be used as $\pi_\sigma(x|x_{t-1})$, where $\sigma$ represents a scale parameter, there exists an optimal scale parameter $\sigma$ such that we obtain an optimal value $a_R^*$. This optimal acceptance rate $a_R^*$ minimizes the correlation among the samples within the chain. Unlike in rejection samplers (see Chaps. 3–4), this optimal rate $a_R^*$ is unknown (it varies depending on the specific problem) and in general differs from 1. Below, we list different scenarios where we can obtain $a_R \approx 1$:

1. When the proposal coincides with the target density, i.e., $\pi(x) \propto p_o(x)$. This is clearly an ideal case, where the MH method is converted into an exact sampler, providing the best possible Monte Carlo performance, i.e., i.i.d. samples from the target $p_o(x)$.
2. When the scale parameter $\sigma$ of the proposal pdf $\pi_\sigma(x|x_{t-1})$ is very small w.r.t. the variance of the target. In this case the MH sampler tends to accept any proposed candidate in order to explore as quickly as possible the state space. The performance is often poor with high correlation among the generated samples.
3. A third scenario is found for certain advanced MCMC techniques (adaptive or not) where, if certain parameters grow to infinity, then $a_R \to 1$. In this case, the

[2]Note that this is no longer a sufficient condition to ensure the ergodicity for adaptive MCMC techniques (where the proposal pdf is adapted online). However, all the techniques described in the sequel satisfy the detailed balance condition, at each iteration $t$.

[3]Since the chain has $p_o$ as invariant pdf, we have $x_{t-1} \sim p_o(x)$ after a "burn in" period. Namely, after a certain number of iterations, we have $(x_{t-1}, z_t) \sim \pi(z_t|x_{t-1})p_o(x_{t-1})$.

performance can be extremely good, providing virtually independent samples, but with an increased computational cost. The key point in these methods is how they handle the trade-off between performance and computational cost.

Other similar scenarios can also exist. In this chapter, we focus on the third case. We will describe different techniques which can yield *asymptotically independent samples* as the proposals approach the target pdf (at the expense of an increase in computational cost).

## 7.3 Independent Generalized MH Methods with Multiple Candidates

In this section, we describe two generalizations of the standard MH scheme, with the common feature that they generate several candidates at each iteration [15]. In both techniques, the next state of the Markov chain is selected, according to certain weights, from a set of candidates drawn from the proposal pdf. The main advantage of this approach is that these methods can explore a larger portion of the sample space, at the expense of a higher computational cost per iteration (as more evaluations of the target are needed at each iteration).

### *7.3.1 Independent Multiple Try Metropolis Algorithms*

In this section, we consider a multiple try Metropolis (MTM) scheme [3, 14, 17, 19, 20], which uses an independent proposal density, i.e., $\pi(x_t|x_{t-1}) = \pi(x_t)$, independently from the previous state. Unlike an MH technique, in an MTM scheme $N$ different candidates are proposed, independently drawn from the proposal $\pi(x_t)$. According to the importance sampling weights[4] [13, 20, 31], one proposed sample is selected as the tentative next element of the chain. Then, the new state is accepted with a suitable probability $\alpha$, specifically chosen to produce an ergodic chain with invariant pdf $p_o(x) \propto p(x)$. Indeed, the kernel of the MTM method satisfies the detailed balance condition (hence, the chain is also reversible). The algorithm consists of the following steps:

1. Choose an initial state $x_0$.
2. For $t = 1, \dots, T$ :

   (a) Draw $N$ candidates, $x^{(1)}, \dots, x^{(N)}$, from $\pi(x)$.

[4]The analytic form of the weights is more general [20]. Indeed, different weights can be used without jeopardizing the ergodicity of the chain. However, here we consider only importance weights, for simplicity.

(b) Select an index $j^* \in \{1, \ldots, N\}$ with probability proportional to the (unnormalized) weights $w(x^{(i)}) = \frac{p(x^{(i)})}{\pi(x^{(i)})}$, with $i = 1, \ldots, N$, and where $p(x) \propto p_o(x)$.

(c) Set $v^{(i)} = x^{(i)}$ for all $i = 1, \ldots, j^* - 1, j^* + 1, \ldots, N$ and $v^{(j^*)} = x_{t-1}$. Namely, the vector $\mathbf{v} = [v^{(1)}, \ldots, v^{(N)}]^\top$ differs with the vector $\mathbf{x} = [x^{(1)}, \ldots, x^{(N)}]^\top$ only in the $j^*$th component.

(d) Set $x_t = x^{(j^*)}$ with probability

$$\alpha_N(x_{t-1}, x^{(j^*)}) = \min\left[1, \frac{\sum_{i=1}^{N} w(x^{(i)})}{\sum_{i=1}^{N} w(v^{(i)})}\right], \tag{7.7}$$

otherwise set $x_t = x_{t-1}$ (with probability $1 - \alpha$).

The underlying idea of the MTM approach is to improve the proposal procedure in order to provide a better exploration of the state space. For instance, suitable larger jumps are facilitated.

Clearly, the performance varies with the number $N$ of candidates. In particular, the correlation among the samples vanishes when greater values of $N$ are used [14, 20]. On the one hand, as $N$ grows, the computational cost also increases since we have to evaluate the target pdf $N$ times per iteration. On the other hand, as $N$ grows, we also have $\alpha_N \to 1$, i.e.,

$$\lim_{N \to +\infty} \alpha_N(x_{t-1}, x^{(j^*)}) = 1, \qquad \forall x_{t-1}, x^{(j^*)} \in \mathcal{D}. \tag{7.8}$$

Indeed, note that $\alpha_N$ in Eq. (7.7) can be rewritten as

$$\alpha_N(x_{t-1}, x^{(j^*)}) = \min\left[1, \frac{\sum_{i=1}^{N} w(x^{(i)})}{\sum_{i=1}^{N} w(x^{(i)}) - w(x^{(j^*)}) + w(x_{t-1})}\right]. \tag{7.9}$$

Since $w(x) \geq 0$, for all $x \in \mathcal{D}$, and the numerator and denominator inside $\alpha_N$ differ only for one element (the remaining $N - 1$ elements coincide), we have $\alpha_N \to 1$ for $N \to +\infty$. This means that as $N$ grows, the probability of accepting a new state becomes greater. Indeed, the candidate $x^{(j^*)}$ is drawn from the set $\{x^{(1)}, \ldots, x^{(N)}\}$, using an improved approximation of the target distribution that we obtain by applying the importance weights $w(x^{(j)}) = \frac{p(x^{(j)})}{\pi(x^{(j)})}$, $j = 1, \ldots, N$ (an exhaustive discussion of this issue can be found in the appendices of [26]). Hence, as $N$ grows, the selected try $x^{(j^*)}$ becomes a better candidate in the sense that it becomes statistically "more similar" to a sample drawn directly from the target $p_o$ [26, Appendix C1]. As a consequence, $\alpha_N$ approaches 1. For $N = 1$ the MTM scheme becomes a standard MH algorithm with an independent proposal pdf $\pi$ and the standard acceptance function $\alpha$ in Eq. (7.1).

### 7.3.2 Ensemble MCMC Method

The *Ensemble MCMC* (EnMCMC) algorithm is a method related to the MTM scheme that has been independently proposed in [30]. A related technique can also be found in [2], as discussed in [22]. The main idea behind EnMCMC is similar to MTM, since different possible candidates are considered at each iteration. The algorithms differ in the way one candidate is chosen and then tested. In the MTM scheme, these two steps are clearly separated in Step 2b and Step 2d of the algorithm in Sect. 7.3.1, respectively. In the EnMCMC method, these two steps are collapsed into a single one. In this section, we consider a special case of EnMCMC using an independent proposal pdf $\pi(x)$ (i.e., independent of the previous state of the chain). The EnMCMC algorithm consists of the following steps:

1. Choose an initial state $x_0$.
2. For $t = 1, \dots, T$ :
   (a) Draw $N$ candidates, $x^{(1)}, \dots, x^{(N)}$, from $\pi(x)$.
   (b) Set $x^{(N+1)} = x_{t-1}$.
   (c) Set $x_t = x^{(j^*)}$ where $j^* \in \{1, \dots, N+1\}$ is drawn according to the normalized weights

$$\alpha_N(x_{t-1}, x^{(j)}) = \frac{w(x^{(j)})}{\sum_{i=1}^{N+1} w(x^{(i)})} = \frac{w(x^{(j)})}{\sum_{i=1}^{N} w(x^{(i)}) + w(x_{t-1})}, \quad j = 1, \dots, N+1,$$

and $w(x^{(i)}) = \frac{p(x^{(i)})}{\pi(x^{(i)})}$ are the importance weights associated to the sample $x^{(i)}$.

Therefore, the $t$th state of the chain is drawn from the set

$$\{x^{(1)}, \dots, x^{(N)}, x^{(N+1)} = x_{t-1}\},$$

according to the associate importance weights, where $N$ elements are independently drawn from $\pi$ at each iteration and the last one is set equal to the previous state $x_{t-1}$.

Let us observe that the EnMCMC algorithm extends the standard MH technique with the Barker's acceptance function [12, 13, 31]. Indeed, with $N = 1$, denoting

$x^{(1)} = x' \sim \pi(x)$, the acceptance function becomes

$$\alpha_N(x_{t-1}, x') = \frac{\frac{p(x')}{\pi(x')}}{\frac{p(x')}{\pi(x')} + \frac{p(x_{t-1})}{\pi(x_{t-1})}}, \tag{7.10}$$

$$= \frac{p(x')\pi(x_{t-1})}{p(x')\pi(x_{t-1}) + p(x_{t-1})\pi(x')} \tag{7.11}$$

that is exactly the classical Barker's acceptance function [13, 31] for an independent proposal pdf. It is possible to show that the detailed balance condition is also satisfied by the EnMCMC algorithm described above (see also the appendix in [22]). Note also that the probability of the chain remaining in the same state $x_{t-1}$,

$$\alpha_N(x_{t-1}, x_{t-1}) = \frac{w(x_{t-1})}{\sum_{i=1}^{N} w(x^{(i)}) + w(x_{t-1})}, \tag{7.12}$$

goes to zero, i.e., $\alpha_N(x_{t-1}, x_{t-1}) \to 0$, as $N \to \infty$. Hence, when $N \to \infty$, the probability of accepting a new state approaches 1 and the correlation among the generated states vanishes, since the approximation of the probability distribution with density $p_o$ is improved and the selected candidate becomes statistically closer to a sample directly drawn from $p_o$ (in the same fashion as in MTM). In fact, as $N \to \infty$ the approximation of the target distribution (via the importance weights) converges, hence the EnMCMC algorithm becomes an independent sampler itself, i.e., not only the correlation but also the statistical dependence of the chain is (asymptotically) removed.

## 7.4 Independent Doubly Adaptive Rejection Metropolis Sampling

In this section, we describe the *Independent Doubly Adaptive Rejection Metropolis Sampling* (IA$^2$RMS) algorithm [21, 23], which relies on a non-parametric construction of the proposal pdf in the same fashion as the Adaptive Rejection Sampling (ARS) and the Adaptive Rejection Metropolis Sampling (ARMS) methods, described exhaustively in Chap. 4. The shape of the non-parametric proposal used in IA$^2$RMS is tailored to the specific target function and its construction relies on a set of support points. The proposal can be shown to converge to the target as the number of support points increases.

Before introducing the IA$^2$RMS algorithm, we first revisit, briefly, the ARS and ARMS methods. Let us also recall that we denote with $\bar{\pi}_t(x)$ and $\pi_t(x)$ the non-normalized and normalized adaptive proposal pdfs at the $t$th iteration, respectively, and with $p(x)$ and $p_o(x)$ the non-normalized and normalized target densities, respectively.

### 7.4.1 Adaptive Rejection Sampling (ARS)

The ARS technique [5, 6], thoroughly described in Chap. 4, is a universal sampling technique which produces independent and identically distributed samples from the target. Hence, it is not an MCMC method, but the way in which the proposal pdf is constructed is tightly related to the IA$^2$RMS approach. Let us consider a set of support points at the $t$th iteration, denoted as

$$\mathcal{S}_t = \{s_1, s_2, \ldots, s_{m_t}\} \subset \mathcal{D},$$

such that $s_1 < \ldots < s_{m_t}$, and let us define $V(x) = -\log p(x)$ and $w_i(x)$ as the tangent line to $V(x)$ at $s_i$ for $i = 1, \ldots, m_t$. Then we can build the piecewise linear (PWL) function,

$$W_t(x) = \max\{w_1(x), \ldots, w_{m_t}(x)\}, \quad x \in \mathcal{D}. \tag{7.13}$$

and, in turn, select the proposal pdf, $\pi_t(x) \propto \bar{\pi}_t(x) = \exp(-W_t(x))$, which consists of exponential pieces in such a way that $W_t(x) \leq V(x)$ (and thus $\bar{\pi}_t(x) \geq p(x)$) when $V(x)$ is convex (i.e., $p(x)$ is log-concave).

Table 7.1 summarizes the ARS algorithm. Note that a new sample is added to the support set whenever it is rejected in the test of Step 4 in the algorithm. The ARS method has the important property that the sequence of proposals always converges to the target pdf. If we denote the $L_1$ distance between $\bar{\pi}_t(x)$ and $p(x)$ as

$$D(\bar{\pi}_t, p) = \int_{\mathcal{D}} |\bar{\pi}_t(x) - p(x)| dx, \tag{7.14}$$

**Table 7.1** Adaptive rejection sampling (ARS) algorithm

**Initialization:**

1. Set $t = 0$ and $n = 0$. Choose an initial set $\mathcal{S}_0 = \{s_1, \ldots, s_{m_0}\}$.

**Iterations (while $n < N$):**

2. Build a proposal, $\bar{\pi}_t(x)$, given a set of support points $\mathcal{S}_t = \{s_1, \ldots, s_{m_t}\}$, according to Eq. (7.13).
3. Draw $x' \sim \pi_t(x) \propto \bar{\pi}_t(x)$ and $u' \sim \mathcal{U}([0, 1])$.
4. If $u' > \frac{p(x')}{\bar{\pi}_t(x')}$, then reject $x'$, update $\mathcal{S}_{t+1} = \mathcal{S}_t \cup \{x'\}$, $m_{t+1} = m_t + 1$ and set $t = t + 1$. Go back to step 2.
5. Otherwise, if $u' \leq \frac{p(x')}{\bar{\pi}_t(x')}$, then accept $x'$, setting $x_n = x'$.
6. Set $\mathcal{S}_{t+1} = \mathcal{S}_t$, $m_{t+1} = m_t$, $t = t + 1$, $n = n + 1$ and return to step 2.

then the ARS method ensures that $D(\bar{\pi}_t, p) = \|\bar{\pi}_t(x) - p(x)\|_1 \to 0$ when $t \to \infty$. This property leads to two important consequences:

1. The acceptance rate,

$$\hat{a}_t = \int \frac{p(x)}{\bar{\pi}_t(x)} \pi_t(x) dx = \frac{c_v}{c_t}, \tag{7.15}$$

   tends to one as $t \to \infty$ (we have denoted $c_v = \int_{\mathcal{D}} p(x)dx$ and $c_t = \int_{\mathcal{D}} \bar{\pi}_t(x)dx$). Typically, $\hat{a}_t \to 1$ very quickly and the ARS scheme becomes virtually a direct i.i.d. sampler after a few iterations.
2. The computational cost remains bounded, as the probability of adding a new support point, $P_t = 1 - \hat{a}_t = \frac{1}{c_t} D(\bar{\pi}_t, p)$, tends to zero as $t \to \infty$.

### *7.4.2 Adaptive Rejection Metropolis Sampling*

Unfortunately, the ARS algorithm can only be applied to log-concave target pdfs (i.e., when $V(x) = -\log p(x)$ is convex). Although several generalizations of ARS have been proposed (cf. [8, 11, 18]), they are still only able to handle specific classes of pdfs. An alternative option is provided by the adaptive rejection Metropolis sampling (ARMS) technique, which combines the ARS method and the MH algorithm [13, 31] (see Sect. 4.6.1). ARMS is summarized in Table 7.2. It performs first a rejection test, and the rejected samples are used to improve the

**Table 7.2** Adaptive rejection Metropolis sampling (ARMS) algorithm

**Initialization:**

1. Set $n = 0$ (this is the chain iteration) and $t = 0$ (this is the algorithm iteration). Choose an initial state $x_0$ and support set $\mathcal{S}_0 = \{s_1, \ldots, s_{m_0}\}$.

**Iterations (while $n < N$):**

2. Build a proposal, $\bar{\pi}_t(x)$, given a set of support points $\mathcal{S}_t = \{s_1, \ldots, s_{m_t}\}$, according to Eq. (7.16).
3. Draw $x' \sim \pi_t(x) \propto \bar{\pi}_t(x)$ and $u' \sim \mathcal{U}([0, 1])$.
4. If $u' > \frac{p(x')}{\bar{\pi}_t(x')}$, then reject $x'$, update $\mathcal{S}_{t+1} = \mathcal{S}_t \cup \{x'\}$, $m_{t+1} = m_t + 1$ and set $t = t + 1$. Go back to step 2.
5. Otherwise, draw $u'' \sim \mathcal{U}([0, 1])$. If $u'' \leq \alpha$, with

$$\alpha = \min \left[ 1, \frac{p(x') \min[p(x_n), \bar{\pi}_t(x_n)]}{p(x_n) \min[p(x'), \bar{\pi}_t(x')]} \right],$$

   then accept $x'$, setting $x_{n+1} = x'$. Otherwise, if $u'' > \alpha$, then reject $x'$, setting $x_{n+1} = x_n$.
6. Set $\mathcal{S}_{t+1} = \mathcal{S}_t$, $m_{t+1} = m_t$, $t = t + 1$, $n = n + 1$ and return to step 2.

proposal pdf, exactly as in ARS. However, unlike ARS, the samples accepted in the rejection test go through an MH test. The MH step removes the main limitation of ARS: requiring that $\bar{\pi}_t(x) \geq p(x)$ $\forall x \in \mathcal{D}$. This allows ARMS to generate samples from a wide variety of target pdfs, becoming virtually a universal sampler.

The choice of the proposal construction approach is critical for the good performance of ARMS [7]. Consider again the set of support points $\mathcal{S}_t = \{s_1, s_2, \ldots, s_{m_t}\}$ and let us define the intervals $\mathcal{I}_0 = (-\infty, s_1]$, $\mathcal{I}_j = (s_j, s_{j+1}]$ for $j = 1, \ldots, m_t - 1$, and $\mathcal{I}_{m_t} = (s_{m_t}, +\infty)$. Moreover, let us denote as $L_{j,j+1}(x)$ the line passing through the points $(s_j, V(s_j))$ and $(s_{j+1}, V(s_{j+1}))$ for $j = 1, \ldots, m_t - 1$. Then, a PWL function $W_t(x) = -\log[\bar{\pi}_t(x)]$ is constructed in ARMS, of the form

$$W_t(x) = \begin{cases} L_{1,2}(x), & x \in \mathcal{I}_0, \\ \min\{L_{1,2}(x), L_{2,3}(x)\}, & x \in \mathcal{I}_1, \\ \varphi_j(x), & x \in \mathcal{I}_j, \\ \min\{L_{m_t-2,m_t-1}, L_{m_t-1,m_t}(x)\}, & x \in \mathcal{I}_{m_t-1}, \\ L_{m_t-1,m_t}(x), & x \in \mathcal{I}_{m_t}, \end{cases} \tag{7.16}$$

where

$$\varphi_j(x) = \min\{L_{j,j+1}(x), \max\{L_{j-1,j}(x), L_{j+1,j+2}(x)\}\},$$

and $j = 2, \ldots, m_t - 1$. Hence, the proposal pdf, $\pi_t(x) \propto \bar{\pi}_t(x) = \exp(-W_t(x))$ is, again, formed by exponential pieces.

It is important to remark that the number of pieces that form the proposal with this construction is larger than $m_t$ in general, since the proposal can be formed by two segments rather than one in some intervals. The computation of intersection points among these two segments is also needed. More sophisticated approaches to build $W_t(x)$ (e.g., using quadratic segments when possible [29]) have been proposed. However, none of them solves the structural problem of ARMS that is briefly described next.

### *7.4.3 Structural Limitations of ARMS*

Unlike ARS, the ARMS algorithm cannot guarantee the convergence of the sequence of proposals to the target, i.e., it cannot be claimed that $D(\bar{\pi}_t, p) \to 0$ as $t \to \infty$ in general. In ARMS, the proposal pdf is updated only when a sample $x'$ is discarded in the rejection test, something that can only happen when $\bar{\pi}_t(x') > p(x')$. On the other hand, when a sample is initially accepted in the rejection test, as it always happens when $\bar{\pi}_t(x') \leq p(x')$, the proposal is never updated. Thus, the satisfactory performance of ARMS depends on two issues:

(a) $W_{t+1}(x)$ should be constructed in such a way that $W_t(x) \leq V(x)$ (i.e., $\bar{\pi}_t(x) \geq p(x)$) inside most of the domain of $\mathcal{D}$, so that support points can be added almost everywhere.

(b) The addition of a support point inside an interval must entail a change of the proposal pdf inside other neighboring intervals when building $W_{t+1}(x)$. This allows the proposal to improve inside regions where $\bar{\pi}_t(x) < p(x)$.

These two conditions lead to unnecessarily complex proposal construction schemes. Furthermore, even if the proposal-building approach fulfills these two requirements [as it happens for the procedure proposed in [7] and described by Eq. (7.16)], the convergence of $\bar{\pi}_t(x)$ to $p(x)$ almost everywhere cannot be guaranteed, due to the fact that support points can never be added inside regions where $\bar{\pi}_t(x) < p(x)$. Indeed, inside some region $\mathcal{C} \subset \mathcal{D}$, where $\bar{\pi}_t(x) < p(x)$, we might obtain a sequence of proposals such that $\bar{\pi}_{t+\tau}(x) = \bar{\pi}_t(x)$ for an arbitrarily large value of $\tau$, or even $\forall \tau \in \mathbb{N}$, i.e., the proposal pdf might never change inside $\mathcal{C} \subset \mathcal{D}$. For further details, see [23].

### 7.4.4 IA$^2$RMS Algorithm

Our aim in this section is to devise a sequence of self-tuning proposals such that $\bar{\pi}_t(x) \to p(x)$, when $t \to \infty$, as fast as possible. Namely, we want to obtain an algorithm having a performance as close as possible to the ARS technique (i.e., ensuring that $D(\bar{\pi}_t, p) \to 0$ as $t \to \infty$ with a bounded computational cost), and the same range of applicability as the ARMS method (i.e., being a universal sampler, able to draw samples virtually from any target pdf). This can be achieved by means of a simple strategy that attains these two goals. This scheme ensures the convergence of the chain to the target distribution and keeps, at the same time, the computational cost bounded. Furthermore, it enables us to completely decouple the adaptation mechanism from the proposal construction, thus allowing simpler alternatives for the latter, as shown in Sect. 7.4.6.

The algorithm is called *independent doubly adaptive rejection Metropolis sampling* (IA$^2$RMS) [21, 23] (for some generalization see also [25]), with the A$^2$ emphasizing that we incorporate an additional adaptive step to improve the proposal pdf, compared to ARMS. The IA$^2$RMS algorithm is summarized in Table 7.3. Initially, IA$^2$RMS proceeds like ARMS, drawing a sample from the current proposal (step 3), performing a rejection test and incorporating rejected samples to the support set (step 4). Then, initially accepted samples go through an MH step to determine whether they are finally accepted or not (step 5.1), as in ARMS. The key improvement w.r.t. ARMS is the introduction of a new test (step 5.2), which allows to add samples (in a controlled way) inside regions of the domain where $\bar{\pi}_t(x) < p(x)$. Therefore, the IA$^2$RMS algorithm guarantees a complete adaptation of the proposal (i.e., $D(\bar{\pi}_t, p) \to 0$ as $t \to \infty$) exactly as in ARS. As a consequence, the correlation among samples is drastically reduced, quickly vanishing to zero, and IA$^2$RMS becomes an exact and direct sampler after some iterations (like ARS and unlike ARMS).

**Table 7.3** IA$^2$RMS algorithm

**Initialization:**

1. Set $n = 0$ (chain iteration) and $t = 0$ (algorithm iteration). Choose an initial state $x_0$ and support set $\mathcal{S}_0 = \{s_1, \ldots, s_{m_0}\}$.

**Iterations (while $n < N$):**

2. Build a proposal, $\bar{\pi}_t(x)$, given the set $\mathcal{S}_t = \{s_1, \ldots, s_{m_t}\}$, using a convenient procedure (e.g. the ones described in [7, 29] or the simpler ones proposed in Sect. 7.4.6).
3. Draw $x' \sim \pi_t(x) \propto \bar{\pi}_t(x)$ and $u' \sim \mathcal{U}([0, 1])$.
4. If $u' > \frac{p(x')}{\bar{\pi}_t(x')}$, then reject $x'$, update $\mathcal{S}_{t+1} = \mathcal{S}_t \cup \{x'\}$, $m_{t+1} = m_t + 1$, set $t = t + 1$, and go back to step 2.
5. Otherwise, if $u' \leq \frac{p(x')}{\bar{\pi}_t(x')}$, then:

   5.1 Draw $u'' \sim \mathcal{U}([0, 1])$. If $u'' \leq \alpha$, with

$$\alpha = \min\left[1, \frac{p(x') \min[p(x_n), \bar{\pi}_t(x_n)]}{p(x_n) \min[p(x'), \bar{\pi}_t(x')]}\right],$$

   then accept $x'$, setting $x_{n+1} = x'$ and $y = x_n$. Otherwise, if $u'' > \alpha$, then reject $x'$, setting $x_{n+1} = x_n$ and $y = x'$.

   5.2 Draw $u''' \sim \mathcal{U}([0, 1])$. If

$$u''' > \frac{\bar{\pi}_t(y)}{p(y)},$$

   then set $\mathcal{S}_{t+1} = \mathcal{S}_t \cup \{y\}$ and $m_{t+1} = m_t + 1$. Otherwise, set $\mathcal{S}_{t+1} = \mathcal{S}_t$ and $m_{t+1} = m_t$.

   5.3 Update $t = t + 1$, $n = n + 1$ and return to step 2.

Finally, let us remark that IA$^2$RMS requires selecting a single set of parameters: the initial set of support points, $\mathcal{S}_0$. After this choice, the algorithm proceeds automatically without any further intervention required by the user. Regarding the robustness of IA$^2$RMS w.r.t. $\mathcal{S}_0$, the only requisite is choosing $m_0 \geq 2$ initial support points where the value of the target is different from zero, i.e., $p(s_i) > 0$ for $i = 1, \ldots, m_0$. Furthermore, if the effective support of the target (i.e., the support containing most of its probability mass) is approximately known, then a good initialization consists of selecting the two points delimiting this support and at least another point inside this support. If the user desires to increase the robustness of IA$^2$RMS, a grid of initial points can be used. This choice speeds up the convergence of the algorithm, but any random selection within the effective support of the target ensures the convergence of IA$^2$RMS.

### 7.4.5 Convergence of the Chain and Computational Cost

The new control test is performed using an auxiliary variable, $y$, which is always different from the new state, $x_{n+1}$. This construction leads to a proposal, $\tilde{\pi}_t(x)$, which is independent of the current state of the chain, $x_n$. Hence, the convergence of the Markov chain to the target density is ensured by Theorem 2 in [10] (see also Theorem 8.2.2 in [12]).[5] The conditions required to apply this theorem are fulfilled as long as $\tilde{\pi}_t(x) \rightarrow p(x)$ almost everywhere. This implies also that $\pi_t(x) \rightarrow p_o(x)$ almost everywhere as $t \rightarrow \infty$. The convergence of $\tilde{\pi}_t(x)$ to $p(x)$ almost everywhere also implies that the probability of adding new support points goes to zero as $t \rightarrow \infty$, thus keeping the computational cost bounded. Note that there is no contradiction between the two previous statements. As more support points are added, $\tilde{\pi}_t(x)$ becomes closer to the target, and this implies a decrease in the probability of adding new support points. However, for $t < \infty$ there is always a non-null (albeit small for large values of $t$) probability of adding new support points to improve the proposal and to make it closer to the target.

The coding and implementation complexity of IA$^2$RMS is virtually identical to ARMS, since all the quantities involved in the ratio of step 5.2 have been previously calculated in steps 4 and 5.1. Thus, no additional evaluation of the proposal and target pdfs is required. Given a specific construction procedure for $\tilde{\pi}_t(x)$, the total number of support points increases w.r.t. ARMS, but it always remains within the same order of magnitude, as can be shown numerically in [23]. Indeed, it is important to emphasize that the number of support points does not diverge: it remains bounded thanks to the two control tests, exactly as in ARS and ARMS, since $\frac{p(x)}{\tilde{\pi}_t(x)} \rightarrow 1$ almost everywhere when $t \rightarrow \infty$.

These properties entail that IA$^2$RMS is drawing samples from the target distribution within a finite number of iterations with a probability arbitrarily close to 1.

### 7.4.6 Examples of Proposal Constructions for IA$^2$RMS

Since IA$^2$RMS improves the adaptive structure of ARMS, simpler procedures can be used to build the function $W_t(x)$, reducing the overall computational cost and the coding effort [21, 23, 24]. A first possibility is to define $W_t(x)$ inside the $i$th interval simply as the straight line $L_{i,i+1}(x)$ going through $(s_i, V(s_i))$ and $(s_{i+1}, V(s_{i+1}))$ for $1 \leq i \leq m_t - 1$ (where $V(x) = -\log[p(x)]$), and extending the straight lines corresponding to $\mathcal{I}_1$ and $\mathcal{I}_{m_t-1}$ towards $\pm\infty$ for the first and last intervals. Formally,

$$W_t(x) = L_{i,i+1}(x), \quad x \in \mathcal{I}_i = (s_i, s_{i+1}], \tag{7.17}$$

[5]Note that, even though the IA$^2$RMS algorithm falls inside the broad category of independent adaptive algorithms, its structure is inspired by [7], not by [10]. Indeed, no RS test is performed in [10] and the construction of the proposals is completely different.

for $1 \leq i \leq m_t - 1$, $W_t(x) = L_{1,2}(x)$ in $\mathcal{I}_0 = (-\infty, s_1]$ and $W_t(x) = L_{m_t-1,m_t}(x)$ in $\mathcal{I}_{m_t} = (s_{m_t}, \infty)$. This is illustrated in Fig. 7.1a. Note that, although this procedure looks similar to the one used in ARMS, as described by Eq. (7.16), it is actually much simpler, since it does not require the calculation of intersection points. Furthermore, an even simpler procedure to construct $W_t(x)$ can be devised from Eq. (7.17): using a piecewise constant approximation with two straight lines inside the first and last intervals. Formally,

$$W_t(x) = \min\{V(s_i), V(s_{i+1})\}, \quad x \in \mathcal{I}_i = (s_i, s_{i+1}], \tag{7.18}$$

for $1 \leq i \leq m_t - 1$, $W_t(x) = L_{1,2}(x)$ in $\mathcal{I}_0 = (-\infty, s_1]$ and $W_t(x) = L_{m_t-1,m_t}(x)$ in $\mathcal{I}_{m_t} = (s_{m_t}, \infty)$. This construction leads to the simplest possible proposal: a collection of uniform pdfs with two exponential tails. Figure 7.1b shows an example of the construction of the proposal using this approach.

Alternatively, we could build the proposal $\bar{\pi}_t(x)$ directly, instead of constructing $W_t(x)$ and setting $\bar{\pi}_t(x) = \exp(-W_t(x))$. Following this approach, we could apply the procedure described in [1] for adaptive trapezoidal Metropolis sampling (ATRAMS), even though the structure of this algorithm is completely different from IA$^2$RMS. In this case, the proposal is constructed using straight lines $\widetilde{L}_{i,i+1}(x)$ passing through $(s_i, p(s_i))$ and $(s_{i+1}, p(s_{i+1}))$, i.e., *directly* in the domain of the target pdf, $p(x)$. Formally,

$$\bar{\pi}_t(x) = \widetilde{L}_{i,i+1}(x), \quad x \in \mathcal{I}_i = (s_i, s_{i+1}], \tag{7.19}$$

for $1 \leq i \leq m_t - 1$, and the tails are two exponential pieces. Figure 7.1c shows an example of a proposal using this approach. Finally, note that Eq. (7.18) would be identical in the pdf domain, since $\exp(\max\{V(s_i), V(s_{i+1})\}) = \max\{p(s_i), p(s_{i+1})\}$. Furthermore, applying Eq. (7.19) directly in the domain of the pdf could yield invalid proposals with $\bar{\pi}_t(x) < 0$ inside some regions. Indeed, although many

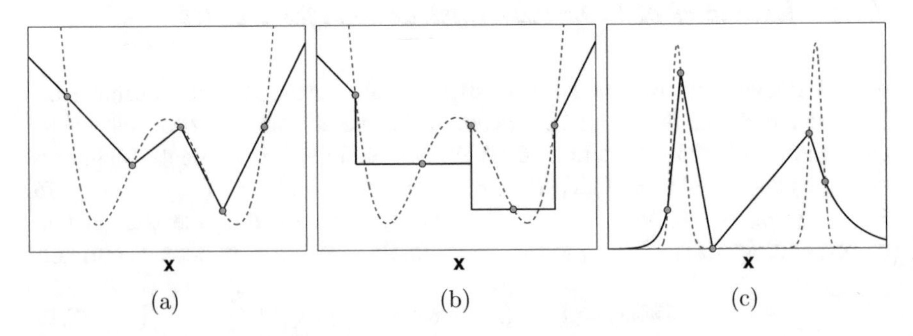


**Fig. 7.1** Examples of proposal constructions ($V(x)$ or $p(x)$ in dashed lines, $W_t(x)$ or $\pi_t(x)$ in solid lines), using five support points and different procedures: **(a)** the procedure described by Eq. (7.17) in the log-domain; **(b)** the procedure described by Eq. (7.18) in the log-domain; **(c)** the procedure described by Eq. (7.19) directly in the pdf's domain

other alternatives can be considered to build the proposal, they have to satisfy the following basic properties:

(1) Valid proposals are always obtained, i.e., $\bar{\pi}_t(x) = \exp(-W_t(x)) > 0$ $\forall x \in \mathcal{D} \subseteq \mathbb{R}$ and $t \in \mathbb{N}^+$.
(2) The sequence of proposals $\{\bar{\pi}_t\}_{t=0}^{\infty}$ tends to $p(x)$ (i.e., $D(\bar{\pi}_t, p) \to 0$) when new support points are added.
(3) Samples from $\pi_t(x) \propto \bar{\pi}_t(x)$ can be efficiently drawn.

The first condition is easily fulfilled in the log-domain, but restricts the use of some constructions in the pdf domain. The second condition is for the performance of the algorithm. The last condition is essential for practical purposes to obtain an efficient algorithm. When the proposal is a piecewise function, as in the four constructions proposed in this section, it demands the ability to compute the area below each piece and to draw samples efficiently from each piece.

## 7.5 Summary

In this chapter, we have described different MCMC techniques that can yield *asymptotically independent samples* at the expense of an increase of the computational cost. First, in Sect. 7.2, we have recalled some basic concepts related to MCMC techniques. In the same section, we have also presented the standard Metropolis-Hastings method. In Sect. 7.3, we have introduced two extensions of the MH algorithm which generate a set of $N$ candidates at each iteration. As the cardinality $N$ of this set grows, the correlation among the generated samples vanishes. Both techniques can be easily used for sampling from multivariate target distributions.

In Sect. 7.4, we have described a different approach (for the purpose of sampling from univariate distributions), introducing the Independent Doubly Adaptive Rejection Metropolis Sampling (IA$^2$RMS) algorithm. This method is based on a non-parametric construction of the proposal pdf in the same fashion as the Adaptive Rejection Sampling (ARS) and the Adaptive Rejection Metropolis Sampling (ARMS) methods, described exhaustively in Chap. 4. The shape of the proposal pdf is adapted using interpolation procedures. As the proposal pdf becomes closer and closer to the target, the correlation among the generated samples decreases.

## References

1. B. Cai, R. Meyer, F. Perron, Metropolis-Hastings algorithms with adaptive proposals. Stat. Comput. **18**, 421–433 (2008)
2. B. Calderhead, A general construction for parallelizing Metropolis-Hastings algorithms. Proc. Natl. Acad. Sci. U. S. A. (PNAS) **111**(49), 17408–17413 (2014)
3. R. Casarin, R.V. Craiu, F. Leisen, Interacting multiple try algorithms with different proposal distributions. Stat. Comput. **23**, 185–200 (2013)

4. D. Gamerman, H.F. Lopes, *Markov Chain Monte Carlo: Stochastic Simulation for Bayesian Inference*. Chapman & Hall/CRC Texts in Statistical Science (Chapman & Hall/CRC, Boca Raton, 2006)
5. W.R. Gilks, Derivative-free adaptive rejection sampling for Gibbs sampling. Bayesian Stat. **4**, 641–649 (1992)
6. W.R. Gilks, P. Wild, Adaptive rejection sampling for Gibbs sampling. Appl. Stat. **41**(2), 337–348 (1992)
7. W.R. Gilks, N.G. Best, K.K.C. Tan, Adaptive rejection metropolis sampling within Gibbs sampling. Appl. Stat. **44**(4), 455–472 (1995)
8. D. Görür, Y.W. Teh, Concave convex adaptive rejection sampling. University College London, Technical Report (2009)
9. W.K. Hastings, Monte Carlo sampling methods using Markov chains and their applications. Biometrika **57**(1), 97–109 (1970)
10. L. Holden, R. Hauge, M. Holden, Adaptive independent Metropolis-Hastings. Ann. Appl. Probab. **19**(1), 395–413 (2009)
11. W. Hörmann, A rejection technique for sampling from T-concave distributions. ACM Trans. Math. Softw. **21**(2), 182–193 (1995)
12. F. Liang, C. Liu, R. Caroll, *Advanced Markov Chain Monte Carlo Methods: Learning from Past Samples*. Wiley Series in Computational Statistics (Wiley, Chichester, 2010)
13. J.S. Liu, *Monte Carlo Strategies in Scientific Computing* (Springer, New York, 2004)
14. J.S. Liu, F. Liang, W.H. Wong, The multiple-try method and local optimization in metropolis sampling. J. Am. Stat. Assoc. **95**(449), 121–134 (2000)
15. L. Martino, A review of multiple try MCMC algorithms for signal processing. Digital Signal Process. **75**, 134–152 (2018)
16. L. Martino, V. Elvira, Metropolis sampling. Wiley StatsRef: Statistics Reference Online (2017)
17. L. Martino, F. Louzada, Issues in the Multiple Try Metropolis Mixing. Comput. Stat. **32**(1), 239–252 (2017)
18. L. Martino, J. Míguez, Generalized rejection sampling schemes and applications in signal processing. Signal Process. **90**(11), 2981–2995 (2010)
19. L. Martino, J. Read, A multi-point Metropolis scheme with generic weight functions. Stat. Probab. Lett. **82**(7), 1445–1453 (2012)
20. L. Martino, J. Read, On the flexibility of the design of multiple try Metropolis schemes. Comput. Stat. **28**(6), 2797–2823 (2013)
21. L. Martino, J. Read, D. Luengo, Independent doubly adaptive rejection Metropolis sampling, in *IEEE International Conference on Acoustics, Speech, and Signal Processing (ICASSP)* (2014)
22. L. Martino, V. Elvira, D. Luengo, J. Corander, F. Louzada, Orthogonal parallel MCMC methods for sampling and optimization. arXiv:1507.08577 (2015)
23. L. Martino, J. Read, D. Luengo, Independent doubly adaptive rejection Metropolis sampling within Gibbs sampling. IEEE Trans. Signal Process. **63**(12), 3123–3138 (2015)
24. L. Martino, H. Yang, D. Luengo, J. Kanniainen, J. Corander, A fast universal self-tuned sampler within Gibbs sampling. Digit. Signal Process. **47**, 68–83 (2015)
25. L. Martino, R. Casarin, D. Luengo, Sticky proposal densities for adaptive MCMC methods, in *IEEE Workshop on Statistical Signal Processing (SSP)* (2016)
26. L. Martino, V. Elvira, D. Luengo, J. Corander, Layered Adaptive Importance Sampling. Stat. Comput. **27**(3) 599–623 (2017)
27. N. Metropolis, S. Ulam, The Monte Carlo method. J. Am. Stat. Assoc. **44**, 335–341 (1949)
28. N. Metropolis, A. Rosenbluth, M. Rosenbluth, A. Teller, E. Teller, Equations of state calculations by fast computing machines. J. Chem. Phys. **21**, 1087–1091 (1953)
29. R. Meyer, B. Cai, F. Perron, Adaptive rejection Metropolis sampling using Lagrange interpolation polynomials of degree 2. Comput. Stat. Data Anal. **52**(7), 3408–3423 (2008)
30. R. Neal, MCMC using ensembles of states for problems with fast and slow variables such as Gaussian process regression. arXiv:1101.0387 (2011)
31. C.P. Robert, G. Casella, *Monte Carlo Statistical Methods* (Springer, New York, 2004)

# Chapter 8
# Summary and Outlook

In this monograph, we have described the theory and practice of pseudo-random variate generation. This is the core of Monte Carlo simulations and, hence, of practical importance for a large number of applications in various fields, including computational statistics, cryptography, computer modeling, games, etc. The focus has been placed on independent and exact sampling methods, as opposed to techniques that produce weighted (e.g., importance sampling) and/or correlated populations (e.g., MCMC).

A number of relevant references can be found in the literature related to these topics [1–4]. However, in this monograph, we have tried to present a comprehensive and unified view of the field of independent random sampling. We have included the most relevant classes of methods and emphasized their generality, as opposed to the common trend of investigating algorithms "tailored" to specific problems. Moreover, we have explored in depth the connections, relationships, and relative merits of the different families of techniques, including systematic comparisons with non-independent samplers, such as MCMC methods and importance samplers. Let us also note that the majority of the contents presented in this monograph correspond to research that has been published during the last decade, especially concerning the various families of adaptive samplers.

Although our main interest when compiling this work was in the theory and methods for independent random sampling, we have made an effort to select application examples that enjoy a clear practical interest. In this respect, we expect that the materials included in this book may be of interest to engineers working in signal processing and statisticians interested in computational methods, but they should also be useful to scientists working in the fields of biology, quantitative finance or physics, where complex models that demand Monte Carlo computations are needed. Matlab code for many of the algorithms and examples presented in this monograph is available online in a companion website.

In the sequel we summarize the main results presented in each chapter and then proceed to expose some promising lines of future research.



L. Martino et al., *Independent Random Sampling Methods*, Statistics and Computing, https://doi.org/10.1007/978-3-319-72634-2_8

In Chap. 2, we have described the so-called *direct methods*: a collection of classical and modern techniques used for random sampling based on suitable transformations and/or specific connections among random variables. All of them assume the availability of a random source with known distribution, and all of them are aimed at producing independent and identically distributed (i.i.d.) samples. Many of them are intrinsically connected, and we have made a special effort to highlight the relationships among different techniques or different categories of methods. Indeed, we have noted that some techniques can be classified within more than one category of algorithms and can be derived in different ways (e.g., the Box-Muller method). These different points of view have been explored and the connections among categories of algorithms have been identified.

In Chaps. 3 and 4, we have presented and discussed the *standard rejection sampling* (RS) and the *adaptive rejection sampling* (ARS) algorithms. The basic RS approach, which is described in Chap. 3, was suggested by John von Neumann as early as in 1946, although it was not published until 1951, and it is a classical technique for universal sampling. In an accept/reject method, each sample is either accepted or rejected by an adequate test of the ratio of the proposal and the target pdfs, and it can be proved that accepted samples are actually distributed according to the target density. The fundamental figure of merit of a rejection sampler is the mean acceptance rate, i.e., the expected number of accepted samples over the total number of proposed candidates. To attain good acceptance rates, adaptive rejection sampling (ARS) schemes, which are the focus of Chap. 4, have been proposed in the literature. These techniques aim at sequentially building proposal functions that become closer and closer to the target pdf as the algorithm is iterated (i.e., as more samples are drawn and more accept/reject tests are carried out). In these two chapters we have presented several variants of standard RS and ARS schemes, together with various recent developments, and included some original material. Again, the connections and relationships among different methods have been highlighted.

In Chap. 5 we have focused on the ratio of uniforms (RoU) method, which is a classical technique that combines both of the approaches of the previous chapters: suitable transformations of the random variables of interest and rejection sampling. Assume that $p(x)$ is the target density from which it is needed to generate samples. The RoU technique aims at calculating a bounded region $\mathcal{A}$ such that points drawn independently and uniformly inside $\mathcal{A}$ yield i.i.d. samples from $p(x)$ in a very straightforward manner. Since uniform samples within $\mathcal{A}$ cannot be obtained usually by direct methods, in practice the RoU approach is often combined with the RS method. First, we have presented the standard RoU technique and some extensions. Then we have focused on adaptive implementations of the RoU method, including some original contributions. The connections of the RoU approach, both with transformation methods and with the accept/reject class of techniques, are highlighted and several application examples have also been provided.

In Chap. 6, we have reviewed some generic or tailored sampling methods to draw samples from multidimensional distributions (i.e., to generate random vector-samples). Most of them are only *partially general*, as they involve a number of constraints on the target distributions. Others are more general, e.g., the multivariate extension of the RoU and ARS techniques discussed in Chaps. 5 and 4, but their computational complexity can be prohibitive. Both theoretical constraints and computational limitations have been explored, and illustrated by way of a few examples.

In Chap. 7, we have introduced another family of samplers, widely used in the literature: Markov chain Monte Carlo (MCMC) algorithms. MCMC methods [5, 6] are Monte Carlo techniques that produce a Markov chain of correlated samples whose stationary distribution is known. From the perspective of this work (the generation of i.i.d. samples), they present two drawbacks: the correlation of the generated samples and the fact that the samples only come exactly from the desired distribution when the chain attains its stationary distribution, a status which is not straightforward to determine. In Chap. 7, we have explored two special classes of MCMC approaches that produce "asymptotically" independent samples. This means that the MCMC sampler tends to become an exact sampler as the number of iterations grows or as a parameter of the algorithm is increased, thus ensuring that the correlation among samples quickly vanishes to zero and the samples generated eventually become i.i.d.

Several extensions of the methods presented herein can be expected in the near future. We expect to see significant developments in two particular directions, namely

- the design of *general* methods for multidimensional random variables, possibly tied to efficient computational methods for Monte Carlo simulation, and
- hybrid techniques, e.g., combinations of independent random samplers with MCMC schemes (as explored in Chap. 7) but also with importance sampling (IS) methods.

Indeed, we view the two areas of work above as complementary. On one hand, it is hard to foresee efficient random sampling methods for large random vectors unless some sort of hybrid algorithms (possibly "almost exact" or asymptotically exact) are developed. On the other hand, many researchers in the computational statistics community are seeking ways to improve the efficiency of MCMC and IS methodologies in high dimensional estimation problems. The adaptive independent samplers described here may possibly play a role as key building blocks to enhance the performance of sophisticated MCMC and IS schemes in such high dimensional settings.

Some relatively straightforward applications of adaptive independent samplers should come about in the short term as these techniques become more popular and, especially, more easily accessible—meaning that easy-to-use, off-the-shelf software becomes available to practitioners working in various problems in engineering (tracking, queueing, code design, etc.), computer science (machine learning and artificial intelligence), or biology (synthetic biology, *in silico* experimentation, etc.) where efficient sampling methods are often used.

## References

1. J. Dagpunar, *Principles of Random Variate Generation* (Clarendon Press, Oxford/New York, 1988)
2. L. Devroye, Random variate generation for unimodal and monotone densities. Computing **32**, 43–68 (1984)
3. J.E. Gentle, *Random Number Generation and Monte Carlo Methods* (Springer, New York, 2004)
4. W. Hörmann, J. Leydold, G. Derflinger, *Automatic Nonuniform Random Variate Generation* (Springer, New York, 2003)
5. F. Liang, C. Liu, R. Caroll, *Advanced Markov Chain Monte Carlo Methods: Learning from Past Samples*. Wiley Series in Computational Statistics (Wiley, London, 2010)
6. C.P. Robert, G. Casella, *Monte Carlo Statistical Methods* (Springer, New York, 2004)

# Appendix A
# Acronyms and Abbreviations

| | |
|---|---|
| a.k.a. | Also known as |
| ARMS | Adaptive Rejection Metropolis Sampling |
| ARoU | Adaptive Ratio of Uniforms |
| ARS | Adaptive Rejection Sampling |
| BR | Band Rejection |
| CARS | Cheap Adaptive Rejection Sampling |
| CCARS | Concave Convex Adaptive Rejection Sampling |
| cdf | Cumulative distribution function |
| DP | Dirichlet Process |
| e.g. | Exempli gratia (for instance). |
| E-IoD | Extended Inverse-of-Density |
| EnMCMC | Ensemble MCMC |
| GARS | Generalized Adaptive Rejection Sampling |
| GBR | Generalized Band Rejection |
| GP | Gaussian Process |
| GRoU | Generalized Ratio of Uniforms |
| IA$^2$RMS | Independent Doubly Adaptive Rejection Metropolis Sampling |
| i.e. | Id est (that is) |
| i.i.d. | Identically and identical distributed |
| IoD | Inverse-of-Density |
| LCG | Linear Congruential Generator |
| MCMC | Markov Chain Monte Carlo |
| MH | Metropolis-Hastings |
| MJP | Markov Jump Process |
| MTM | Multiple Try Metropolis |
| PARS | Parsimonious Adaptive Rejection Sampling |
| pdf | Probability density function |
| PMMLCG | Prime Modulus Multiplicative Linear Congruential Generator |
| PRN | Pseudo Random Number |



L. Martino et al., *Independent Random Sampling Methods*, Statistics and Computing, https://doi.org/10.1007/978-3-319-72634-2

| | |
|---|---|
| PRNG | Pseudo Random Number Generator |
| RC | Rejection Control |
| RoU | Ratio of Uniforms |
| RS | Rejection Sampling |
| r.v. | Random variable |
| SDE | Stochastic Differential Equation |
| TDR | Transformed Density Rejections |
| TRM | Transformed Rejection Method |
| VDR | Vertical Density Representation |
| WP | Wiener Process |
| w.r.t. | With respect to |

# Appendix B
# Notation

## B.1 Vectors, Points, and Intervals

Scalar magnitudes are denoted using regular face letters, e.g., $x$, $X$, while vectors are displayed as boldface letters, e.g., $\mathbf{x}$, $\mathbf{X}$. The scalar coordinates of a vector in $n$-dimensional space are denoted with square brackets, e.g., $\mathbf{x} = [x_1, \ldots, x_n]$. Often, it is more convenient to interpret $\mathbf{x}$ as a point in the space. When needed, we emphasize this representation with the alternative notation $\mathbf{x} = (x_1, \ldots, x_n)$.

We use a similar notation for the intervals in the real line. Specifically, for two boundary values $a \leq b$, we denote $[a, b] = \{x \in \mathbb{R} : a \leq x \leq b\}$ for a closed interval, while

$$(a, b] = \{x \in \mathbb{R} : a < x \leq b\}, \quad [a, b) = \{x \in \mathbb{R} : a \leq x < b\},$$

are half-open intervals and finally $(a, b) = \{x \in \mathbb{R} : a < x < b\}$ is an open interval.

## B.2 Random Variables, Distributions and Densities

We indicate random variables (r.v.) with uppercase letters, e.g., $X$, $\mathbf{X}$, while we use lowercase letters to denote the corresponding realizations, e.g., $x$, $\mathbf{x}$. Often, when we draw a collection of samples of a r.v., we use the superscript notation $x^{(i)}$, $\mathbf{x}^{(i)}$ where $i$ indicates the sample number.

We use lowercase letters, e.g., $q(\cdot)$, to denote the probability density function (pdf) of a random variable or vector, e.g., $q(y)$ is the pdf of $Y$. The conditional pdf of $X$ given $Y = y$ is written $p(x|y)$. The cumulative distribution function (cdf) of a r.v. $X$ is written as $F_X(\cdot)$. The probability of an event, e.g., $X \leq x$, is indicated as $\text{Prob}\{X \leq x\}$. In particular, $F_X(a) = \text{Prob}\{X \leq a\}$.



L. Martino et al., *Independent Random Sampling Methods*, Statistics and Computing, https://doi.org/10.1007/978-3-319-72634-2

The target pdf from which we wish to draw samples is denoted as $p_o(x)$ while $p(x)$ is a function proportional to $p_o(x)$, i.e., $p(x) \propto p_o(x)$.

The uniform distribution in an interval $[a, b]$ is written $\mathcal{U}([a, b])$. The Gaussian distribution with mean $\mu$ and variance $\sigma^2$ is denoted $\mathcal{N}(\mu, \sigma^2)$. The symbol $\sim$ means either that a r.v. $X$ or a sample $x'$ has the indicated distribution, e.g., $X \sim \mathcal{U}([a, b])$ and $x' \sim \mathcal{N}(\mu, \sigma^2)$, or that a sample $x'$ has a particular pdf, e.g., $x' \sim p_o(x)$. Finally, $\mathcal{N}(x; \mu, \sigma^2)$ represents a Gaussian pdf with mean $\mu$ and variance $\sigma^2$.

## B.3 Sets

Sets are denoted with calligraphic uppercase letters, e.g., $\mathcal{R}$. The support of the r.v. of interest $X$ is denoted as $\mathcal{D} \subseteq \mathbb{R}$ [i.e., $\mathcal{D}$ is the domain of the target pdf $p_o(x)$]. In some cases, without loss of generality, we may consider $\mathcal{D} = \mathbb{R}$ for convenience. When needed, we denote with $\mathcal{C}$ the support of auxiliary variables.

Finally, we write the indicator function on the set $\mathcal{S}$ as $\mathbb{I}_{\mathcal{S}}(x)$. It takes value 1 if $x \in \mathcal{S}$ and 0 otherwise, i.e.,

$$\mathbb{I}_{\mathcal{S}}(x) = \begin{cases} 1 \text{ if } x \in \mathcal{S} \\ 0 \text{ if } x \notin \mathcal{S} \end{cases}. \tag{B.1}$$

## B.4 Summary of Main Notation

- $p_o(x)$ : (normalized) target density.
- $p(x)$ : target function proportionally to $p_o(x)$ .
- $\pi(x)$: proposal density.
- $x$ or $X$: scalar magnitudes are denoted using regular face letters.
- $\mathbf{x}$ or $\mathbf{X}$: vectors are displayed as boldface letters.
- $\mathbf{x} = [x_1, \ldots, x_n]$: the scalar coordinates of a vector in $n$-dimensional space are denoted with square brackets.
- $\mathbf{x} = (x_1, \ldots, x_n)$: point in the space $\mathbb{R}^n$.
- $X$: scalar random variable.
- $x^{(i)}$: $i$th sample.
- $q(\cdot)$: (lowercase letter) probability density function (pdf) of a random variable or vector.
- $p(x|y)$: the conditional pdf of $X$ given $Y = y$.
- $F_X(x)$: cumulative density function of $X$.
- Prob$\{\cdot\}$: the probability of an event.
- $\mathcal{U}([a, b])$: uniform distribution between $a$ and $b$.
- $\mathcal{N}(\mu, \sigma^2)$: Gaussian distribution of mean $\mu$ and variance $\sigma^2$.
- $\mathcal{N}(x; \mu, \sigma^2)$: Gaussian density of mean $\mu$ and variance $\sigma^2$.

# Appendix C
# Jones' RoU Generalization

The analysis of the proof of GRoU in Sect. 5.2 suggests further generalizations of the GRoU technique [1, 2]. Let us consider the generic transformation of variables $(v, u) \rightarrow (x, z)$,

$$\begin{cases} x = t(v, u) \\ z = u \end{cases}, \tag{C.1}$$

where $t(v, u)$ is a transformation of $u$ and $v$ which is invertible with respect to the variable $v$, i.e., such that we can write

$$v = f(x, u). \tag{C.2}$$

Therefore, the entire inverse transformation can be expressed as

$$\begin{cases} v = f(x, z) \\ u = z \end{cases}. \tag{C.3}$$

Moreover, let us denote with $r(x, z)$ a function with the following three properties:

1. The first derivative of $r(x, z)$ w.r.t. $z$ is equal to $f(x, z)$, i.e.,

$$f(x, z) = \frac{\partial r(x, z)}{\partial z} \rightarrow r(x, z) = \int f(x, z) dz, \tag{C.4}$$

2. The first derivative of $r(x, z)$ w.r.t. $x$ is zero when $z = 0$, i.e.,

$$\left[ \frac{\partial r(x, z)}{\partial x} \right]_{z=0} = 0. \tag{C.5}$$


L. Martino et al., *Independent Random Sampling Methods*, Statistics and Computing, https://doi.org/10.1007/978-3-319-72634-2

3. We also need $\frac{\partial r(x,z)}{\partial x}$ to be invertible in $z$.

With the definitions and assumptions above, we can enunciate the following theorem.

**Theorem C.1** *Let $u$ and $v$ be uniformly distributed over*

$$\mathcal{A} = \Big\{(v,u) : 0 \leq u \leq \eta\big(t(v,u)\big)\Big\}. \tag{C.6}$$

*If $\eta(x)$ is given such that*

$$p_o(x) \propto \left[\frac{\partial r(x,z)}{\partial x}\right]_{z=\eta(x)}, \tag{C.7}$$

*then $x = t(v,u)$ has density $p_o(x)$.*

*Proof* The vector $(v,u)$ is distributed uniformly on $\mathcal{A}$. Hence, the joint pdf $q(x,z)$ of the vector $(x,z)$ is

$$q(x,z) = \frac{1}{|\mathcal{A}|}|J^{-1}| \quad \text{for} \quad 0 \leq z \leq \eta(x), \tag{C.8}$$

where $x = t(v,u)$ [see Eq. (C.1)] and $|\mathcal{A}|$ indicates the measure of $\mathcal{A}$. With $J^{-1}$, we denote the Jacobian of the inverse transformation

$$J^{-1} = \det\begin{bmatrix} f_x(x,z) & f_z(x,z) \\ 0 & 1 \end{bmatrix} = f_x(x,z), \tag{C.9}$$

where $f_x(x,z) = \frac{\partial f(x,z)}{\partial x}$. Therefore,

$$q(x,z) = \begin{cases} \frac{1}{|\mathcal{A}|} f_x(x,z) & \text{for} \quad 0 \leq z \leq \eta(x) \\ 0 \text{ otherwise.} \end{cases} \tag{C.10}$$

Marginalizing the joint density $q(x,z)$ w.r.t. $z$, we obtain

$$\begin{aligned} \int_{-\infty}^{+\infty} q(x,z)dz &= \int_0^{\eta(x)} \frac{1}{|\mathcal{A}|} f_x(x,z)dz = \frac{1}{|\mathcal{A}|}\int_0^{\eta(x)} \frac{\partial f(x,z)}{\partial x}dz, \\ &= \frac{1}{|\mathcal{A}|}\frac{\partial}{\partial x}\int_0^{\eta(x)} f(x,z)dz = \frac{1}{|\mathcal{A}|}\left[\frac{\partial r(x,z)}{\partial x}\right]_{z=0}^{z=\eta(x)} \\ &= \underbrace{\frac{1}{|\mathcal{A}|}\left[\frac{\partial r(x,z)}{\partial x}\right]_{z=\eta(x)}}_{p_o(x)} - \underbrace{\frac{1}{|\mathcal{A}|}\left[\frac{\partial r(x,z)}{\partial x}\right]_{z=0}}_{0} = p_o(x). \end{aligned}$$

where we have interchanged the symbols of integral and derivative and we have used the Eqs. (C.5)–(C.7). □

## C.1 Possible Choices of $t(v, u)$

A possible family of functions $t(v, u)$, studied in [2], is

$$x = t(v, u) = a^{-1}\left(\frac{v - b(u)}{\dot{g}(u)}\right), \tag{C.11}$$

where $g(u)$ and $a(x)$ are each differentiable and invertible functions, with $g(0) = 0$. Moreover, $b(u)$ is another generic function of $u$. In this case, we have the function

$$\eta(x) = g^{-1}\left(-c_A\frac{p_o(x)}{\dot{a}(x)}\right)$$

where $c$ is a positive constant and $\dot{a} = \frac{da}{dx}$. Hence, the region $\mathcal{A}$ is defined as

$$\mathcal{A} = \left\{(v, u) : 0 \leq u \leq g^{-1}\left[-c_A\left[\frac{p_o}{\dot{a}}\right]\left(a^{-1}\left(\frac{v - b(u)}{\dot{g}(u)}\right)\right)\right]\right\} \tag{C.12}$$

where the functions $p_o(x)$ and $\dot{a}(x)$ are both evaluated in $x = a^{-1}\left(\frac{v-b(u)}{\dot{g}(u)}\right)$.

The form of $t(v, u)$ in Eq. (C.11) is composed by a monotonic transformation $a^{-1}$ combined with a generalized "location" function $b(u)$ and a "scale" term $\dot{g}(u)$. Now, for the sake of simplicity, we set $c_A = 1$ and analyze some specific cases below.

1. *$a(x)$ and $g(u)$ identity functions, and $b(u)$ zero:* In this case

$$\mathcal{A} = \mathcal{A}_0 = \{(v, u) : 0 \leq u \leq p_o(v)\}$$

with $x = u$, i.e., we find the definition of the area below the target pdf $p_o(x)$. We come back to the fundamental theorem of simulation in Sect. 2.4.3.

2. *$a(x)$ and $g(u)$ identity functions:* If $b(u) = k$, where $k$ is a constant, this case corresponds only to a shift of the target pdf $p_o(x)$, with

$$\mathcal{A} = \{(v, u) : 0 \leq u \leq p_o(v - k)\}$$

and $x = v - k$. The density $p_o(x)$ could be "relocated" by an arbitrary function $b(u)$,

$$\mathcal{A} = \{(v, u) : 0 \leq u \leq p_o(v - b(u))\}$$

with $x = v - b(u)$. In [1], the authors study the special case where $b(u) = -u$ and $x = v + u$. In [3, 4], the authors observe that this relocation can ease the achievement of higher acceptance rates in an RS scheme using the RoU method.

3. *$a(x)$ identity and $b(u)$ zero:* This is the case in Sect. 5.4, i.e.,

$$\mathcal{A} = \mathcal{A}_g = \left\{(v, u) : 0 \leq u \leq g^{-1}\left[p\left(\frac{v}{\dot{g}(u)}\right)\right]\right\}$$

and $x = v/\dot{g}(u)$.

4. *$a(x)$ identity function:* Starting from the generalization in Sect. 5.4 we add the "relocation" function $b(u)$, i.e.,

$$\mathcal{A} = \left\{(v, u) : 0 \leq u \leq g^{-1}\left[p\left(\frac{v - b(u)}{\dot{g}(u)}\right)\right]\right\},$$

and we have to take $x = (v - b(u))/\dot{g}(u)$.

5. *$g(u)$ identity function, $b(u)$ zero:* In this case, the region is defined as

$$\mathcal{A} = \left\{(v, u) : 0 \leq u \leq \left[\frac{p_o}{\dot{a}}\right]\left(a^{-1}(u)\right)\right\},$$
$$= \left\{(v, u) : 0 \leq u \leq \frac{p_o(a^{-1}(v))}{\dot{a}(a^{-1}(v))}\right\},$$

and $x = a^{-1}(v)$ is then distributed as $p_o(x)$. The set $\mathcal{A}$ can be rewritten as

$$\mathcal{A} = \left\{(v, u) : 0 \leq u \leq p_o(a^{-1}(v))\frac{da^{-1}}{dv}\right\},$$

where we can notice that it is that case of a monotonic transformation $a(x)$ of the variable $X \sim p_o(x)$. Indeed, the r.v. $V = a(X)$ has density $q(v) = \frac{da^{-1}}{dv}p_o(a^{-1}(v))$. Therefore, in this case the RoU method is equivalent to the r.v. transformation $V = a(X)$: we first generate a sample $v' \sim q(v)$ and then take $x' = a^{-1}(v')$.

## References

1. G. Barbu, On computer generation of random variables by transformations of uniform varaibles. Soc. Sci. Math. R. S. Rom. Tome 26 **74**(2), 129–139 (1982)
2. M.C. Jones, A.D. Lunn, Transformations and random variate generation: generalised ratio-of-uniforms methods. J. Stat. Comput. Simul. **55**(1), 49–55 (1996)
3. A.J. Kinderman, J.F. Monahan, New methods for generating student's t and gamma variables. Computing **25**(4), 369–377 (1980)
4. J.C. Wakefield, A.E. Gelfand, A.F.M. Smith, Efficient generation of random variates via the ratio-of-uniforms method. Stat. Comput. **1**(2), 129–133 (1991)

# Appendix D
# Polar Transformation

Given a hyper-sphere $\mathcal{B}_\rho$ of radius $\rho$ in $\mathbb{R}^n$, a point $\mathbf{x} = [x_1, \ldots, x_n]^\top$ on the boundary of this hyper-sphere, we can be represented using $\rho$ and $n-1$ angles $\theta_1, \ldots, \theta_{n-1}$, namely

$$\begin{cases} x_1 &= \rho \sin\theta_1 \sin\theta_2 \ldots \sin\theta_{n-2} \sin\theta_{n-1}, \\ x_2 &= \rho \sin\theta_1 \sin\theta_2 \ldots \sin\theta_{n-2} \cos\theta_{n-1}, \\ x_3 &= \rho \sin\theta_1 \sin\theta_2 \ldots \sin\theta_{n-3} \cos\theta_{n-2}, \\ &\vdots \\ x_{n-2} &= \rho \sin\theta_1 \sin\theta_2 \cos\theta_3, \\ x_{n-1} &= \rho \sin\theta_1 \cos\theta_2, \\ x_n &= \rho \cos\theta_1. \end{cases} \tag{D.1}$$

As we have seen in Chap. 2 (and we recall in Sect. 6.7), in this case we have to compute the determinant of the Jacobian matrix of the corresponding transformation, i.e.,

$$|\mathbf{J}| = \rho^{n-1} (\sin\theta_1)^{n-2} (\sin\theta_2)^{n-3} \cdots \sin\theta_{n-2}.$$

Thus, the choice of a "direction" uniformly in $\mathbb{R}^n$ is equivalent to generating random angles $\theta_1, \ldots, \theta_{n-1}$ according to

$$h(\theta_1, \ldots, \theta_{n-1}) = \prod_{i=1}^{n} h_i(\theta_i) \propto (\sin\theta_1)^{n-2} (\sin\theta_2)^{n-3} \cdots \sin\theta_{n-2}, \tag{D.2}$$

with $0 < \theta_i < \pi$, $i = 1, \ldots, n-2$ and $0 < \theta_{n-1} < 2\pi$. Then, we can observe that

$$\begin{aligned} \Theta_{n-1} &\sim \mathcal{U}([0, 2\pi]), && 0 < \theta_{n-1} < 2\pi, \\ \Theta_i &\sim h_i(\theta_i) \propto (\sin\theta_i)^{n-1-i}, && 0 < \theta_i < \pi, \quad i = 1, \ldots, n-2. \end{aligned}$$


L. Martino et al., *Independent Random Sampling Methods*, Statistics and Computing, https://doi.org/10.1007/978-3-319-72634-2

Therefore, the previous sampling method for drawing uniform points inside the hyper-sphere $\mathcal{B}_r$ (i.e., with radius $r$) could be rewritten as following:

1. Draw $\rho' \sim q(\rho) \propto \rho^{n-1}$ with $\rho \in (0, r]$.
2. Draw $\theta_i' \sim h_i(\theta_i) \propto (\sin\theta_i)^{n-1-i}$, with $0 < \theta_i < \pi$ for $i = 1, \ldots, n-2$, and $\theta_{n-1}' \sim \mathcal{U}([0, 2\pi])$.
3. Set $x_i'$, $i = 1, \ldots, n$, as in Eq. (D.1).

Printed in the United States
By Bookmasters